Proceedings of the Specialist Meeting on
Personal Dosimetry and Area Monitoring Suitable
for Radon and Daughter Products

Compte rendu d'une réunion de spécialistes sur
la dosimétrie individuelle et la surveillance de l'atmosphère
en ce qui concerne le radon et ses produits de filiation

Elliot Lake, Canada
4-8 Oct. 1976

NUCLEAR ENERGY AGENCY
ORGANISATION FOR ECONOMIC CO-OPERATION AND DEVELOPMENT

AGENCE POUR L'ÉNERGIE NUCLÉAIRE
ORGANISATION DE COOPÉRATION ET DE DÉVELOPPEMENT ÉCONOMIQUES

The Organisation for Economic Co-operation and Development (OECD) was set up under a Convention signed in Paris on 14th December, 1960, which provides that the OECD shall promote policies designed:
- to achieve the highest sustainable economic growth and employment and a rising standard of living in Member countries, while maintaining financial stability, and thus to contribute to the development of the world economy;
- to contribute to sound economic expansion in Member as well as non-member countries in the process of economic development;
- to contribute to the expansion of world trade on a multilateral, non-discriminatory basis in accordance with international obligations.

The Members of OECD are Australia, Austria, Belgium, Canada, Denmark, Finland, France, the Federal Republic of Germany, Greece, Iceland, Ireland, Italy, Japan, Luxembourg, the Netherlands, New Zealand, Norway, Portugal, Spain, Sweden, Switzerland, Turkey, the United Kingdom and the United States.

The OECD Nuclear Energy Agency (NEA) was established on 20th April 1972, replacing OECD's European Nuclear Energy Agency (ENEA) on the adhesion of Japan as a full Member.

NEA now groups all the European Member countries of OECD and Australia, Canada, Japan, and the United States. The Commission of the European Communities takes part in the work of the Agency.

The objectives of NEA remain substantially those of ENEA, namely the orderly development of the uses of nuclear energy for peaceful purposes. This is achieved by:
- *assessing the future role of nuclear energy as a contributor to economic progress, and encouraging co-operation between governments towards its optimum development;*
- *encouraging harmonisation of governments' regulatory policies and practices in the nuclear field, with particular reference to health and safety, radioactive waste management and nuclear third party liability and insurance;*
- *forecasts of uranium resources, production and demand;*
- *operation of common services and encouragement of co-operation in the field of nuclear energy information;*
- *sponsorship of research and development undertakings jointly organised and operated by OECD countries.*

In these tasks NEA works in close collaboration with the International Atomic Energy Agency, with which it has concluded a Co-operation Agreement, as well as with other international organisations in the nuclear field.

© OECD, 1977
Queries concerning permissions or translation rights should be addressed to:
Director of Information, OECD
2, rue André-Pascal, 75775 PARIS CEDEX 16, France.

L'Organisation de Coopération et de Développement Économiques (OCDE), qui a été instituée par une Convention signée le 14 décembre 1960, à Paris, a pour objectif de promouvoir des politiques visant :
- à réaliser la plus forte expansion possible de l'économie et de l'emploi et une progression du niveau de vie dans les pays Membres, tout en maintenant la stabilité financière, et contribuer ainsi au développement de l'économie mondiale ;
- à contribuer à une saine expansion économique dans les pays Membres, ainsi que non membres, en voie de développement économique ;
- à contribuer à l'expansion du commerce mondial sur une base multilatérale et non discriminatoire, conformément aux obligations internationales.

Les Membres de l'OCDE sont : la République Fédérale d'Allemagne, l'Australie, l'Autriche, la Belgique, le Canada, le Danemark, l'Espagne, les États-Unis, la Finlande, la France, la Grèce, l'Irlande, l'Islande, l'Italie, le Japon, le Luxembourg, la Norvège, la Nouvelle-Zélande, les Pays-Bas, le Portugal, le Royaume-Uni, la Suède, la Suisse et la Turquie.

L'Agence de l'OCDE pour l'Energie Nucléaire (AEN) a été instituée le 20 avril 1972, en remplacement de l'Agence Européenne pour l'Énergie Nucléaire de l'OCDE (ENEA) à la suite de l'adhésion du Japon en tant que Membre de plein exercice.

L'AEN groupe à présent tous les pays européens Membres de l'OCDE ainsi que l'Australie, le Canada, les États-Unis et le Japon. En outre la Commission des Communautés Européennes participe également aux travaux de l'Agence.

Les objectifs de l'AEN restent pour la plupart les mêmes que ceux de l'ENEA et concernent la promotion du développement harmonieux des utilisations pacifiques de l'énergie nucléaire. Elle entreprend à cet effet :
- *d'évaluer le rôle futur de l'énergie nucléaire dans la réalisation du progrès économique et d'encourager la coopération entre les gouvernements en vue de son développement optimal ;*
- *de promouvoir une harmonisation des politiques et pratiques réglementaires des gouvernements dans le domaine nucléaire, en particulier pour la protection de la santé et la sécurité, la gestion des déchets radioactifs, la responsabilité civile et l'assurance en matière nucléaire ;*
- *d'établir des prévisions sur les ressources, la production et la demande d'uranium ;*
- *d'assurer le fonctionnement de services communs et d'encourager la coopération dans le domaine de l'information nucléaire ;*
- *de patronner des entreprises de recherche et de développement organisées et exploitées en commun par des pays Membres de l'OCDE.*

Pour remplir ces fonctions, l'AEN travaille en étroite collaboration avec l'Agence Internationale de l'Énergie Atomique (avec laquelle elle a conclu un accord de coopération) ainsi qu'en liaison avec d'autres organisations internationales dans le domaine nucléaire.

© OCDE, 1977
Les demandes de reproduction ou de traduction doivent être adressées à :
M. le Directeur de l'Information, OCDE
2, rue André-Pascal, 75775 PARIS CEDEX 16, France.

Foreword

Within the OECD Nuclear Energy Agency's programme, it was foreseen that topics relating to radiation and environmental protection matters in mining and milling operations would be studied. In order to prepare such a programme, and ad hoc Expert Group met during April 1976 and proposed as first steps the setting up of an exchange of information system and the organisation of a Specialist Meeting on Personal Dosimetry and Area Monitoring suitable for Radon and Daughter Products.

Following an invitation from the Canadian authorities, that meeting took place at Elliot Lake (Ontario), from 4th to 8th October 1976. The meeting dealt with questions relating to personal dosimetry techniques as well as the monitoring of the mine atmosphere with reference to radon and radon daughters. The Proceedings reproduce the specialists' papers as well as the Panels' discussions.

Avant-propos

Dans le cadre du programme de l'Agence de l'OCDE pour l'Energie Nucléaire, il avait été prévu d'étudier les problèmes de radioprotection et de protection de l'environnement liés aux opérations d'extraction et de traitement du minerai d'uranium. Afin d'élaborer un programme dans ce domaine, un Groupe d'experts ad hoc s'est réuni en avril 1976 et a proposé, dans un premier temps, d'instituer un système d'échange d'informations et d'organiser une Réunion de spécialistes sur la dosimétrie individuelle et la surveillance de l'atmosphère en ce qui concerne le radon et ses produits de filiation.

Sur invitation des autorités canadiennes, cette réunion a eu lieu à Elliot Lake (Ontario), du 4 au 8 octobre 1976. La réunion a porté sur des questions relatives aux techniques de dosimétrie individuelle, ainsi qu'à la surveillance de l'atmosphère dans les mines, eu égard au radon et aux produits de filiation du radon. Le présent compte rendu reproduit les exposés des spécialistes, ainsi que les discussions des tables rondes.

Table of Contents

Table des matières

Session I   - INTRODUCTORY PAPERS
Séance I   - COMMUNICATIONS D'INTRODUCTION

        Chairman - Président : P.E. HAMEL (Canada)

RADON AND ITS HAZARDS
R.M. Fry, Australia 13

INTERPRETATION OF MEASUREMENTS IN URANIUM MINES : DOSE EVALUATION AND BIOMEDICAL ASPECTS
W. Jacobi, Federal Republic of Germany 33

RESULTATS BIOLOGIQUES EXPERIMENTAUX ET RELATION DOSE-EFFET DU RADON AVEC SES PRODUITS DE FILIATION
J. Chameaud, R. Perraud, J. Lafuma, R. Massé,
J. Chrétien, France 49

SOCIETAL ASPECTS OF HAZARDS IN URANIUM MINES
D.J.M. Ham, Canada 57

STATEMENT OF IAEA ACTIVITIES IN THE FIELD OR RADIATION PROTECTION IN MINING AND MILLING OF NUCLEAR MATERIALS AND PERSONAL AND AREA MONITORING IN GENERAL
J.U. Ahmed, IAEA 67

Session II   - PERSONAL DOSIMETRY
Séance II   - DOSIMETRIE INDIVIDUELLE

        Chairman - Président : J. PRADEL (France)

MEASUREMENT OF EMPLOYEES' INDIVIDUAL CUMULATIVE EXPOSURES TO RADON DAUGHTERS AS PRACTICED IN THE UNITED STATES
R.L. Rock, United States 71

MOD WORKING LEVEL DOSIMETER
A.J. Breslin, United States 91

MEASUREMENT AND RECORDING OF OCCUPATIONAL RADIATION EXPOSURES FOR URANIUM MINERS
D. Grogan, Canada 97

FIELD TEST OF MOD WORKING LEVEL DOSIMETERS
G.R. Yourt, Canada 101

APPAREILS DE MESURE EN CONTINU DE LA CONCENTRATION DES EMETTEURS α DE LA CHAINE DE U 238 (POUSSIERES A VIE LONGUE ET DESCENDANTS DU RADON) EN SUSPENSION DANS L'AIR
J. Pradel, Ph. Duport, P. Zettwoog, France 117

A COMBINED PERSONAL SAMPLER FOR DUST AND RADON-DAUGHTER EXPOSURE IN MINES
G. Knight, R.A. Washington, W.M. Gray, Canada 127

CODE FOR RADIATION EXPOSURE IN ONTARIO MINES
W.A. Bardswich, Canada 135

PERSONAL DOSIMETRY IN THE NINGYO-TOGE MINE
Y. Kurokawa, K. Nakashima, Y. Kitahara, R. Kurosawa, Japan 137

HAZARDS TO WORKERS FROM RADON AND RADON DAUGHTER PRODUCTS IN TUNNELS AND DRIFTS OF SWITZERLAND
E. Kaufmann, Switzerland 145

PROBLEMES PRATIQUES RENCONTRES DANS LA DETERMINATION DES DOSES α INHALEES PAR LE PERSONNEL DES MINES D'URANIUM
J. Pradel, Y. François, P. Zettwoog, France 149

IN VIVO MEASUREMENTS AND BIOASSAY FOR Pb 210, AS AN INDICATOR OF CUMULATIVE EXPOSURE OF URANIUM MINERS TO RADON DAUGHTERS
C. Pomroy, Canada 155

Session III   - <u>PANEL DISCUSSION ON PERSONAL DOSIMETRY</u>
Séance III   - <u>TABLE RONDE SUR LA DOSIMETRIE INDIVIDUELLE</u>

    Chairman - Président : R.M. FRY (Australia)
Summary of discussion, R.M. Fry 159

Session IV   - <u>AREA MONITORING</u>
Séance IV   - <u>SURVEILLANCE DE L'ATMOSPHERE</u>

    Chairman - Président : D. ROSS (United States)

TWO AREA MONITORS WITH POTENTIAL APPLICATION IN URANIUM MINES
A.J. Breslin, United States 165

OPERATIONAL RADIATION PROTECTION IN UK MINING
M.C. O'Riordan, United Kingdom 171

PROTECTION OF WORKERS IN RADON-RICH ATMOSPHERES : THE MANDATE FOR QUICK DETERMINATION OF RADON-DAUGHTER CONCENTRATIONS AND A SOLUTION
J.D. Shreve, Jr., United States 181

SUPERVISION OF RADON DAUGHTER EXPOSURE IN MINES IN SWEDEN
J.O. Snihs, H. Ehdwall, Sweden 191

SOME DATA OF PERSONAL AND AREA MONITORING IN THE SWEDISH URANIUM MINE, RANSTAD
P.O. Agnedal, Sweden 199

AREA MONITORING IN THE NINGYO-TOGE MINE
Y. Kurokawa, K. Nakashima, Y. Kitahara, R. Kurosawa, Japan 201

MEASUREMENTS OF RADON DAUGHTER CONCENTRATIONS IN STRUCTURES BUILT ON OR NEAR URANIUM MINE TAILINGS
F.F. Haywood, G.D. Kerr, W.A. Goldsmith, P.T. Perdue, J.H. Thorngate, United States 219

NIVEAUX RENCONTRES DANS LA CONTAMINATION DE L'AIR EN DESCENDANTS DU RADON DANS LES REGIONS URANIFERES ET AU VOISINAGE DES INSTALLATIONS D'EXTRACTION ET DE TRAITEMENT DE MINERAI
J. Pradel, Ph. Duport, P. Zettwoog, France 231

PROBLEMS IN AREA MONITORING FOR RADON DAUGHTERS
R.A. Washington, J.L. Horwood, Canada 239

PRESENT PRACTICES OF THE DEPARTMENT OF NATIONAL HEALTH AND WELFARE FOR THE AREA MONITORING OF RADON AND DAUGHTER PRODUCTS
H. Taniguchi, Canada 251

LIMITATION OF EXPOSURE TO THE COMBINED RISKS FROM RADON DAUGHTERS, GAMMA RADIATION AND OTHER RADIOLOGICAL HAZARDS
W.R. Bush, Canada 255

ASSESSMENT OF AIRBORNE RADIOACTIVITY IN ITALIAN MINES
G. Sciocchetti, Italy 259

CONTINUOUS AIR MONITORING USING BETA-GAMMA DETECTION FOR AREA MEASUREMENTS OF RADON DAUGHTERS
H.M. Johnson, Canada 271

Session V   - PANEL DISCUSSION ON AREA MONITORING

Séance V   - TABLE RONDE SUR LA SURVEILLANCE DE
              L'ATMOSPHERE

  Chairman - Président : W. JACOBI (Federal Republic of Germany)
    Summary of discussion, W. Jacobi 275

Session VI - SUMMARY, CONCLUSIONS AND RECOMMENDATIONS

Séance VI   - RESUME, CONCLUSIONS ET RECOMMANDATIONS

          Chairman - Président : P.E. HAMEL (Canada)
    Discussion, P.E. Hamel 281

LIST OF PARTICIPANTS - LISTE DES PARTICIPANTS 311

Session I
Introductory papers

Chairman - Président
P.E. HAMEL
(Canada)

Séance I
Communications d'introduction

RADON AND ITS HAZARDS

R. M. Fry
Australian Atomic Energy Commission
Sydney, Australia

1. INTRODUCTION

I propose to outline something of what is known about radon and its occurrence, the relation between exposure to its daughter products and dose to the lung, and the consequences of such exposures for man. I cannot pretend that this will be a critical review of the knowledge in this complex field. As I understand my role, it is to introduce the subject of this Specialist Meeting with a broad overview of some of the basic facts and outstanding problems of radon risk assessment so that the papers of the invited speakers and your own contributions to the techniques of personnel and area monitoring for radon and its daughter products, can be seen in perspective.

2. THE SOURCE OF RADON

Radon, ^{222}Rn, is the heaviest of the noble gases, with a density 7.7 times that of air. Its only source is the decay of its parent ^{226}Ra, one curie of which produces about 2μCi of ^{222}Rn per second. Uranium, and consequently ^{226}Ra concentrations in soils and rocks are around 1pCi/g, so that on average each gram of the surface layers of the earth is a perpetual source of some 2×10^{-18} Ci of ^{222}Rn per second.

Being an inert gas, and having a relatively long half-life (3.82 days) some of this radon will escape from the particles in which it is produced and diffuse into interstices in the soil whence it can migrate to the surface of the earth. The rate of exhalation of radon to the atmosphere is a complex function of the nature of the soil and of meteorological conditions and has been studied for many years. A review of radon migration in the ground will be found in Tanner [1], and of the effects of atmospheric variables on ^{222}Rn flux and exhalation, in Megumi and Mamuro [2], Kraner et al [3], and Harley [4]; the literature is extensive. A representative figure for ^{222}Rn emanation rate from the earth's surface is about 0.5pCi m^{-2} sec^{-1}. This leads to a rate of ^{222}Rn injection into the world's atmosphere of about 2.4×10^9 Ci/y and an equilibrium content of 3.6×10^7 Ci (Wilkening et al [5], Harley [4], Israel [6]).

A special category of radon exhalation studies of relevance to uranium mining and milling is the theory and measurement of ^{222}Rn emanation from rocks and minerals, especially of uranium bearing minerals in situ, and from uranium mill tailings piles. Rate of emanation from uranium ores, which is a determining factor in the levels of radon found in uranium mines, depends on the mineral, the type of rock, its porosity and permeability, the size of the particles in which the uranium is incorporated and the degree of weathering of the rock; it is not directly related to the richness of the orebody (IAEA [7]).

Emanation from tailings piles, which may contain 100 to 1000 pCi/g of ^{226}Ra, depends very much on the moisture content of the tailings, the relative disposition of the slimes and sand-like fraction, and the nature and moisture content of any inactive soil cover (Sears et al [8]). A useful rough guide for the emanation rate from a dry pile is 1pCi ^{222}Rn m^{-2} sec^{-1} for each 1pCi ^{226}Ra/g of tailings (Swift et al [9]).

The actual volume of ^{222}Rn in a uranium ore body is extremely small. One curie of ^{222}Rn occupies only 6.6×10^{-4} cm^3 at NTP. Thus in 1000 tonnes of ore containing 1 tonne of uranium, and consequently 0.3Ci of ^{226}Ra and therefore also 0.3Ci of ^{222}Rn, there is only about 2×10^{-4} cm^3 of radon.

3. RADON IN THE ATMOSPHERE

Outside

Radon and its daughter products constitute the major part of the naturally occurring background radioactivity in the air of the lower atmosphere. Their concentration varies at any one place diurnally and seasonally depending on meteorological conditions. These affect not only the amount of mixing and dilution that occurs in the atmosphere, but also, as mentioned above, the rate of emanation of radon from the earth's surface. Radon levels will, for example, build up near the surface under inversions, and decrease under unstable conditions. Radon levels over the ocean are much less than over land (about two orders of magnitude less) because of the lower radium content of the oceans. For this reason radon concentrations in the atmosphere at coastal sites are dependent on whether the wind is blowing from the land or the sea. The concentration of radon in ground level air out of doors is of the order of 0.1 pCi/l (average range 0.04 - 0.4 pCi/l; UNSCEAR [10]).

Because of turbulent diffusion in the atmosphere and radioactive decay, the air concentration of radon decreases with height. Integration of this variation of radon concentration with height to the top of the atmosphere leads to a total radon content which agrees well with the observed average emanation rate from the earth's surface of about 0.5 pCi m^{-2} sec^{-1} (Israel [6]).

Indoors

Radon levels inside buildings are usually higher than in the outside air, particularly if the radon can accumulate in poorly ventilated spaces. Indoor concentrations depend upon the rate at which radon emanates from the walls and floor, which in turn is dependent upon the porosity and ^{226}Ra content of the building materials, upon the ventilation rate in the room, meteorological variables and, to a minor extent, upon the radon concentration in the outside air. In some areas where the radon content of drinking water is significant radon may be released into houses from tap water; natural gas used for cooking and domestic heating may be an additional minor source (Barton et al [11], Gesell [12], Johnson et al [13]).

Indoor levels of radon may be a factor of 10 or more higher than outside (UNSCEAR [10]). Gemesi [14] found an average value of 3.05 pCi/l for radon in apartments compared with 0.2 pCi/l out of doors. Levels may be much higher again in poorly ventilated basements; for example a mean level of 6.0 pCi/l was found in the HASL basement compared with a mean outdoor city level of 0.17 pCi/l (George [15]). A representative figure for dwellings in the USA would appear to be about 0.5 pCi/l.

In Mines

Not surprisingly, radon concentrations can reach very much higher levels than these in poorly ventilated underground uranium mines. Median radon levels measured in the 1940's in some US mines having only natural draft ventilation were 2,500 pCi/l in Colorado and 5,000 pCi/l in Utah (Holaday et al [16]).

Measurements made in the late 1930's in the uranium mines at Schneeberg and Jachymov ranged between 50 and 5,000 pCi/l (Schraub [17]). At this time some effort had been made to reduce radon levels by forced ventilation of the mines, but estimates of levels before ventilation was introduced go as high as 15,000 pCi/l.

It is well known that high radon levels can also arise in non-uranium underground mines (Snihs [18], Strong et al [19]). Average concentrations in excess of 1,000 pCi/l have been measured in poorly ventilated areas in a Newfoundland mine producing fluospar where uranium was present in only trace amounts in the ore (FRC [20]). The radon entered the mine dissolved in ground water. At equilibrium, radon will be present in water at about a quarter the concentration that it will have in air in contact with it; the radon is readily released when the water is exposed, especially if it is aerated by flowing or splashing (Holaday [24]).

4. RADON DAUGHTER PRODUCTS

The "Working Level"

It is now known that the radon level in an atmosphere expressed in terms of the activity of ^{222}Rn per unit volume is not well correlated with the radiological consequences of its inhalation. The radiation dose to the respiratory tract due to radon is negligibly small compared with that delivered by its associated short-lived daughter products. Of course if radon were always in radioactive equilibrium with its decay products in the atmosphere, and these were always in the same physical state, radon concentration would be a good index of its radiological hazard. This however is not the case. The short-lived decay products may exist as free ions or free atoms (or molecules, if they react with some constituent of the atmosphere); or as the nuclei of small aggregations of molecules; or they may be attached to aerosol particles. The distribution of the daughter products between these various states, and the size distribution of the various fractions, will clearly affect the subsequent behaviour of the daughter products, both in the atmosphere and in the lung. A knowledge of these distributions as they occur in various atmospheres is therefore essential to an understanding of radon dosimetry and for the development of measuring techniques which will give a good physical correlative of radiological dose.

The concept of the "Working Level" (WL) was introduced to provide a better physical measure of the hazard of radon exposure. The most important contributors to the dose to the lung, and in particular to the cells lining the bronchi, are the two alpha emitting radon daughters, RaA and RaC'. The dose contribution from the beta particles and gamma rays emitted by RaB and RaC is negligible, whilst the long-lived daughters, RaD, RaE and RaF are eliminated from the bronchi before any significant number of disintegrations of these isotopes has occurred. What one wishes to know therefore is the total potential alpha energy associated with the short-lived daughter products in a given quantity of air.

The basis of the definition of the working level is the potential alpha energy associated with 100 pCi of ^{222}Rn in radioactive (secular) equilibrium with its short-lived decay products; this is 1.3×10^5 Mev.

The assumption is then made that any combination of RaA, RaB, RaC, RaC' which will result in an ultimate emission of 1.3×10^5 Mev of alpha energy has the same potential hazard as the equilibrium combination, if distributed in the same volume of air. Note that it is assumed that the potential hazard is independent of the details of how disequilibrium may be achieved, and independent also of the distribution of the decay products among the various aerosol states possible in the atmosphere. If the volume of air is taken as one litre, the new unit, the "Working Level" (WL) emerges. Thus one WL is defined as "any combination of the short-lived decay products of radon (RaA, RaB, RaC and RaC') in one litre of air which will result in the ultimate emission by them of 1.3×10^5 MeV of alpha particle energy.

1 WL is equivalent to 100 pCi/l of radon only in the one special case when the radon is in equilibrium with its short-lived decay products. To express a radon concentration measured in pCi/l, in WL's, it is necessary to know the state of disequilibrium between the radon and its short-lived daughters under the conditions obtaining at the time. In general, concentration in $WL's = F \times \left(\frac{\text{concentration in pCi/l}}{100}\right)$ where F is a factor, variously called the "disequilibrium factor" or "working level ratio", which is a measure of the lack of equilibrium between the radon and its short-lived daughters.

F is variable in practice and rarely has a value of one. Because of the half-lives of RaB and RaC it would take about 3 hours for initially fresh radon to come into equilibrium with its short-lived daughters. Any process which removes the daughter products, particularly movement and dilution of the air, will prevent equilibrium being attained and ensure that F remains less than 1. Even in a space where the ventilation rate was as low as 1 change per hour, F would be kept to about 0.7; it would drop to about 0.4 at 2 changes per hour.

A WL is a measure of a concentration. To obtain an accumulated or integrated exposure, it must be multiplied by the time of exposure. Integrated exposures are usually measured in terms of the "Working Level Month" (WLM). Inhalation of air with a concentration of 1 WL of radon daughters for the working hours in a month (taken as 170 h) results in a cumulative exposure of 1 WLM. Note that breathing rate, which varies markedly with physical activity, is not specified in the definition of the WLM.

The Physical State of the Decay Products

^{222}Rn is formed by the emission of a 4.78 MeVα particle from the nucleus of a ^{226}Ra atom. At the moment of its formation it will therefore have a recoil energy of some 86 keV and will be stripped of some of its outer electrons. Because of its inert gas structure it should readily become neutralized and behave as a "gas" atom. (The partial pressure of radon at a concentration of 100 pCi/l is less than 10^{-16} atmospheres). For this reason radon is not retained on particulate filters or electrostatic precipitators and must be collected as a gas, liquefied (its boiling point is -62°C), or adsorbed onto an appropriate medium.

The situation is different with the subsequent decay products of radon. ^{222}Rn decays according to the following well known scheme:

$$^{222}Rn \xrightarrow[3.82 \text{ days}]{5.49 \text{MeV}\alpha} {}^{218}Po \text{ (RaA)} \xrightarrow[3.05 \text{ min}]{6.00 \text{MeV}\alpha} {}^{214}Pb \text{ (RaB)}$$

$$\xrightarrow[26.8 \text{ min}]{\sim 0.7 \text{MeV}\beta} {}^{214}Bi \text{ (RaC)} \xrightarrow[19.7 \text{ min}]{\sim 2 \text{MeV}\beta} {}^{214}Po \text{ (RaC')}$$

$$\xrightarrow[164 \mu \text{ sec}]{7.69 \text{MeV}\alpha} {}^{210}Pb \text{ (RaD)} \xrightarrow[21y]{\sim 0.2 \text{MeV}\beta}$$

The recoil energy of ^{218}Po, 106keV, will again ensure the stripping of orbital electrons but because of the nature of the polonium atom it will be formed as a positively charged ion, which, if created in the atmosphere, can be expected to retain its charged state with some stability. That the daughter products of radon were positively charged was well known to early investigators of the radium decay chain. The collection of "active deposit" on negatively charged plates and wires became a common technique in the study of radon decay products. The fate of these ions in the atmosphere is most important for the subsequent behaviour of ^{218}Po and the other short-lived decay products of radon, in the air, and in the body, if they are inhaled.

A number of events may overtake the newly created ^{218}Po ion in the atmosphere:

. It may remain free as a positively charged ^{218}Po atom until it decays radioactively. This is only likely in an atmosphere with a low negative ion and low aerosol concentration, though if these conditions do obtain the mean life of positive ions in air can be of the order of tens of minutes. It should be noted however that the presence of radon and its daughters will itself contribute significantly to the production of negative ions in the atmosphere. (Raabe [21]).

. It may become the nucleus of a stable cluster of polar molecules. Diffusion measurements on ^{218}Po in air indicate that this is likely, and that some fraction of the airborne polonium behaves as if it were comprised of ^{218}Po atoms surrounded by some six water molecules (Raabe [21]).

. The free ions, or much more likely, the water-atom clusters, attach themselves to aerosol particles. These are often electrostatically charged or have a negatively polarized surface which facilitates adsorption of positive ions or the clusters of polar molecules.

A similar fate awaits the newly created daughter of ^{218}Po except that its behaviour will depend to some extent upon whether its parent is attached to an aerosol particle or not. The recoil energy of ^{214}Pb is about 112 keV which should be sufficient, if the recoil is in the right direction, to detach the ^{214}Pb ion from the aerosol particle. The recoil ion could, on the other hand, be driven further into the particle, or, if the aerosol diameter is less than its range, penetrate it altogether. The range of the recoiling ^{214}Pb ion in air is about 60μm, or about 0.04μm in a particle of density 2g/cm^3, though this will depend upon the charge state of the ion on formation, which is not well known. This effect must be taken into account in assessing the relative distribution of radon daughters between attached and unattached states in the atmosphere. It may also be important in the measurement of radon daughters since it could lead to losses of RaB from the collected active deposit (Jonassen et al. [22]).

The Attached and Unattached Fractions

A number of studies, both theoretical and experimental, have been made on the state of radioactive equilibrium of the short-lived daughter products in various atmospheres and of their distribution within the various possible particulate fractions indicated above. In particular one wishes to know the number concentration in the unattached and attached state, and the size distribution of the aerosols to which the daughters are attached. This will depend upon the concentration and size distribution of the inactive aerosols in the atmosphere, which will in turn depend upon air movements and ventilation rates and the presence of boundary surfaces on which the aerosols may deposit. The state of disequilibrium between radon in the atmosphere and its airborne daughters will be determined by this complex interaction of processes.

It is not possible here to do more than mention a few characteristic results of these studies.

- In the free atmosphere and indoors the size distribution of the active aerosol is relatively constant; particle diameter ranges between about 0.002 to 0.4μm. About half of the ^{214}Pb and ^{214}Bi activity is on particles less than about 0.08μm diameter (IAEA [23]). Measured activity median diameters can however vary markedly depending on the method of measurement (see George [15]). The mean diameter of the radioactive aerosol in mines is not much greater than in the atmosphere (IAEA [23]).

- The number concentration of the aerosol is of the order of 10^4/cm^3 in the open air away from city traffic, 10^4 to 10^5/cm^3 indoors and may be in excess of 10^5/cm^3 in mines. These numbers are very variable and depend significantly on ventilation rate, and in the case of mines, upon the mining operation. (George [15], Holaday [24], Blanc et al [25] Fusamura et al [26]).

- The unattached fraction has a diffusion coefficient of about 0.054 cm^2/s (Chamberlain et al [27]) and is identified with the cluster of molecules mentioned above (Raabe [21]). These have a mean life before attachment to aerosols of about 90 seconds in the open air and 2-4 seconds in a mine atmosphere (Jacobi [28]).

- The unattached fraction decreases with increasing aerosol concentration and increases with ventilation rate. Since ventilation may also reduce aerosol concentration, ventilation may be responsible for a marked increase in the unattached fraction (Jacobi [28]). Under medium-ventilation conditions in a mine the unattached fraction for RaA is estimated to be about 1-3% and for RaB, 0.2-0.5%, which corresponds to measurements made in some French mines (see Jacobi [28]). Recent measurements by George [15] outdoors and indoors in the vicinity of New York gave mean unattached fractions for ^{218}Po between 0.045 and 0.094, the highest value being found in the country out of doors where the concentration of dust particles was lowest.

- The state of disequilibrium between radon and its short-lived decay products in the atmosphere is a complex function of the rate of attachment of unattached ions and clusters to aerosols and to boundary surfaces, and to ventilation rate. The degree of disequilibrium is enhanced by ventilation. Disequilibrium factors calculated by Jacobi agree well with values measured in mines (Jacobi [28]).

George et al [29] have published a detailed study of the aerosol concentration, uncombined RaA fraction, radon levels and daughter product disequilibrium in a number of US uranium mines.

The number and size distribution of the short-lived radon daughters between the unattached and attached aerosol fractions discussed above, is the source term for assessing the behaviour of these particulates when they are inhaled, and consequently, the dose delivered to critical portions of the lung.

5. THE RISK FROM EXPOSURE TO RADON

It is now well established that lengthy exposure to high levels of radon in mine atmospheres can lead to an increased incidence of lung cancer in the miners. This has been observed in a number of the mines mentioned above, where radon levels in excess of some 1000pCi/l were common. The question that must be answered for protection purposes is: what is the maximum integrated exposure to radon in the circumstances encountered in a mine that will not lead to an unacceptable increase in the risk of a miner contracting lung cancer?

Since the primary standards of radiological protection are expressed as allowable doses of radiation, the conventional approach to the problem would be to calculate the radiation dose to the organ at risk, in this case the lung. The recommended maximum permissible dose rate to the lungs of radiation workers is 15 rem/y. One would like therefore to be able to calculate the limiting concentration of radon that would lead to a lung dose of 15 rem/y in exposed workers. Such calculations have been tried but, in the light of present knowledge, cannot give a satisfactory answer. Instead, observations on the incidence of lung cancer in miners, as a function of exposure, have been used to determine acceptable maximum limits of exposure. Although definitive calculation of dose is not practical, the attempts have been instructive, and it will be worthwhile describing briefly the theoretical approach.

The Dosimetric Approach

The principal workers associated with the development of these dosimetric models are Altshuler et al [30], Jacobi [31], [28], [32], Haque et al [33] and Harley et al [34]. The work is reviewed in Parker [35], UNSCEAR [10], IAEA [23] and FRC [20]. The models begin with the characterisation of the radon daughter aerosol as discussed above. When this aerosol is inhaled the unattached fraction, having a high mobility, is deposited most efficiently in the upper respiratory tract, whilst the attached fraction tends to be deposited deeper in the lung. The more sophisticated models attempt to calculate in some detail the pattern of deposition in the various regions of the lung as a function of breathing rate, depth of breathing and aerosol particle size distribution. The effect of mouth breathing and nose breathing is also taken into account. The Landahl lung model, the ICRP lung model (Jacobi [32]) and the geometry of the Weibel lung model (Harley et al [34]) have been used to calculate fractional retention in the various airway regions.

The steady state distribution of radioactivity within the regions of the lung is determined by the rate of clearance of the deposited material and its translocation from one region of the lung to another, and the rate of decay of the various daughter fractions. Clearance and translocation within the tracheo-bronchial tree is due to the movement of the mucous sheath. Mucous transit times in the various regions of the lung can be estimated, leading to a fractional retention distribution pattern of the radioactive aerosol. Transit times vary from a few minutes in the main bronchi and secondary bronchioles to hours within the terminal bronchioles (Harley et al [34]). These transit times are however long enough to ensure that most of the short-lived ^{222}Rn daughters will decay before they leave the lung. Material within the pulmonary region is not of course removed by ciliary action. It may accumulate in situ or be transferred to blood or to the lymphatic system but, whatever its precise fate, its biological half-life can be taken as 10 hours or more, so that the short-lived ^{222}Rn daughters can be considered to decay where they are deposited.

Before the equilibrium distribution of radioactive material can be converted to a biologically meaningful absorbed dose rate, the relevant biological target must be identified. This is considered, in all models, to be the nuclei of the basal cells of the bronchial epithelium. These are situated some 7μm above the basement membrane. Since the dose to these nuclei would, on the basis of these models, be due almost entirely to α particles from the ^{218}Po and ^{214}Po daughters, the thickness of intervening material between the active deposit and the nuclei is critical. This is assumed to be 7μm for the mucous layer, plus 7μm for the serous fluid enveloping the cilia, plus the thickness of the epithelium above the basal cell nuclei. The latter is very variable between different regions of the respiratory tree, and within any particular region of the bronchus. In general the thickness of the epithelium decreases as the branches get smaller and may vary by ±50% about the median thickness in any one region. (Gastineau et al [36]). The median thickness of the layer to be traversed is about 40μm for the epithelium above the target nuclei plus 15μm for the mucous and serous layer; with the variability indicated above this could range between about 30μm and 90μm. Given that the ranges of the alpha particles from ^{218}Po and ^{214}Po are about 45μm and 70μm respectively (Harley et al [34]) it is clear that if the active particles are resting on the upper surface of the mucous layer, many target cells will not be irradiated at all. Gastineau et al [36] estimate that two thirds of the basal cells of the lobar

- 20 -

and segmental bronchi are within the range of the ^{214}Po alphas, but only one fifth of these cells in the same region can be reached by the alphas from ^{218}Po.

Various assumptions have been made about the distribution of the activity within the mucous-serous layer. Altshuler et al [30] assume it rests on the mucous surface. Jacobi [31] assumes that the ^{218}Po decays while on the mucous surface but that subsequent daughters are distributed in a concentration gradient that varies from a maximum at the mucous surface to zero at the boundary of the epithelium; while Harley et al [34] consider all the alpha activity to be distributed homogenously within the 15μm thick mucous-serous layer and claim some experimental support for this.

Dose Conversion Factors

With such a number of model parameters and exposure conditions to be quantified it is not surprising that the relationship between exposure and dose found by various investigators varies over a wide range of values. All however agree that the maximum dose is delivered to the basal cells of the segmental bronchi. Dose conversion factors, expressed as rads to the basal cells of the critical lung region per WLM of exposure, assessed by various workers are summarized in Table 1. (See also Johnson et al [13]).

The estimated dose conversion factor is seen to vary between about 0.2 and 10 rads/WLM depending on the degree of disequilibrium, the aerosol characteristics and the lung model used. Quite critical in each model is the depth assumed for the biological target. In reviewing the situation, the BEIR Committee in 1972 considered that a value of 1 rad/WLM was probably close to an upper limit for a reasonably uniform dose to the basal cell layer of the epithelium of the larger bronchi; and perhaps 0.5 rad/WLM was an appropriate figure for miners, who were likely to have chronic bronchitis and therefore an increased thickness of the mucous layer and of the epithelium (BEIR [37]).

There is of course a further step to be taken in the dosimetric approach to the assessment of the biological hazards of radon exposure and that is the conversion of the estimated absorbed dose into a dose equivalent. The end point is bronchogenic cancer and the quality factor for this for alpha particles is unknown. The conventional value for the quality factor for alpha particles is 10; there is some evidence that 3 may be more appropriate in this context (Johnson et al [13]). The models all agree in predicting that the most heavily irradiated region of the lung is the region where the primary lung cancer in miners is thought to originate.

Although the status of the theoretical approach cannot be considered satisfactory, these studies have contributed much to our understanding of the processes involved, provided guidance to areas requiring additional research and enabled the relative importance of the many different biological and physical parameters to be assessed.

Deficiencies in the Working Level Concept

One conclusion emerges from these studies; the Working Level, though a considerable improvement upon the mere measurement of radon activity concentration, is still deficient as a physical correlative of biological hazard. As defined, the WL is independent of the distribution of the potential alpha emission amongst the attached and unattached fraction, and of the manner

TABLE 1: DOSE CONVERSION FACTORS FOR RADON-222 DAUGHTERS

(Dose to nuclei of basal cells of the epithelium of critical bronchial region)

Exposure Conditions			Atmosphere Particle size & concentration	Lung Model Breathing rate Nuclei depth	Dose factor rad/WLM*	Ref.
Daughter distn. Rn:RaA:RaB:RaC		F				
10 : 10 : 6 : 4 25% free RaA		0.57	Indoors, 1 change/h 0.09µm $10^4/cm^3$	Landhal 14ℓ/min 30µm	3.0	[31] 1964
10 : 10 : 10 : 10 25% free RaA		1.0	Outdoors 0.09µm $10^4/cm^3$		2.7	
Average conditions 1 - 2% free RaA			Clean Air High conc.	New ICRP 20ℓ/min	0.38-1.0 0.08-0.13	[32] 1972
10 : 9 : 6 : 4 9% free RaA		0.58	Mine 40%>0.1µm	Landhal 15ℓ/min 36µm mouth breathing nose breathing	3.5 1.4	[30] 1964
10 : 9 : 5 : 3.5 35% free RaA		0.49	Indoors 0.05µm $3x10^4/cm^3$	Weibel 15ℓ/min (mouth) 30µm 60µm	9.9 1.4	[33] 1967
10 : 10 : 10 : 10 4% free RaA		1.0	Mine 0.3µm	Weibel 15ℓ/min 22µm	0.24	[34] 1972
10 : 6 : 3 : 2 4% free RaA		0.29			0.36	
10 : 10 : 0 : 0 4% free RaA		0.1			1.1	
10 : 9 : 6 : 4 9% free RaA		0.58	Mine 40%>0.1µm	(Altshuler deposit) nose breathing 22µm	1.6	

*rad/WLM = $\dfrac{\text{dose rate (rad/y)}}{\text{Rn conc. (pCi/}\ell\text{)}}$ x $\dfrac{100 \text{ (pCi/}\ell\text{)}}{\text{F (WL)}}$ x $\dfrac{170 \text{ (h/month)}}{\text{exposure time (h/y)}}$

(Not normalized for breathing rates)

in which disequilibrium is achieved. Jacobi [32] has pointed out, as have Morken [38] and others, that the bronchial dose can vary by an order of magnitude for a given inhaled potential alpha energy. This is illustrated by the dose conversion factor assessed by Harley et al [34] and shown in Table 1 which is 0.24 for an equilibrium mixture of daughter products (F = 1) and 1.1 for the same concentration of RaA unaccompanied by RaB or RaC (F = 0.1). Harley and Harley [39] have shown that the relative contribution of ^{218}Po to the calculated dose to the basal cells is considerably larger than its contribution to the WL. The relative contributions of RaA, RaB and RaC' activity to the WL are RaA:RaB:RaC' = 1:5.2:3.8. However on the basis of the Harley and Pasternack model the relative contributions to dose become 1:1.4:1.2 for 0.2μm particles, and 1:1.7:1.6 for 0.3μm particles.

The aerosol content of the air and the ventilation rate in the area are two important parameters that need to be defined in addition to the potential alpha energy content of the atmosphere.

The Epidemiological Approach

An enhanced incidence of lung cancer has been observed in underground miners from a number of regions and attempts have been made to relate excess cancers to cumulative radon exposure. These investigations have been reviewed in a number of places. (UNSCEAR [10] Vol. II, BEIR [37], Jacobi [40] [41], Snihs [18]).

The most thorough study of this kind is one carried out by the US National Institute for Occupational Safety and Health (Lundin et al [42]) and deals with the incidence of lung cancer in underground uranium miners from the Colorado Plateau. The sample consists of 3,366 white and 780 non-white (99% Navajo Indians) underground uranium miners who had one or more months of underground uranium employment before 1st January 1964. The study considers deaths from all causes in this sample in the period July 1950 through September 1968. Total number of deaths in the white miner group was 437 compared with only 277 that would be expected in a group with the same age distribution and smoking habits. (Most of these miners were smokers; the interaction between smoking and radon exposure is discussed later.) The excess deaths were largely due to cancer of the respiratory tract and accidents. A total of 70 deaths from lung cancer were observed against an expected 12.

Estimates were made of the cumulative exposure of each miner in terms of WLM. No excess cancer was observed among miners who received less than about 120 WLM. The group is still under study and additional respiratory cancer cases are continuing to appear. Between September 1968 and December 1971, a further 37 deaths from lung cancer were recorded, plus an additional 8 cases that were still alive in February 1973. As well, an additional 22 cases of probable lung cancer - 9 deaths and 13 still alive - exist among the 3,366 sample (Archer et al [43]). The newly reported cases have occurred in the higher exposure categories and the statement that no excess cancers are observed among miners in the less than 120 WLM category still holds.

These more recent data were included in a review of the NIOSH study carried out in 1972 by the US Advisory Committee on the Biological Effects of Ionizing Radiations (BEIR [37]). The results of this review are shown in Figure 1 where the excess number of lung cancer cases above expectation (expressed as cases per year per million persons exposed) are plotted against

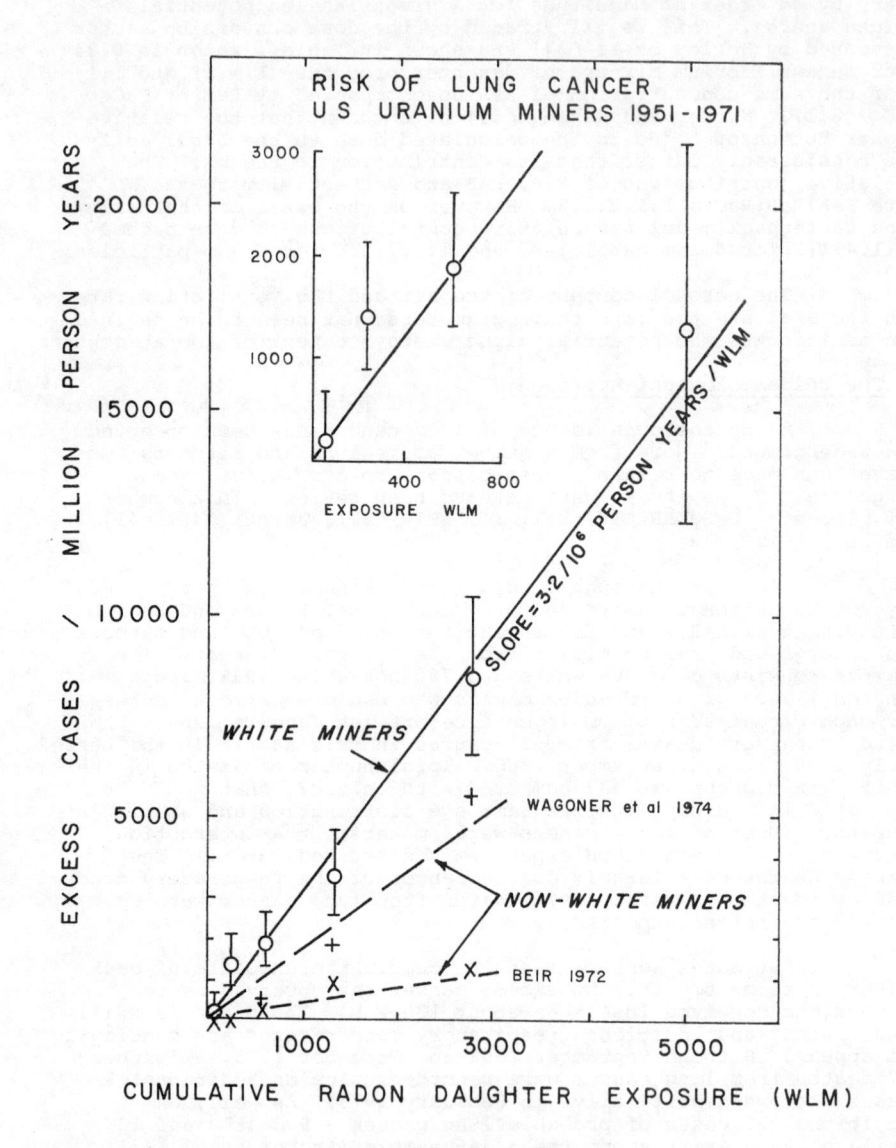

Figure 1

Exposure-response data for lung cancer in US uranium miners
(BEIR [37], Wagoner et al [44]),
Insert: Lowest exposure range for white miners.

cumulative exposure to radon daughters. A fairly clear linear relationship between exposure and lung cancer incidence is emerging, at least at the higher exposure levels, indicating an absolute risk co-efficient of about 3.2 cases per year per million smoking miners per WLM. It should be pointed out, however, that neither these data, nor those obtained from similar studies, are adequate to establish the linearity of the exposure-response curve at the lowest exposures (less than 100 WLM) and that a curvilinear relationship could fit the data just as well.

If one assumes that the exposure-response curve remains linear to the lowest exposures, and that this risk continues on average for 30 years, the total number of excess cases that might be expected in a population of a million smoking miners after exposure to 1 WLM, is about 100. Thus, if the exposure is limited to 4 WLM/y, a maximum permissible exposure rate to radon daughters now adopted in a number of countries, the risk amongst miners who smoke is about 400 cases per million per year.

Of the 780 non-white miners examined, observed deaths (72) to September 1968 were actually less than the total expected deaths (90) due to a deficit in heart disease among the miners; lung cancer incidence was not significantly different from the control group. However more recent data available to the BEIR committee in 1972, included additional lung cancer cases, and a linear response-exposure relationship could be demonstrated for the non-white miners as well. These data are also plotted in Figure 1.

Effect of Smoking

Most of the Indians were non-smokers, or smoked only the odd cigarette per day, and it was tempting to interpret the relative slopes of these two exposure-response curves as indicating the enhanced risk of radon exposure amongst those who smoke. Wagoner et al [44] have warned against this. Still more recent data (to December 1973) show 11 lung cancer deaths among the Indian miners against an expected 2.58 and their distribution in the cumulative exposure categories has raised the slope of the non-white exposure-response curve significantly, though it does not yet equal that for the white miners. See Figure 1.

The relationship between cigarette smoking and lung cancer induction by exposure to radon daughters is still not clear, but smoking alone cannot account for more than a fraction of the total number of cancers in uranium miners. Evidence is now accumulating that smoking may act as a "promoter"; that is, it accelerates the induction of the lung cancer after the first exposure to the carcinogen proper. This is supported by the observation that the latent period for the induction of cancer appears to be six or seven years less for smokers than for non-smokers. The average induction period amongst the Erz Mountain miners who did not smoke cigarettes was 20 to 21 years, and is 19 years among the American Indians, compared with an average of 14 years among the smoking US miners (Archer et al [45], Wagoner [44].

Absolute Risk Coefficients

Lung cancer absolute risk coefficients for exposure to radon daughters estimated from the US and other studies are as follows:

US white uranium miners $\quad 3.2 \times 10^{-6}$ per WLM year (BEIR [37])

Newfoundland fluorspar
miners $\quad 8.0 \times 10^{-6}$ per WLM year (BEIR [37])

Czechoslovak uranium
miners $\quad 8.0 \times 10^{-6}$ per WLM year (Sevc et al [46])

(assuming the mean period of risk for the exposed group is 20 years).

Swedish non-uranium
miners $\quad 3.4 \times 10^{-6}$ per WLM year (Snihs [18])

Although these risk coefficients may not be strictly comparable because of differences in smoking habits - the fluospar miners are said to be heavier smokers than the US miners and no information is given on the Czech or Swedish miners - age pattern of mining experience and the length of the observation of the groups compared with the total period at risk following radon exposure, they do appear remarkably consistent and it is tempting to take an average. The mean risk coefficient is 5 or 6×10^{-6} per WLM year though if properly weighted, a lower value would probably be a more justifiable representation of the available data.

If the mean value for the period over which the risk remains elevated following exposure is taken as 30 years, the absolute risk coefficient becomes about 200×10^{-6} per WLM. This is 200×10^{-6} per rad if a dose conversion factor of 1 rad/WLM is assumed, or 400×10^{-6} per rad if, as BEIR suggests, 0.5 rad/WLM is a more appropriate factor for miners. BEIR further suggests an RBE of 10 for α irradiation of the lung which would give a risk coefficient of either 20×10^{-6} per rem or 40×10^{-6} per rem.

These risk coefficients are not inconsistent with those estimated from lung cancer induction in spondylitic patients irradiated with X-rays and in the survivors of Hiroshima irradiated with both neutrons and γ-rays (UNSCEAR [10], BEIR [37]).

The risk coefficients are also compatible with the natural incidence of lung cancer in man. If the average radon level inside dwellings in the USA is of the order of 0.002 WL (ie about 0.5 pCi/l with F = 0.5), continuous exposure at this level would be an exposure rate of about 0.1 WLM/y. With a risk coefficient of 200×10^{-6} per WLM, natural radon background could account for some 20 cases of lung cancer per year per 10^6 people. Annual incidence of lung cancer in the USA among non-smokers is about 130 per million, about half of which may be primary lung cancers originating in the bronchi.

Protection Standards for Exposure to Radon Daughters

On the basis of the epidemiological studies a number of countries are adopting a maximum permissible exposure rate to radon daughters for miners of 4 WLM/y. This may be justified on two grounds:

- Assuming that the observed relationship between exposure and risk is linear to the lowest exposures and has no threshold, the risk associated with 4 WLM/y is "acceptable" compared both with other risks to which miners are exposed, and with risks found acceptable in other industries; and

- The epidemiological studies have shown no significant excess of lung cancers in miners receiving exposures of 120 WLM or less. At the 4 WLM/y limit, it would take 30 years to accumulate an exposure of 120 WLM.

With an absolute risk coefficient of 200×10^{-6} per WLM, the lung cancer risk associated with 4 WLM/y is 800×10^{-6} per year, or 800 deaths per million per year (d/M/y). This would be 400 d/M/y on the basis of experience with the Colorado uranium miners and would of course be much lower still if the exposure-response relationship were curvilinear at low exposures. This risk may be compared with the risk associated with other occupational radiological protection standards and with fatality rates in other industries.

The absolute risk for the induction of all types of cancers associated with whole body irradiation is assumed to be about 6×10^{-6} per rem per year (BEIR [37]). (The fatal cancer risk rate may be somewhat less than this.) Assuming a 30 year mean period of elevated risk following irradiation, the corresponding lifetime risk becomes about 200×10^{-6} per rem. The risk associated with the occupational maximum permissible whole body dose rate of 5 rem/y is thus 1000 d/M/y which is consistent with the assessed risk of exposure to 4 WLM/y.

Fatality rates due to accidents and occupational disease vary widely from one industry to another. The following figures are taken from a recent survey by Pochin [47]:

Industry	Fatality Rate d/M/y
Trades, light manufacturing	50
Metal manufacture	120
Construction	300 to 800
Quarrying	400
Mining	600 to 1000

In some specialist industries (deepsea fishing, commercial flying, and occupations involving exposure to toxic materials) the rates may be considerably in excess of 1000 d/M/y. In the particular case of pneumonoconiosis and silicosis amongst miners, fatality rates as high as 5000 d/M/y were being reported as late as 1970. On the basis of the linear hypothesis, the proposed standard of 4 WLM/y would therefore appear to place uranium mining among the more hazardous industries in terms of potential risk of lung cancer induction. It should be remembered that the radon daughter exposure standard is a limit and current radiological protection philosophy requires that operational exposure rates be kept as far below this limit as is reasonably achievable. Actual mean exposure rates to uranium miners should therefore be less than 4 WLM/y. This should certainly be achievable in open cut mines.

It is also unlikely that many uranium miners will continue mining uranium for 30 years, the period required to accumulate 120 WLM at the maximum permissible rate of exposure. It is true that no excess lung cancers have been demonstrated in miners receiving less than 120 WLM and it is possible that an "effective" threshold exists for the induction of lung cancer by alpha irradiation. There is some evidence to support this.

A cumulative exposure of 120 WLM in miners would be about 60 rads, assuming, with BEIR, a dose conversion factor of 0.5 rad/WLM. The lowest cumulative dose at which lung cancer has been observed in animal experiments involving the inhalation of alpha-emitters is 70 rads. This was observed in rats exposed to a sodium chloride aerosol containing ^{210}Po; the absorbed dose was assessed by averaging the emitted alpha energy over the mass of the lung. The lowest beta dose associated with an observed increase in lung cancer in rats is 600 rads. This work is reviewed by Bair [48]. Bair is not prepared to accept that these and other experiments demonstrate the existence of a true threshold, but he points out that it is possible that at low doses of radiation, the latent period for the manifestation of a cancer could exceed the normal life-span of the individual. One could then have an "effective" or "practical" threshold dose. Bair cites his own work on dogs which had inhaled ^{239}Pu to support this possibility. The BEIR committee states that an inverse relationship between cumulative exposure and the latent period for cancer after initial exposure, is observable in the US Colorado data but that the effect is not very striking at the present time [37].

6. CONCLUSION

Although the consistency of the above observations appear to establish beyond doubt a relationship between radon exposure and lung cancer induction, it should be pointed out that the reality of a direct casual relationship has been questioned.

There is little - until recently some would say no - evidence from animal experiments that radon and its decay products can induce lung cancer. A brief review of these experiments is given by Morken [49]. In 1973 he said that though work with other alpha-emitting isotopes has been successful in producing neoplasia in the lung, "a large number of animal experiments over the past fifty years have been unable to confirm that radon or its decay-products can cause lung cancer." Very high doses of radon daughter products have been given to mice, dogs and rats and while extensive injury of the bronchial tree was apparent, the lesions did not lead to bronchial cancer. Recent studies by French workers however disprove this. Chameaud and his colleagues [50] have been able to induce bronchopulmonary cancers in rats due to the inhalation of radon and its daughters and have demonstrated a definite increase in the frequency of cancer induction as a function of cumulative exposure. The risk coefficient (initial slope) for bronchogenic cancers in the rats is found to be about 65×10^{-6} per WLM; and for both bronchogenic and pulmonary cancers, about 180×10^{-6} per WLM.

Because of the difficulty in obtaining experimental confirmation of the carcinogenicity of radon daughters, and because of what appeared to be a clear enhancement of the lung cancer risk in humans by smoking, it has been thought that other toxic constituents of a mine atmosphere might, either acting alone, or more probably acting synergistically with the alpha irradiation, play a part in the induction of lung cancer in miners. The epidemiological evidence appears to be against this. Underground mining does not in itself lead to an increase in lung cancer risk; the risk only appears heightened in those mines where high levels of radon occur regardless of the mineral mined. Nevertheless some non-radioactive constituents of the mine atmosphere are potential carcinogens and a knowledge of such components as heavy metal and mineral dusts (including the long lived alpha emitters in the uranium decay chain) diesel fumes and the by-products of explosives would appear necessary for a complete description of the toxicological characteristics of a mine atmosphere.

REFERENCES

[1]   Tanner, A.B. Radon migration in the ground: A review.
      The Natural Radiation Environment, Adams, J.A.S. and Lowder,
      W.M. (Editors), University of Chicago Press (1964).

[2]   Megumi, K. and T. Mamuro. Radon and thoron exhalation from
      the ground. Journal of Geophysical Research, 78, 1804 (1973).

[3]   Kraner, H.W., G.L. Schroeder and R.D. Evans. Measurements
      of the effects of atmospheric variables on radon-222 flux
      and soil-gas concentrations. The Natural Radiation Environment,
      Adams, J.A.S. and Lowder, W.M. (Editors), University of
      Chicago Press (1964).

[4]   Harley, J.H. Environmental radon. Noble Gases, Stanley, R.E.
      and Moghissi, A.A. (Editors) CONF-730915, US Environmental
      Protection Agency (1973).

[5]   Wilkening, M.H., W.E. Clements and D. Stanley. Radon-222
      flux measurements in widely separated regions. Natural
      Radiation Environment II, Adams, J.A.S., Lowder, W.M.
      and Gesell, T.F. (Editors) CONF-720805, Rice University,
      Houston, Tex. (USA) (1972).

[6]   Israel, H. Radioactivity of the atmosphere. Compendium of
      Meteorology, Malone, T.F. (Editor) American Meteorological
      Society (1951).

[7]   IAEA. Radon in uranium mining. Proceedings of a Panel,
      Washington, D.C., 4-7 September 1973. IAEA, Vienna (1975).

[8]   Sears, M.B., Blanco, R.E., Dahlmar, R.C., Hill, G.S., Ryan,
      A.D. and Witherspoon, J.P. Correlation of radioactive waste
      treatment costs and the environmental impact of waste
      effluents in the nuclear fuel cycle for use in establishing
      "as low as practicable" guides - Milling of uranium ores.
      ORNL-TM-4903, Vol. 1 p. 144 (1975).

[9]   Swift, J.J., Hardin, J.M. and Calley, H.W. Potential
      radiological impact of airborne releases and direct gamma
      radiation to individuals living near inactive uranium
      mill tailings piles. U.S. Environmental Protection Agency
      (1976).

[10] UNSCEAR Ionizing radiation: Levels and effects Vol. 1: Levels.
      United Nations (1972).

[11] Barton, C.J., Moore, R.E. and Rohmer, P.S. Contribution of
      radon in natural gas to the dose from airborne radon-daughters
      in homes. Noble Gases, Stanley, R.E. and Moghissi, A.A.
      (Editors) CONF-730915 US Environmental Protection Agency
      (1973).

[12] Gesell, T.F. Some radiological aspects of radon-222 in
      liquefied petroleum gas. Ibid (1973).

[13] Johnson, R.H. Bernhart, D.B. Nikon, N.S. and Calley, H.W.,
      Assessment of potential radiological health effects from
      radon in natural gas. US Environmental Protection Agency
      (1973).

[14] Gemesi, J. Szy, D. and Toth, A. Radon-222 content in the internal atmosphere of Hungarian residential buildings. Natural Radiation Environment II. Ibid (1972).

[15] George, A.C. Indoor and outdoor measurements on natural radon and radon decay products in New York City air. Natural Radiation Environment II. Ibid (1972).

[16] Holaday, D.A. and Doyle, H.N. Environmental studies in uranium mines. Proc. Symposium on Radiological Health and Safety in Mining and Milling of Nuclear Materials, IAEA, Vienna (1964).

[17] Schraub, A. Introductory remarks concerning health and safety in Uranium mines. Proc. Symposium on Radiological Health and Safety in Mining and Milling of Nuclear Materials. Ibid (1964)

[18] Snihs, J.O. The approach to radon problems in non-uranium mines in Sweden. Proc. Third International Congress of the International Radiation Protection Association, CONF-730907, USAEC (1974).

[19] Strong, J.C. Laidlow, A.J. and O'Riordan, M.C. Radon and its daughters in various British mines NRPB-R39, UK National Radiation Protection Board (1975).

[20] Federal Radiation Council Guidance for the control of radiation hazards in uranium mining Report No. 8 Revised US Federal Radiation Council (1967).

[21] Raabe, O.G. Concerning the interactions that occur between radon decay products and aerosols. Health Phys. $\underline{17}$, 177 (1969).

[22] Jonassen, N. and McLaughlin, J.P. The effect of RaB recoils losses on radon daughter measurements. Health Phys. $\underline{30}$, 234 (1976).

[23] IAEA Inhalation risks from radioactive contaminants. Technical Report Series No. 142, IAEA, Vienna (1973).

[24] Holaday, D.A. Evolution and control of radon daughter hazards in uranium mines. NIOSH Technical Information. US Dept. Health, Education and Welfare (1974).

[25] Blanc, D., Fontan, J., Chapuis, A., Billard, F., Madelaine, G. and Pradel, J. Dosage du radon et de ses descendants dans une mine d'uranium. Répartition granulométrique des aérosol radioactifs. Assessment of Airborne Radioactivity, IAEA, Vienna (1967).

[26] Fusamura, N., Kurosawa, R. and Misawa, H. The measurement of radioactive gas and dust in uranium mines. Health Phys. $\underline{10}$, 909 (1964).

[27] Chamberlain, A.C. and Dyson, E. The dose to the trachea and bronchi from the decay products of radon and thoron. AERE Report HP/R 1737, UKAEA (1956).

[28] Jacobi, W. Activity and potential α-energy of ^{222}radon - and ^{222}radon-daughters in different air atmospheres. Health Phys. $\underline{22}$, 441 (1972).

[29] George, A.C. and Hinchcliffe, L. Measurement of uncombined radon daughters in uranium mines. Health Phys. 23, 791 (1972).

[30] Altshuler, B., Nelson, N. and Kuschner, M. Estimation of lung tissue dose from the inhalation of radon and its daughters. Health Phys. 10, 1137 (1964).

[31] Jacobi, W. The dose to the human respiratory tract by inhalation of short-lived ^{222}Rn- and ^{220}Rn- decay products. Health Phys. 10, 1163 (1964).

[32] Jacobi, W. Relations between the inhaled potential α-energy of ^{222}Rn- and ^{220}Rn-daughters and the absorbed α-energy in the bronchial and pulmonary region. Health Phys. 23, 3 (1972).

[33] Haque, A.K.M.M. and Collinson, A.J.L. Radiation dose to the respiratory system due to radon and its daughter products. Health Phys. 13, 431 (1967).

[34] Harley, N.H. and Pasternack, B.S. Alpha absorption measurements applied to lung dose from radon daughters. Health Phys. 23, 771 (1972).

[35] Parker, H.M. The dilemma of lung dosimetry. Health Phys. 16, 553 (1969).

[36] Gastineau, R.M., Walsh, P.J. and Underwood, N. Thickness of bronchial epithelium with relation to exposure to radon. Health Phys. 23, 857 (1972).

[37] BEIR The effects on populations of exposure to low levels of ionizing radiation. Report of the Advisory Committee on the Biological Effects of Ionizing Radiations, National Academy of Sciences. National Research Council (1972).

[38] Morken, D.A. The relation of lung dose rate to working level. Health Phys. 16, 796 (1969).

[39] Harley, J.H. and Harley N.H. Permissible levels for occupational exposure to radon daughters. Health Phys. 27, 593 (1974).

[40] Jacobi, W. Relation between cumulative exposure to radon-daughters, lung dose, and cancer risk. Noble Gases CONF-730915 ibid (1973).

[41] Jacobi, W. Lung cancer risk from inhalation of radon-222 products. Biophysik, 10, 103 (1973).

[42] Lundin, F.E., Wagoner, J.K. and Archer, V.E. Radon daughter exposure and respiratory cancer, quantitative and temporal aspects. National Institute for Occupational Safety and Health, Joint Monograph No. 1, US Department of Health, Education and Welfare (1971).

[43] Archer, V.E. and Wagoner, J.K. Lung cancer among uranium miners in the United States. Health Phys., 25, 351 (1973).

[44] Wagoner, J.K., Archer, V.E. and Gillam, J.D. Mortality of American Indian miners. Proc. XI International Cancer Congress, Florence, 1974; Vol. 3. Cancer Epidemiology, Environmental Factors. Exerpta Medica, Amsterdam, Elsevier, N.Y. (1975).

[45] Archer, V.E., Wagoner, J.K., Hyg, S.D. and Lundin, F.E. Uranium mining and cigarette smoking effects on man. Journal of Medicine, 15, 204 (1973).

[46] Sevc, J. and Placek, V. Lung cancer risk in relation to long-term exposure to radon daughters. Health Physics Problems of International Contamination, Proc. 2. Europ. Rad. Protection Congress, Budapest (1972).

[47] Pochin, E.E. Occupational and other fatality rates. Community Health, 6, 2 (1974).

[48] Bair, W.J. Inhalation of radionuclides and carcinogenesis. Inhalation Carcinogenesis. AEC Symposium Series 18. CONF-691001, USAEC (1970).

[49] Morken, D.A. The biological effects of radon on the lung. Noble Gases CONF-730915 ibid (1973).

[50] Chameaud, J., Perraud, R., Masse, R., Nenot, J.C. and Lafuma J. Cancers du poumon provoqués chez le rat par le radon et ses descendants a diverses concentrations. Int. Symp. on Biological Effects of Low Level Radiation Pertinent to Protection of Man and his Environment. Vol. II. IAEA, ibid (1976).

INTERPRETATION OF MEASUREMENTS IN URANIUM MINES:
DOSE EVALUATION AND BIOMEDICAL ASPECTS

W. Jacobi

Institut für Strahlenschutz
Gesellschaft für Strahlen- und Umweltforschung mbH
München - Neuherberg (Federal Republic of Germany)

INTRODUCTORY REMARKS

Measurements in a radiation protection programme have in general the following principal objectives. They should guarantee that the level of radiological safety meets appropriate standards, as they are recommended by the ICRP and layed down in national regulations. In addition they should give an early warning when unexpected changes of the radiation level are occurring. Finally these measurements should provide information how the protection system can be improved and the exposure reduced.

The appropriate basic standards of radiation protection are given by the system of dose limits to organs and tissues of the human body. Consequently it is the ideal objective of measurements in a radiation protection programme to allow an interpretation which gives the mean dose to relevant organs and tissues.

However, these organ doses cannot be measured directly and the relationship between measurable quantities and the organ doses are complex and in many cases still uncertain. We are confronted with this interpretation problem not only in monitoring programmes for the inhalation or ingestion of radionuclides but also in case of external radiation measurements. This means that the final objective of a radiation protection programme is in practice only achievable by the introduction of simplifying interpretation models. These models should reach the form of standardized methods for the practical application in radiation protection.

In uranium mines the main radiation hazard to workers is the irradiation of the lung by inhalation of the short-lived daughters of ^{222}Rn (Radon). In this case the choice of practicable and reasonable models for the interpretation of measurements of Rn and its daughters in air meets special problems: The complexicity of the geometry and the activity atmosphere in a mine, the technical difficulties of personnel Radon-dosimetry, and the problems involved in the estimation of lung dose from inhaled Rn-daughters.

In this paper the problems in the interpretation of measurements in terms of lung dose and the corresponding biomedical aspects are discussed and the existing concepts of interpretation models are outlined. The interpretation precedes in the three steps which are shown in the block diagram in figure 1.

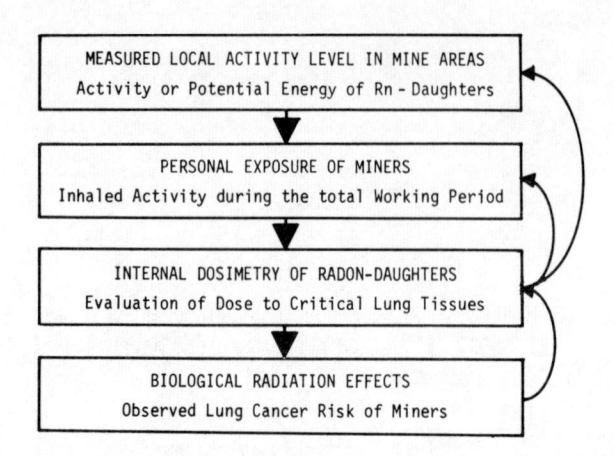

```
┌─────────────────────────────────────────────┐
│     MEASURED LOCAL ACTIVITY LEVEL IN MINE AREAS │
│     Activity or Potential Energy of Rn-Daughters│
└─────────────────────────────────────────────┘
                      ▼
┌─────────────────────────────────────────────┐
│          PERSONAL EXPOSURE OF MINERS        │
│   Inhaled Activity during the total Working Period │
└─────────────────────────────────────────────┘
                      ▼
┌─────────────────────────────────────────────┐
│       INTERNAL DOSIMETRY OF RADON-DAUGHTERS │
│       Evaluation of Dose to Critical Lung Tissues │
└─────────────────────────────────────────────┘
                      ▼
┌─────────────────────────────────────────────┐
│          BIOLOGICAL RADIATION EFFECTS       │
│       Observed Lung Cancer Risk of Miners   │
└─────────────────────────────────────────────┘
```

Figure 1: Sequence of steps in the interpretation of measured air activity in U-mines

ESTIMATION OF THE INDIVIDUAL RADON-EXPOSURE

The activity level in the air of a mine varies considerably with time and place. It depends on the U-content of the cutted ore, its emanation efficiency, the working methods and the working activity, the mine geometry and the residence time of the air in the mine areas and its variation with ventilation. Simplified theoretical models for the estimation of the resulting activity distribution in a mine have been proposed [1-5]. However, the applicability of these theoretical models is restricted and must be checked by measurements.

The first step in the interpretation of these measurements concerns the correlation between the measured values of Rn- or Rn-daughter activity in the mine air and the cumulated individual Rn-exposure of miners, which can be defined as the total activity or potential energy inhaled during the working period. The accuracy of this correlation depends on the type of measurements.

Continuous measuring methods.
Taking into account the variation of the activity level with time and place in a mine the best information about the individual exposure can be obtained by continuous, personnel air-samplers. This method would be adequate to the methods of personnel dosimetry for external radiation.

In the last years several types of personnel Rn-monitors have been developed which make use of the filter method and a built-in integrating α-detector, for example an α-track etch foil. Under the normal conditions in mines the measured number of α-particles is with sufficient accuracy directly proportional to the integral α-activity or α-energy of the Rn-daughters in the sampled air, integrated over the whole sampling period of normally 8 hours per day.

For the interpretation in terms of inhaled potential activity or α-energy two assumptions have to be made:

(1) The sampling is representative of the breathing air of the worker
(2) The breathing rate of the worker is constant and corresponds to the breathing rate v_b = 20 l/min of the ICRP-reference worker.

With these assumptions the total inhalation intake I of the worker during the whole working period T can be derived from the measured number N_α of α-particles by the formula

$$I(T) = k \cdot \frac{v_b}{v_s} \cdot N_\alpha$$

where v_s means the sampling rate and k the calibration factor of the instrument.

Instead of personnel Rn-monitors local monitors can be used, which sample and measure continuously the air activity at representative places of working areas in the mine. A battery-operated, integrating instrument of this type has been developed by us and is described elsewhere [6, 7]. The instrument has a direct reading of the detector response, which is proportional to the time integral of the potential α-energy concentration in the sampled air.

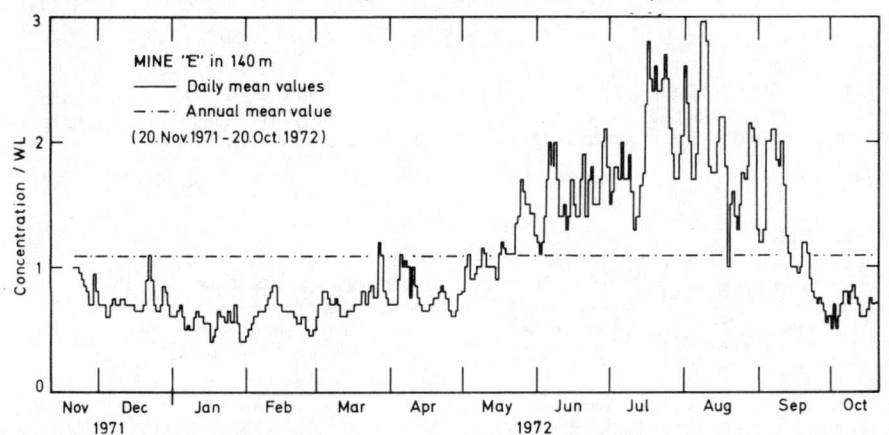

Figure 2: Variation of the continuously measured Rn-daughter activity in a working area of a fluorspar mine in East Bavaria (Nov. 1971 - Oct. 1972)

Figure 2 shows for example the daily mean values of Rn-daughter activity in WL-units, which were measured with this instrument in the working area of a fluorspar mine in Bavaria. This figure gives an impression about the variation of the daily exposures at this place during one year.

Spot check-measurements.
Currently the air monitoring in most mines restricts to the measurement of spot checks of small air samples which are taken more or less frequently at characteristic places in the mine. Either the activity of the Radon-gas is measured afterwards with a scintillation flask or an ionization-chamber, or the Rn-daughters are sampled with the filter method. In the so-called instant WL-meters the filter activity is directly measured and indicated in WL-units, or the KUSNETZ-method with its variations is used.

The main problem in the interpretation of such spot-check-measurements concerns the question: Are the measured activity values in such short-time samples representative for the long-term individual exposure of the miners? Comparative measurements with continuously operating instruments, like personnel air-samplers, are necessary to answer this question. From my knowledge such simultaneous measurements are so far not available. At the present time we have no other choice than to answer the question mentioned above with yes. But we must admit that this assumption can lead in some cases to large errors. In future investigations this assumption should be thoroughly tested, to derive a more realistic and safe model for the interpretation of such spot-check measurements in terms of the individual inhalation exposure of miners.

The fraction of free Rn-daughter atoms.
The routine monitoring in U-mines is restricted to the measurement of the activity or potential α-energy of Radon and its daughters in the air. An additional parameter, which has an influence on lung dose, is the physical state of the inhaled radioactive atoms. After their formation the daughter atoms are rapidly attaching

to dust particles in the air. Therefore, most of the inhaled activity is carried by such particles. However, some daughter atoms, especially RaA-atoms, are inhaled in form of free, unattached atoms, whose deposition pattern in the lung is quite different from that of the aerosol fraction. For the interpretation of activity measurements in terms of lung dose, additional information about the fraction of free daughter atoms is required.

From the coagulation theory follows that the activity fraction f_a of free RaA-atoms decreases with increasing dust concentration n corresponding to the formula [4, 8]:

$$f_a (\%) \approx \frac{100}{1 + \frac{k \cdot n}{\lambda_{RaA} + \lambda_v}}$$

In this equation means λ_{RaA} the decay constant of RaA, λ_v the air exchange rate in the considered air volume by ventilation, and k the coagulation constant of free RaA-atoms with aerosol particles, which is about 0,01 cm³/h.

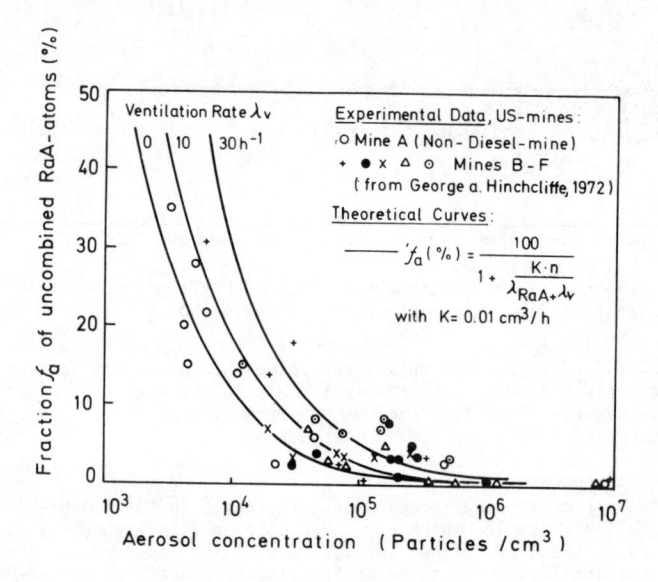

Figure 3: Relationship between the unattached fraction of RaA-atoms and the dust concentration in uranium mines

Comparative, simultaneous measurements of the free RaA-fraction and dust concentration have been performed by GEORGE and HINCHCLIFFE [9] in 5 U-mines in USA. In figure 3 the measured values are compared with the theoretical curves for different ventilation constants λ_v. Within the limits of errors the agreement between experimental data and the theory is satisfactory. It follows from this figure that in areas with a dust concentration of more than about 10⁴ particles/cm³ the free RaA-fraction is rather small and lower than 10%. Such dust concentration can be expected in most working areas, where ore is cutted or loaded. Only in areas with low dust content, especially in mines with no diesel engines, sometimes rather high f_a-values up to 30 - 50% were measured.

For RaB and RaC the fraction of free atoms is lower than for RaA [9, 10]. This means that the free atom-fraction f_p of the total potential α-energy of Rn-daughters will be smaller than the activity fraction f_a of free RaA-atoms alone.

RELATION BETWEEN INHALED ACTIVITY AND LUNG DOSE

The second phase in the interpretation and judgement of measurements in U-mines concerns the estimation of the inhalation dose to the lung, in which the short-lived Rn-daughters are deposited and enriched. We have several objectives in mind:

(1) We have to confirm that the basic dose limit of 15 rem per year for the occupational exposure of the lung will not be exceeded.
(2) We want to know the main physical parameters of the mine atmosphere which have an influence on the lung dose distribution.

On the basis of this knowledge we can judge retrospectively the suitability of the air monitoring programme and of the measured quantities. In addition, information for the improvement of the ventilation system can be obtained.

Principles of lung dose evaluation.
In figure 4 the steps involved in the evaluation of the lung dose are outlined. At first the deposition of Rn-daughters in the different lung regions must be considered. After their deposition the radioactive atoms are involved in the translocation and clearance processes which are running off in the deposition region. Due to these clearance mechanisms only a fraction of the deposited activity is retained in the lung and decays there. From the resulting number of decays and the decay energy the absorbed energy in the considered lung region can be estimated.

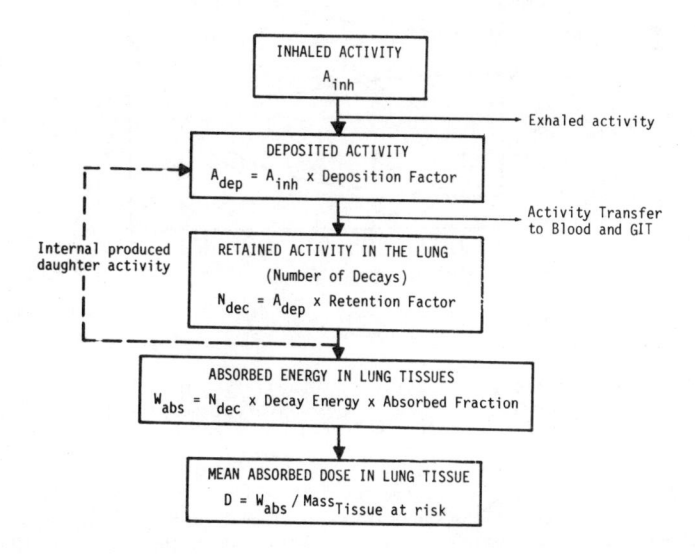

```
           ┌─────────────────────────┐
           │    INHALED ACTIVITY     │
           │         A_inh           │
           └─────────────────────────┘
                       │         ─────► Exhaled activity
                       ▼
           ┌─────────────────────────┐
    ┌ ─ ─ ►│    DEPOSITED ACTIVITY   │
    │      │ A_dep = A_inh x Deposition Factor │
    │      └─────────────────────────┘
    │                  │         ─────► Activity Transfer
    │                  ▼                 to Blood and GIT
    │      ┌─────────────────────────┐
Internal produced │ RETAINED ACTIVITY IN THE LUNG │
daughter activity │    (Number of Decays)   │
    │      │ N_dec = A_dep x Retention Factor │
    └ ─ ─ ─└─────────────────────────┘
                       ▼
           ┌─────────────────────────┐
           │ ABSORBED ENERGY IN LUNG TISSUES │
           │ W_abs = N_dec x Decay Energy x Absorbed Fraction │
           └─────────────────────────┘
                       ▼
           ┌─────────────────────────┐
           │ MEAN ABSORBED DOSE IN LUNG TISSUE │
           │  D = W_abs / Mass_Tissue at risk │
           └─────────────────────────┘
```

Figure 4: Steps involved in the lung dose evaluation of inhaled Radon-daughters

The inhalation of Rn-daughters leads to a rather inhomogeneous distribution of the activity and the absorbed energy in the lung, especially in the tracheobronchial region (TB-region). However, in the national regulations which are based on the recommendations of ICRP only the mean lung dose is limited. On the other side, lung cancer among U-miners is presumed to originate in the bronchial region.

On the basis of this experience it seems reasonable to inquire in addition for the mean dose to the total bronchial region and the basal cells of the bronchial epithelium, where the bronchogenic lung cancer is originating. This mean dose is given by the ratio of the absorbed energy to the mass of the considered target region in the lung. In the following a model for the interpretation of measurements in terms of dose and the resulting dose/exposure-ratios are given.

Deposition and retention model for inhaled Rn-daughters.

For purposes of radiation protection a task group of ICRP has developed a new model for the deposition and retention of inhaled radioactive aerosols in the lung [11]. This model will be used for the future ICRP-recommendations on limits of intake by inhalation. It seems reasonable to apply this model for the internal dosimetry of radon daughters.

For nose breathing the resulting effective deposition factor of Rn-daughters in the naseopharyngeal (NP), the tracheobronchial (TB) and pulmonary (P) region of the lung are given in figure 5 as function of the free-atom fraction f of the inhaled activity.

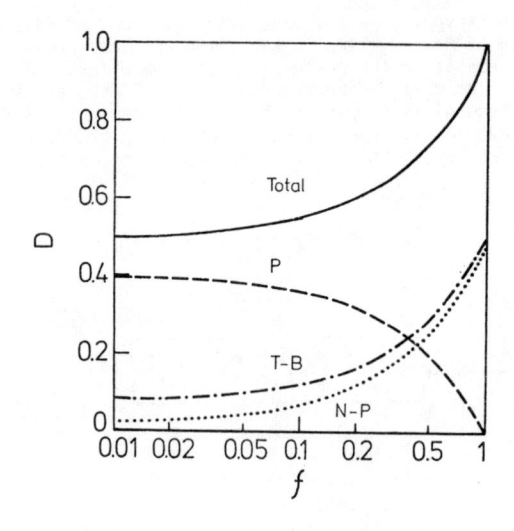

Figure 5: Effective deposition probability D of a Rn- or Tn-daughter nuclide in the different respiratory regions as function of its free-atoms fraction f in the inhaled air

For low f-values the deposition pattern is determined by the radioactive aerosol particles, which are mainly deposited in the deep lung (P-region). Only about 8% of these particles are deposited in the bronchi (TB-region). Under these conditions results a deposition probability of about 50% in the total lung. This value lies at the upper limit of the measured total deposition values for mouth breathing of Rn-daughters in different mine atmospheres. With increasing fraction f of free atoms or ions the deposition in the bronchial region increases, because the free, unattached atoms are efficiently removed in the upper respiratory airways. Consequently the deposition in the pulmonary region decreases with increasing free atom fraction.

Animal experiments and retention studies in humans indicate that the elimination rate of deposited Rn-daughters from the lung to the blood is significantly higher than the elimination rate of the dust particles, to which the radioactive atoms are attached. This suggests that the adsorbed radioactive atoms must be able to leave their carrier particles by desorption or solution soon after their deposition. This means that the retention model for Rn-daughters should be similar to that for inhaled soluble materials.

Therefore, the proposed new ICRP-lung model for soluble aerosols (retention class D) seems to be a reasonable basis to estimate the retention of Rn-daughters in the different lung regions. In figure 6 the compartment diagram of this retention model is given, showing the different pathways (a - f) of elimination. For each pathway the biological half-life time T and the fraction f of the deposited material which is eliminated on this pathway is given. In the tracheobronchial region (TB) the deposited radioactive atoms or particles are transported upwards in the mucus sheet of the bronchi by the action of ciliary movement, and are swallowed after reaching the gut. Another fraction is directly absorbed into the blood similar like in the pulmonary region (P-region).

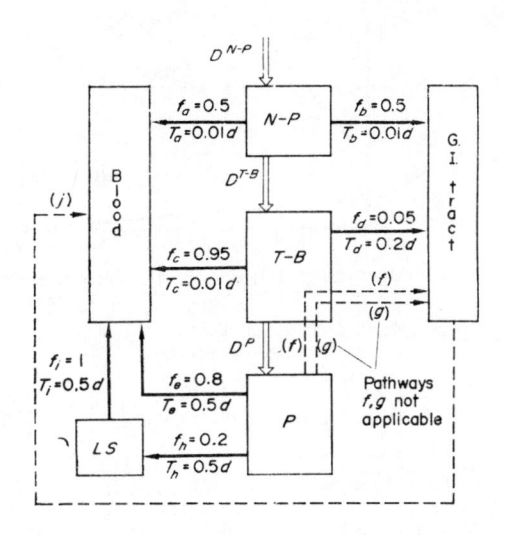

Figure 6: The new ICRP-lung retention model for soluble aerosols (class D)

In the pulmonary region the biological half-life time of 0,5 days is large compared with the radioactive half-life time of the short-lived ^{222}Rn-daughters. This means that nearly 100% of the Rn-daughter activity deposited there will decay in this region. In the tracheobronchial region the elimination rate is considerably higher. For this region the fraction of the deposited potential energy wich is absorbed in this region results to about 80% for RaA, 40% for RaB and 50% for RaC.

Dose to the bronchial and pulmonary region.

On the basis of the described new ICRP-model for the deposition and retention of aerosols in the human lung the mean α-dose to the different lung regions of this model can be calculated. As mentioned this dose depends on the fraction f of daughter atoms inhaled in the form of free, unattached atoms or ions. For this purpose, in most current studies the activity fraction f_a of free RaA-atoms alone was considered. The dose estimation on the basis of the new lung model however indicates that the deciding parameter is the fraction f_p of the total potential α-energy of the radon-daughter mixture which is inhaled in form of free atoms or ions.

In figure 7 the resulting ratio between the absorbed α-dose in rads and the inhalation exposure in WLM-units is given as function of this parameter f_p for the tracheobronchial (TB) and the pulmonary (P) region of the lung. The curves were calculated with a breathing rate of 20 l/min and a mass of 40 g and 960 g for the TB- and P-region, respectively, as they were recommended by ICRP for the reference worker [12].

Figure 7: Mean α-dose per WLM in the tracheobronchial and pulmonary region by inhalation of ^{222}Rn-daughters as function of the unattached fraction of the potential α-energy; calculated from the new ICRP-lung model, class D

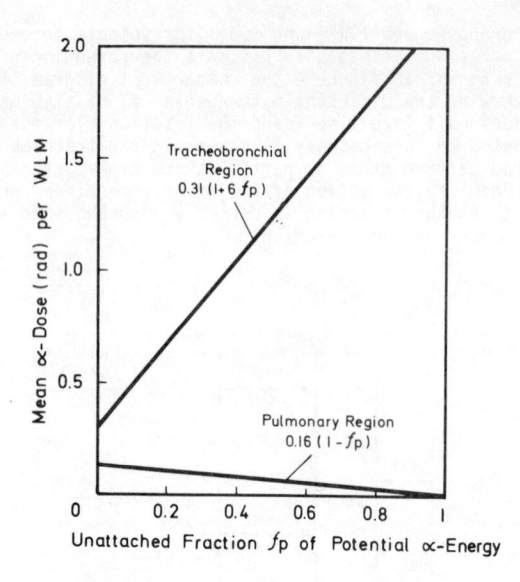

Figure 7 shows that for a low free-atom fraction f_p the mean bronchial dose will be about a factor 2 - 3 higher than the mean pulmonary or lung dose. However, with increasing free-atom fraction f_p the bronchial dose per WLM inhaled increases linear and the pulmonary dose decreases. In either case the bronchial region will be the critical region of the lung, when we assume the same radiosensitivity for both lung regions. The following simple relationships between the α-dose D in rads and the inhalation exposure E in WLM-units can be derived:

Bronchial region $\quad D_{TB} / E \approx 0.31 (1 + 6 f_p)$ rad / WLM

Pulmonary region $\quad D_P / E \approx 0.16 (1 - f_p)$ rad / WLM

As mentioned earlier (see figure 3) the available experimental data indicate that the activity fraction f_a of free RaA-atoms will probably be lower than 10% or 0,1 in most mine areas. The free-atom fraction f_p of the total Rn-daughter mixture will be smaller than this value. Under these conditions a ratio of

$$0.3 - 0.5 \text{ rad} \triangleq 3 - 5 \text{ rem per WLM}$$

should be used for the interpretation of the inhalation exposure of miners in terms of the mean dose to the bronchial region, assuming a quality factor QF = 10 for the α-radiation.

However, it should be kept in mind that the free atom fraction increases with increasing ventilation rate and decreasing aerosol production in a working area. We have estimated the influence of this effect on the rad/WLM-ratio on the basis of a box model, taking into account the removal of dust activity by ventilation and wall deposition [13].

Figure 8:

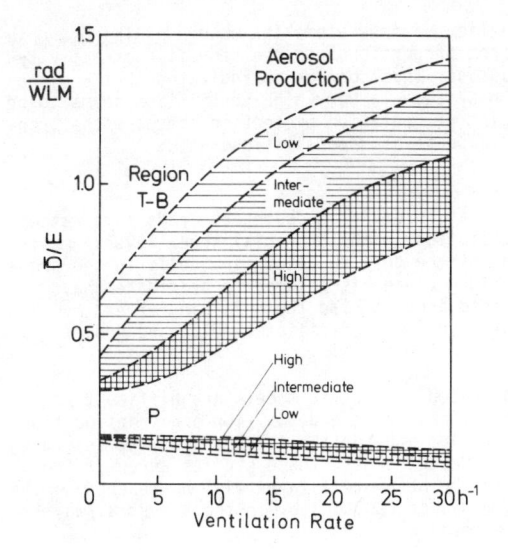

Estimated influence of ventilation rate and aerosol production on the dose / exposure-ratio

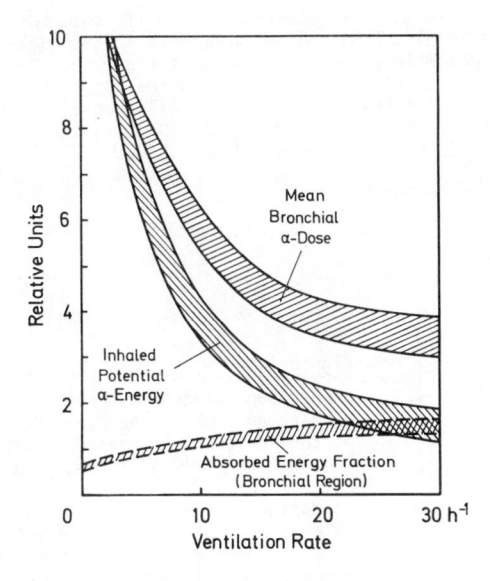

Figure 9:

Estimated relative change of exposure and dose with ventilation in a working area at constant radon and dust production

Figure 8 shows the resulting variation of the rad / WLM-ratio with the ventilation rate λ_V, which should be roughly expected in mine areas with a high, intermediate and low rate of aerosol production. This figure indicates clearly that especially in high-ventilated areas with low dust production the mean bronchial dose can reach about 1 rad or 10 rem per WLM.

This increase of the bronchial dose per WLM with the rate of ventilation is of practical importance. It cancels out to some extent the dose reduction due to the decreasing concentration of Rn-daughters, which we obtain by an enhanced ventilation. Figure 9 shows in a relative scale this expected influence of ventilation on the inhaled potential energy of Rn-daughters in the air and the corresponding mean bronchial dose.

Due to the increasing fraction of free atoms the change of the bronchial dose with increasing ventilation is not proportional to the reduction of activity or potential energy in air by ventilation. This conclusion indicates that - disregarding the costs for the ventilation system - a very high ventilation seems to be also not reasonable from the standpoint of radiation protection, because the bronchial dose reachs a nearly constant value at high ventilation rates.

Dose to the basal cells of the bronchi.

It was already mentioned that the inhalation of Radon-daughters leads to a rather inhomogeneous activity and dose distribution in the bronchial tree. Several estimations of this dose distribution and of the dose to basal stem cells in the bronchial epithelium, where bronchogenic lung cancer is assumed to originate, have been made [14 - 17]; the models take into account also the shielding effect by the mucus sheet and the rather insensitive layer of ciliated cells and goblet cells on the bronchial epithelium.

The results of these complicated dosimetric models are difficult to compare because partly different assumptions for the deposition and translocation of Rn-daughters on the bronchial epithelium and for the thickness of the shielding layer over the basal cells were made. Nevertheless, these studies agree in the conclusion, that the maximum α-dose in the basal cell layer will be reached in the segmental-subsegmental bronchi (lung model of LANDAHL) or in the 4. - 9. airway generation (lung model of WEIBEL), respectively [18].

The calculated absolute dose values are however, quite different, due to the different assumptions in these models. Critical parameters in these models are especially the depth of the basal cell layer below the mucus surface and the uptake and distribution of radon daughters in the mucus sheet and the upper cell layers of the bronchial epithelium. The mean depth of the basal cells ranges probably from about 50 - 100 μm in the trachea and main bronchi, 30 - 60 μm in the segmental-subsegmental bronchi and 20 - 40 μm in the terminal bronchi.

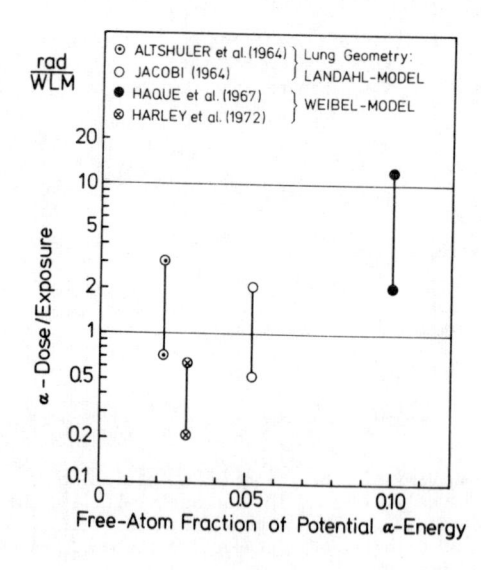

Figure 10:

Comparison between the mean α-doses per WLM to the basal cells of the segmental-subsegmental bronchi, derived from different dosimetric models. The results were normalized to nose breathing at a rate of 20 l/min; the α-dose was averaged over a depth of 30 - 60 μm below the mucus surface

In figure 10 the variation range of the α-dose per WLM in 30 - 60 μm depth in the segmental-subsegmental bronchi is given as it can be derived from the different models. The calculated values are plotted as function of the unattached fraction f_p of the total potential α-energy in the inhaled air and are normalized for nose breathing at a rate of 20 l/min.

The dose/exposure-ratios from the different models cover a range from about 0.2 - 10 rad/WLM. The more recent models [17] indicate, that under normal mining conditions ($f_p < 0.1$) the mean α-dose in the basal cell layer of this critical bronchial region lies probably at the lower end of this range. Under these conditions the reference factor for the conversion of exposure to the mean dose in the basal cell layer of the segmental-subsegmental bronchi should be chosen in the range of 0.3 - 1 rad ≅ 3 - 10 rem per WLM. A comparison with the previously derived conversion factor for the dose to the total bronchial region leads to the conclusion, that the mean α-dose to the basal cell layer in the most exposed bronchial region is probably not considerably higher than the mean α-dose to the total bronchial tree [18].

This conclusion solves one of the main problems in the application of the ICRP-dose limit of 15 rem/year. Taking into account a conversion factor in the range of 0.3-1 rad/WLM, which range covers the results for both reference regions, this dose limit yields an exposure limit of 2 - 5 WLM per year. The exposure limit of 0.3 WL or 4 WLM per year, which has been adopted for uranium miners in USA and other countries falls within the range derived from this dose concept. However, as pointed out, this limit should be reduced in highly ventilated, dust-free working areas, when the fraction of free daughter atoms exceeds considerably 10%.

THE LUNG CANCER RISK OF URANIUM MINERS

The final question in the interpretation of measurements in U-mines concerns the biomedical consequences of the Rn-exposure of miners. Just about hundred years ago - in 1879 - HÄRTING and HESSE [19] have recognized that the specific lung disease among ore miners in the Schneeberg-Joachimsthal region - the so-called "Schneeberger Bergkrankheit" - was identical with lung cancer. In those days nearly 50% of the miners in this region died from lung cancer, whose real cause was unknown.

But only fifteen years ago the epidemiological studies of the US-Public Health Service among U-miners in the Colorado-plateau revealed the evidence of an increased indicence of lung cancer associated with the high Rn-level in these mines [20]. Additional information about the influence of Rn-exposure on lung cancer risk was obtained, especially from studies among U-miners in the CSSR [24] and non-uranium miners in Canada [25], England [26] and Sweden [27]. On the basis of these data the relationship between the Rn-exposure of miners and the expected additional risk of lung cancer can be estimated.

Figure 11 shows the observed excess lung cancer (LC) mortality among the study group of about 3400 white U-miners in the Colorado-plateau [21 - 23]. During the total observation period from 1952 - 1971 about 38 000 person-years at risk have been reached in this group. A significant increase of LC-risk was observed above a comulative exposure of about 100 WLM. From the given 95% confidence limits follows that the data do not allow a quantitative distinction between a linear and a nonlinear relationship between LC-risk and cumulative Rn-exposure in WLM-units. Assuming a linear relationship, a mean annual incidence rate of

$$3{,}2 \pm 0{,}8 \text{ excess LC-cases} / 10^6 \text{ person} \cdot \text{years per WLM}$$

can be derived from the given regression line.

In figure 12 the final data about the LC-risk among U-miners in the CSSR during the obervation period from 1948 - 1973 are presented, which were published recently [24]. In contrary to the US-data for the CSSR-miners only the relative numbers of observed and expected LC-cases, normalized to 1 000 miners are available, but not the absolute number of miners and LC-cases. However, the observation period and the total number of about 35 000 person · years at risk for the CSSR-group are not very different from the data for the US-group. This supports the assumption that the absolute number of U-miners in the CSSR-group is comparable with that in the US-group.

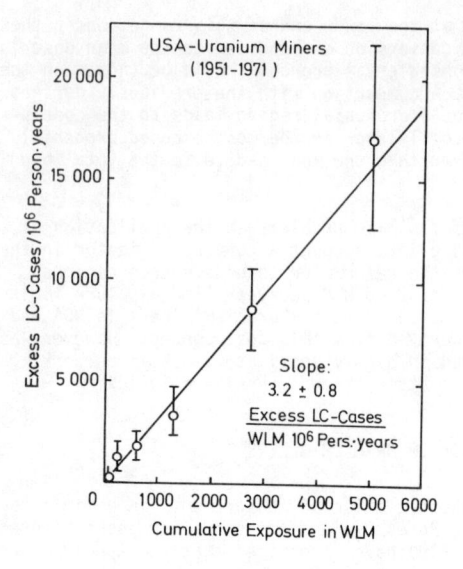

Figure 11:

Observed mean excess incidence rate of lung cancer among uranium miners in USA [23]

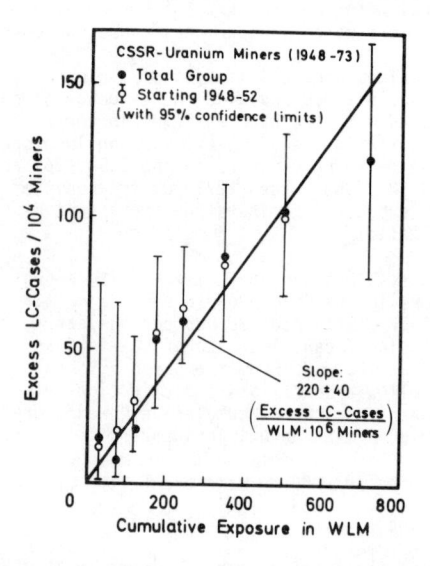

Figure 12:

Observed excess lung cancer incidence among uranium miners in the CSSR [24]

Similar like in the US group a significant increase of the absolute and relative LC-risk among the CSSR-miners was only obtained for an exposure higher than about 100 WLM. However a linear exposure-risk relationship cannot be excluded, leading to a risk coefficient of

220 ± 40 excess LC-cases/10⁶ miners per WLM

for the CSSR-miners, which started underground work in 1948-52. Taking into account the exposure period of 21 - 25 years a mean annual LC-rate of about

10 ± 3 excess LC-cases/10⁶ miners · year per WLM

can be derived, which is about a factor 3 higher than the observed excess rate among the Colorado-miners.

In the following table the observed mean annual rates of LC-risk per WLM of Rn-exposure for all groups of miners are summarized, including non-uranium miners in Canada, Sweden and the United Kingdom. However these risk values consider only the LC-cases observed during the given observation period. As the mean latency of lung cancer among miners is about 20 years, additional cases should be expected in the future years; therefore, the total risk during the life-time of miners will be higher. In the last column of the table this total life-time risk is estimated from the observation period of each group.

Table 1: Observed mean annual rate of lung cancer incidence and

Tabelle 1: Observed mean annual rate of lung cancer incidence and estimated life-time risk per unit of radon-exposure among different groups of miners

Group of miners	Observation Period (years)	Excess LC - incidence per 10^6 Persons	
		Observed mean rate [1] LC-cases / year·WLM	Life-time risk [2] LC-cases / WLM
U-miners (Colorado / USA)	1952 - 71 = 20 years	3.2 ± 0.8	120 ± 50
U-miners (CSSR)	1948 - 73 = 26 years	10 ± 3	300 ± 100
Fluorspar-miners (Newfoundland/Can.)	1952 - 68 = 17 years	2.2	ca. 100
Non-U-miners (Sweden)	1961 - 68 = 8 years	3.4	ca. 150
Iron miners (U.K.)	ca. 15 years	6.0	ca. 200

[1] during the observation period
[2] extrapolated, assuming a mean latency period of 20 years

The resulting mean risk values are in the range from

100 - 300 excess LC-cases/10^6 miners per WLM

Taking into account the difficulties and different methods of Rn-exposure measurement and evaluation in the ascertainment of lung cancer, as well as the different living habits of the studied groups of miners the estimated risk values can be regarded as reasonably consistent. It should be also mentioned that cigarette smoking may have a cocarcinogenic effect on the formation of lung cancer by inhaled Rn-daughters.

Under these circumstances it seems reasonable to propose as reference value for the interpretation of radiation hazards in U-mining and for the assessment of Rn-exposure limits the following risk value:

200 LC-cases/10^6 persons per WLM

Applying a conversion factor of 0,5 rad or 5 rem per WLM for the critical tissue in the bronchial region of the lung this corresponds to

40 LC-cases/10^6 persons per rem

This value is in good agreement with the estimated life-time risk of lung cancer from external neutron irradiation, which was derived from the observations among the atomic bomb survivors in Hiroshima [23].

To judge the practical consequences of this risk estimate it should be emphasized that it was derived on the assumption of a linear exposure-risk relationship from groups of miners which were exposed to high radon levels. For example the mean exposure was about 800 WLM in the US-group and about 300 WLM in the CSSR-group of U-miners. A significant increase of LC-incidence at low exposure - below 100 WLM - could not be ascertained.

At the annual exposure limit of 4 WLM a miner would reach a maximum life-time exposure of 120 - 160 WLM, if he receives the annual limit each year during a working period of 30 - 40 years. If we assume that a linear relationship between exposure and LC-risk is valid for this low exposure range, it results a maximum occupational LC-risk for this worker of about 0,02 - 0,03 or 2 - 3%. This is about 1/10 of the normal cancer risk from other causes observed among the male population in Germany. For the total group of miners the occupational LC-risk will be considerably smaller, because the average exposure in this group is lower than the exposure limit.

FINAL CONCLUSIONS

In this review paper the main problems were described, with which we are confronted in the interpretation of measurements in uranium mines in terms of individual exposure, lung dose and lung cancer risk from radon-daughters. To solve these problems concepts and models of interpretation were outlined. From this analysis the following general conclusions can be drawn:

(1) Uranium miners belong probably to those few groups of radiation workers which are exposed to a relatively high radiation risk. A regular medical supervision of the miners is necessary to ensure an early recognition of radiation effects in the lung.

(2) The monitoring programme in an uranium mine should be adequate to this risk. It should finally guarantee the same level of radiological safety which we have reached in other fields of radiation protection.

(3) To reach this goal methods and techniques for the measurement of Rn and its daughters should be improved; especially instruments for the long-term measurement of the cumulative individual exposure are required.

(4) For the routine monitoring in mines it seems sufficient to measure the total potential α-energy of radon-daughters or the equivalent equilibrium activity of radon, if the fraction of unattached radon-daughters is low. However, additional informations about this fraction and the disequilibrium ratios between radon-daughters in representative working areas of mines are required for the final judgement of the radiotoxicity of the mine atmosphere.

(5) For the practical application in mines standard models for the interpretation of measurements in terms of individual exposure and lung dose should be developed.

Finally it will be necessary to reconsider the current limits for the radon-exposure of miners on the basis of the new system of dose limits which will be recommended by ICRP in the near future.

REFERENCES

[1] HOLADAY, D.A., D.E. RUSHING, R.D. COLEMAN, P.F. WOOLRICH, H.L. KUSNETZ, W.F. BALE: Control of Radon and daughters in uranium mines and calculations of biological effects; Publ. Health Service Publ. No. 494, 1957

[2] WRENN, Mc D.E., J.C. ROSEN, W.R.v. PELT: Steady state solutions for the diffusion equations of Rn-222 daughters; Health Phys. 16 (1969), 647-656

[3] BRESLIN, A.J., A. GEORGE, M. WEINSTEIN: Investigation of the radiological characteristics of uranium mine atmospheres; US Health and Safety Lab. Rep 220

[4] JACOBI, W.: Activity and potential α-energy of Rn-222 and its daughters in different air atmospheres; Health Phys. 22 (1972), 441-450

[5] ROLLE, R.: Radon daughters and age of ventilation air; Health Phys. 23 (1972), 118-120

[6] HAIDER, B., W. JACOBI: Entwicklung von Verfahren und Geräten zur langzeitigen Radon-Überwachung im Bergbau; Research Rep. BMBW-FB-K-72-14, August 1972

[7] HAIDER, B., W. JACOBI: A new monitor for long-term measurement of radon daughter activity in mines; Sympos. on Radiat. Protection in Mining and Milling of Uranium and Thorium; Bordeaux / France, 9.-11. Sept. 1974

[8] RAABE, O.G.: Concerning the interactions that occur between radon decay products and aerosols; Health Phys. 17 (1969), 177-186

[9] GEROGE, A.C., L. HINCHCLIFFE: Measurements of uncombined radon daughters in uranium mines; Health Phys. 23 (1972), 791-803

[10] MERCER, T.T.: Unattached radon decay products in mine air; Health Phys. 28 (1957), 158-161

[11] ICRP-Task Group on Lung Dynamics: Deposition and retention models for internal dosimetry of the human respiratory tract; Health Phys. 12 (1966), 173-207; model revised at the ICRP-meeting in Oxford, 1969

[12] Report of the ICRP-Task Group on Reference Man; ICRP-Publ. 23, Pergamon Press, 1975

[13] JACOBI, W.: Relation between the inhaled potential α-energy of ^{222}Rn- and ^{220}Rn-daughters and the absorbed α-energy in the bronchial and pulmonary region; Health Phys. 23 (1972), 3-11

[14] ALTSHULER, B., N. NELSON, M. KUSCHNER: Estimation of lung tissue dose from the inhalation of radon and daughters; Health Phys. 10 (1964), 1137-1161

[15] JACOBI, W.: The dose to the human respiratory tract by inhalation of short-lived ^{222}Rn- and ^{220}Rn-decay products; Health Phys. 10 (1964), 1163-1174

[16] HAQUE, A.K.M.M., A.J.L. COLLINSON: Radiation dose to respiratory system due to radon and its daughter products; Health Phys. 13 (1967), 431-444

[17] HARLEY, N.H., B.S. PASTERNACK: Alpha absorption measurements applied to lung dose from radon daughters; Health Phys. 23 (1972), 771-782

[18] WALSH, P.J.: Radiation dose to the respiratory tract of uranium miners - a review of literature; Environmental Research 3 (1970), 14-36

[19]   HAERTING, F.H., W. HESSE: Der Lungenkrebs, die Bergkrankheit in den Schnee-
        berger Gruben; Vierteljahresschrift f. gerichtl. Medizin u. öffentl. Gesund-
        heitswesen 30 (1879), 296

[20]   Governors Conference: Health hazards in uranium mines; Publ. Health Service
        Publ. No. 843, Washington D.C., 1961

[21]   LUNDIN, F.E., J.K. WAGONER, V.E. ARCHER: Radon daughter exposure and respi-
        ratory cancer, quantitative and temporal aspects; Nat. Inst. f. Occup.
        Safety a. Health - Nat. Inst. f. Environmental Health Services, Joint
        Monograph. No. 1, 1971

[22]   ARCHER, V.E., J.K. WAGONER, F.E. LUNDIN: Lung cancer among uranium miners
        in the United States; Health Phys. 25 (1973), 351-372

[23]   BEIR-Report: The effects on populations of exposure to low levels of
        ionizing radiation; Nat. Acad. of Sciences - Nat. Res. Council,
        Washington D.C., 1972

[24]   ŠEVC, J., E. KUNZ, V. PLAČEK: Lung cancer in uranium mines and long-term
        exposure to radon daughter products; Health Phys. 30 (1+76), 433-437

[25]   DE VILLIERS, A.J., J.P. WINDISH: Lung cancer in a fluorspar mining community;
        I. Radiation, dust and mortality experience; Brit. J. Ind. Med. 21 (1964), 94

[26]   BOYD, J.T., R. DOLL, J.S. FOULDS: Cancer of the lung in iron ore (haematite)
        miners; Brit. J. Ind. Med. 27 (1970), 97

[27]   SNIHS, J.O.: The approach to radon problems in non-uranium mines in Sweden;
        Proceedings 3. Int. IRPA-Congress, Washington 1973; CONF-730907 (1974), 900-911

RESULTATS BIOLOGIQUES EXPERIMENTAUX ET RELATION DOSE-EFFET
DU RADON AVEC SES PRODUITS DE FILIATION

CHAMEAUD J., PERRAUD R.
COGEMA - Service Médical - Branche Mines
B.P. N° 1 - 87640 - RAZES (France)

LAFUMA J., MASSE R.
C.E.A. - Département de Protection
C.E.N de FONTENAY-AUX-ROSES
FONTENAY-AUX-ROSES (France)

CHRETIEN J.
Laboratoire de Pathologie pulmonaire
Clinique de Pneumologie
Hopital LAENNEC - PARIS

INTRODUCTION

Lorsqu'on reprend l'historique de l'étiologie des cancers du poumon dans les mines d'uranium, on constate qu'il a fallu très longtemps pour établir une relation de cause à effet entre l'augmentation du nombre de cancers parmi les mineurs et l'exposition au radon et à ses descendants.
Bien que les enquêtes épidémiologiques de ces dernières années aient mis nettement ce phénomène en évidence, nombreux étaient ceux qui pensaient, encore récemment, que ce gaz jouait seulement un rôle de cofacteur pour d'autres nuisances de la mine, et la fumée de cigarette en particulier.
C'est en définitive l'induction de cancer bronchique chez l'animal par l'inhalation de radon et de ses descendants qui a apporté la preuve indiscutable de son action carcinogène.
Si de la même façon les études biologiques nous permettaient de mieux cerner les phénomènes d'induction du cancer, de relations dose-effet, surtout pour les doses faibles, elles présenteraient un intérêt majeur.
En effet, malgré les très remarquables travaux qui ont été faits dans le domaine des modèles physiques (1) (2) (3), leur interprétation devra toujours surmonter l'imprecision de nos connaissances de certains paramètres biologiques tels que, par exemple, les phénomènes de l'épuration pulmonaire ou l'appréciation de la barrière du mucus, sans oublier la grande complexité des phénomènes cellulaires, tissulaires et généraux, qui interviennent dans l'apparition d'un cancer.
Pour les enquêtes épidémiologiques (4) (5) (6) (7), même parfaitement conduites comme l'ont été celles de ces dernières années, il est permis de critiquer la précision des données physiques : soit, par exemple, que les prélèvements de l'ambiance de travail ne sont pas assez fréquents et nombreux,

soit que la mesure elle-même reste imprécise ou enfin à cause de la difficulté que l'on rencontre lorsqu'on veut reconstituer, même approximativement, le passé professionnel d'un mineur. Or, c'est justement la précision des données physiques et professionnelles qui est importante, car ce qui nous intéresse c'est le bas de la courbe exprimant la fréquence en fonction de l'exposition totale, et beaucoup moins le pourcentage d'augmentation de cancer pour les doses élevées.

Les appareils de mesure portatifs nous apporteront cette précision en nous permettant d'avoir une meilleure connaissance de la dose totale inhalée. Mais ces résultats ne seront pas exploitables avant de nombreuses années et même alors, dans notre cas particulier, nous aurons toujours beaucoup de difficultés à constituer des groupes statistiquement comparables.

Même à supposer que l'exposition des mineurs ait été très sous-estimée dans les études épidémiologiques, la marge de sécurité entre la dose inhalée par un mineur dans sa vie de travail, dans une mine correctement protégée, et la dose qui marque le début de la zone évidente de risque, est faible. Il convient donc de préciser encore nos connaissances dans la relation dose-effet à ces niveaux de contamination respiratoire. L'expérimentation animale devrait pouvoir nous aider à atteindre ce but.

Pour ces raisons, il parait intéressant de passer en revue les résultats que nous avons obtenus sur le rat en pathologie expérimentale et de les confronter aux données humaines afin de savoir ce qu'ils peuvent nous apporter et si nous pouvons envisager de les extrapoler à l'homme.

Dans ce but, nous examinerons et commenterons les effet provoqués par les inhalations de radon et de ses produits de filiation communs à l'homme et à l'animal, les effets rencontrés chez l'animal et qui n'ont pas été constatés chez l'homme, puis nous parlerons des expériences en cours.

MATERIEL ET METHODE -

Le rat a été choisi comme l'animal d'expérience car il permet de constituer des groupes assez nombreux et aussi parce qu'il ne fait pratiquement jamais de cancer du poumon spontanément. C'est un point important lorsqu'on sait que la souris, par exemple, développe très facilement une variété de cancer, le carcinome alvéolaire, qu'on ne rencontre presque jamais chez l'homme.

Les animaux ont été exposés au radon à différents équilibres avec ses produits de filiation de 1 à 100 %, à différentes concentrations de 100 à 12000 W.L. pendant quelques heures par jour réparties sur 3 ou 4 mois (8) (9). Nous avons utilisé pour cela des cuves d'inhalation de faible volume dans les premières expériences et de 10 m^3 maintenant, ce qui permet une expérimentation sur des lots importants (10). Nous ne reviendrons pas sur le protocole de chacune de ces expériences qui avaient des buts différents, nous ne tiendrons compte que de l'exposition totale au radon, exprimée en W.L.M.

La méthodologie a été toujours la même. Les animaux, sacrifiés, soit systématiquement, soit lorsqu'ils présentent des signes de mort imminente, sont exploités après leur mort. Les poumons, fixés en bloc, sont coupés en tranches de 10 microns, et lorsqu'une anomalie est repérée, un examen anatomopathologique est entrepris. Le dépistage des lésions est donc d'une grande sureté.

RESULTATS

EFFETS COMMUNS à l'HOMME et à l'ANIMAL

Effet du Radon sur le processus silicotique -

Les premières expériences ont débuté en 1963.

Dans cette première période il ne s'agissait pas d'induire des cancers mais de savoir si le radon pouvait aggraver le processus silicotique comme certains, alors, le pensaient (11). Les rats étaient exposés à des inhalations de cristobalite, puis de radon, à 2500 W.L.M. au total.

Nous n'avons pas, chez l'animal, constaté d'aggravation des silicoses, ce qui a été confirmé par les faits en pathologie humaine puisque jusqu'à présent on n'a pas relaté dans les mines d'uranium, même riches en quartz, l'apparition de silicose d'une particulière gravité. Nous n'avons pas non plus constaté de cancer chez les rats qui étaient sacrifiés tôt, vers le 6ème mois.

Cancers expérimentaux -

Les premiers cancers ont été obtenus en 1969, simultanément chez des rats qui avaient préalablement été exposés à du cérium stable (12) qui a la propriété d'induire des lésions précancéreuses, et chez deux rats silicotiques conservés jusqu'à 18 mois. Dans les deux cas un cofacteur était associé au radon, mais par la suite des cancers furent provoqués uniquement après des inhalations de radon et de ses descendants (13).

Types de lésions -

Près de 3000 animaux ont maintenant été utilisés, l'examen histologique systématique des poumons a permis de constater chez les rats exposés environ un millier de lésions bénignes ou malignes dont 250 cancers bronchopulmonaires et pas un seul cancer chez les témoins. Ces lésions sont identiques à celles qu'on rencontre chez l'homme ; les lésions bénignes sont soit des métaplasies épidermoïdes, soit des adénomatoses de types divers, les tumeurs bénignes des adénomes alvéolaires ou bronchioloalvéolaires. Pour les cancers il s'agit surtout de carcinomes épidermoïdes, d'adénocarcinomes, de carcinomes bronchioloalvéolaires et d'autres formes plus rares (14). Nous n'avons pas provoqué de cancer à petites cellules, bien qu'il soit assez fréquent chez l'homme. La cellule qui est vraisemblablement à l'origine de ce type de lésion, la cellule de KULTSCHITZKI, est très faiblement représentée dans les bronches des rongeurs, c'est sans doute l'explication de cette anomalie (15) qui représente la seule disparité entre les lésions malignes de l'homme et de l'animal.

Temps de latence -

Les lésions bénignes précèdent dans leur apparition les cancers qui, eux, surviennent de 1 an à 2 ans après le début de l'exposition à des doses totales inhalées de 250 à 15000 W.L.M. Cette période qui s'écoule entre le début de l'exposition et la première manifestation de la maladie cancéreuse est appelée temps de latence ; il est, compte tenu de la durée de vie des espèces, comparable chez le rat et chez l'homme - respectivement 1 à 2 ans et 20 ans et plus. Plus la dose délivrée est forte, plus le temps de latence est court et plus la dose totale inhalée est faible, plus il est long. Pour de faibles expositions par exemple - 250 W.L.M - les cancers sont apparus au bout de 2ans. Cette notion de variabilité du temps de latence en fonction de l'exposition ne pourra être vérifiée chez l'homme que lorsque les enquêtes épidémiologiques auront été exploitées entièrement.

Fréquence -

Le nombre de cancers provoqués est d'autant plus grand que la dose totale délivrée est importante, ceci jusqu'à 10.000 W.L.M ; au dessus de ce chiffre la durée de vie des animaux est trop raccourcie pour que la maladie ait le temps d'apparaitre, lorsque les séances d'inhalation sont trop prolongées ou la concentration trop élevée il en va de même.

Si nous comparons cette fréquence des cancers bronchopulmonaires expérimentaux en fonction de la dose totale inhalée et la fréquence des cancers du poumon des mineurs telle qu'elle ressort des enquêtes épidémiologiques, on constate que les données expérimentales et les données épidémiologiques s'intercalent parfaitement. L'enquête américaine montre une fréquence légèrement plus faible mais son exploitation n'est pas encore terminée. On se situerait par contre au niveau de la fréquence tchèque en considérant que les cofacteurs de la mine peuvent avoir une influence synergique (16).

DISCUSSION -

Sur le plan qualitatif, (c'est-à-dire l'anatomo-pathologie des lésions, le type de cancer) comme sur le plan quantitatif, (durée du temps de latence, fréquence en fonction de l'exposition au risque) nous voyons donc que les données humaines et expérimentales sont parfaitement superposables. Ceci nous autorise à penser que l'expérimentation sur le rat est un excellent modèle biologique.

On peut se poser la question de savoir pourquoi le radon serait le seul concérogène pulmonaire qui nous permette de tirer de telles conclusions. Il y a à celà trois raisons essentielles.

- Il s'agit d'un gaz rare dépourvu d'action toxique surajoutée, ce qui permet de soumettre durant la même fraction de vie un gramme de tissu pulmonaire de rat à une activité cumulée, comparable à celle que subit un gramme de tissu pulmonaire d'un mineur d'uranium. Si on a rarement pu induire de cancer du poumon d'autres cancérogènes gazeux ou suspectés comme tels, c'est à cause de leur toxicité chimique. Pour la fumée de cigarette en particulier, il faudrait, pour remplir ces conditions, faire fumer pendant six mois à un rat les milliers de cigarettes qu'un homme fume en 20 ans, ce qui est parfaitement impossible, les animaux mouraient très rapidement d'oxycarbonémie aigüe.

- La répartition homogène du gaz autorise, d'autre part, à tirer une relation dose-effet, ce n'est pas le cas lorsqu'on a affaire à l'agression ponctuelle d'un implant bronchique.

- Ajoutons pour terminer que le radon et ses descendants sont les seuls emetteurs α pour lesquels on ait une référence humaine telles que les enquêtes épidémiologiques dans les mines.

EFFETS RENCONTRES UNIQUEMENT CHEZ L'ANIMAL

L'inhalation de ce gaz a chez l'animal d'autres conséquences qui n'ont pas été décrites chez l'homme. Leur gravité est fonction là encore de la dose totale inhalée.

Il s'agit d'abord d'un raccourcissement de la durée de vie que nous avons constaté chaque fois que nous avons laissé vivre les animaux, c'est-à-dire lorsque nous ne les avons pas sacrifiés systématiquement. Ce raccourcissement est très élevé pour des doses inhalées fortes, ainsi pour 10.000 W.L.M. 50 % des rats sont morts 15 mois après le début de l'expérience, alors que la demi-mortalité des témoins est de 2 ans. Pour 7000 W.L.M. la demi-mortalité est à 18 mois. Nous n'avons pas encore de chiffres pour les doses faibles. Phénomène curieux, cette mortalité précoce n'intéresse pas seulement les rats cancéreux, qui meurent d'ailleurs rarement d'une complication de leur maladie, mais aussi ceux qui n'ont pas de cancer.

Quelques animaux, dans chaque expérience et pour la même dose cumulée, meurent rapidement, cancéreux ou non, les autres résistent plus ou moins longtemps. Tout se passe comme si chaque individu portait en lui génétiquement sa propre capacité de restauration cellulaire ou de résistance à l'agression, le temps de latence serait ainsi fonction non seulement de l'intensité de l'agression cumulée mais aussi de ce facteur génétique individuel.

On remarque aussi au niveau du tissu pulmonaire, même pour des doses faibles, l'installation d'une pneumonie interstitielle fibrogène (14), qui peut sans doute pour une part contribuer à raccourcir la vie des animaux. Cette lésion est difficile à mettre en évidence chez l'homme ; elle entre dans le cadre de l'étiologie des insuffisances respiratoires, pathologie classique du mineur, souvent très invalidante, c'est pourquoi il est important de savoir à partir de quelle agression elle apparait.

Expériences en cours -

Les résultats que nous avons obtenus jusqu'à présent l'ont été avec des expositions cumulées fortes, bien que comparables à celles auxquelles ont été soumis certains mineurs. Ils ont permis de mettre en évidence une identité de réponse chez l'homme et l'animal.

L'essentiel de notre programme concerne maintenant les réponses aux faibles doses - 100 à 200 W.L.M - c'est-à-dire très proches de celles qu'un mineur peut normalement inhaler dans sa vie. Une de ces expériences comprend 450 animaux, 300 soumis à des inhalations de radon, 150 servant de témoins ; elle a débuté il y a 15 mois.

Un autre domaine nous a paru intéressant, c'est celui des cofacteurs. Nous avons entrepris pour cela l'étude de l'effet associé au radon et à ses descendants, du SO^2 et de la fumée de tabac. Cette dernière expérience n'est pas encore terminée mais l'exploitation des premiers résultats nous permet de penser que le tabac a plutôt un effet additif, aggravant l'évolution des cancers. Nous n'avons pas pour l'instant noté d'augmentation de leur fréquence. Si ce résultat était confirmé il serait, aussi, en accord avec les données humaines (17).

Encore un point mérite actuellement notre attention, c'est l'influence que peut avoir l'âge auquel débute l'exposition (6).

Enfin nous avons commencé l'étude expérimentale du facteur qui nous semble le plus important, la sensibilité individuelle.

L'exploitation de ces différentes expériences demandera encore de nombreux mois, car un des reproches qu'on peut formuler à l'égard de l'expérimentation biologique c'est la lenteur, d'ailleurs toute relative quand on sait que les enquêtes épidémiologiques demandent un demi siècle avant d'être totalement exploitées.

CONCLUSION

Nous avons passé en revue l'essentiel des résultats que nous avons obtenus depuis une quinzaine d'années en pathologie expérimentale sur les effets du radon inhalé chez le rat.

La concordance entre ces résultats et les données épidémiologiques et pathologiques humaines nous permettent de penser que nous avons là un excellent modèle expérimental. Il devrait nous apporter, dans les années à venir, les précisions qui nous manquent dans la connaissance de l'action des faibles doses de radon associé ou non à d'autres nuisances. La facilité d'obtention de ces cancers pulmonaires, comparables à ceux de l'homme, la reproductivité des résultats, font, de plus, de ce modèle un outil précieux en cancérologie fondamentale.

REFERENCES

1. ALTSHULER B., NELSON N., and KUSCHNER M. - Estimation of lung tissue dose from the inhalation of radon and daughters - Health. Phys. Pergamon Press 1964, vol. 10., p. 1137-1161.

2. JACOBI W. - Relations between the inhaled potential energy of ^{222}Rn and ^{220}Rn daughters and the absorbed energy in the bronchial and pulmonary region. Health Physics Pergamon Press 1972, vol.23, pp. 3.11.

3. HAQUE A.K.M.M. and COLLINSON A.J.L. - Radiation dose to the respiratory system due to radon and its daughter products. Health physics, Pergamon Press 1967, vol. 13, pp. 431-443.

4. LUNDIN F.E., WAGONER J.K., ARCHER C.E - Radon daughter exposure and respiratory cancer quantitative and temporal aspects, NIOSH and NIEHS Joint Monograph N°1, NTIS, SPRINGFIELD (1971).

5. MULLER J. and WHEELER W.C. - Causes of death in Ontario uranium miners. Radiation Protection in mining and milling of uranium and thorium - International Labour Office, CH-1211 Geneva 22, SWITZERLAND.

6. SEVC J. and KUNZ E. - Lung cancer in uranium miners and Long-Term exposure to radon daughter products - Institute of Hygiene and Epidemiology, Department of Radiation Hygiene, 100 42 Praha 10.

7. SNIHS J.O. - The approach to radon problems in non-uranium mines in Sweden. National Institute of Radiation Protection. Stockholm 1973.

8. CHAMEAUD J., PERRAUD R., LAFUMA J., MASSE R., PRADEL J. - Lesions and lung cancers induced in rats by inhaled radon 222 at various equilibriums with radon daughters. Experimental lung cancer, Carcinogenesis and Biossays (KARBE E., PARK J., Eds). Springer-Verlag, BERLIN (1974) 411.

9. CHAMEAUD J., PERRAUD R., MASSE R., NENOT J.C., LAFUMA J. - Cancers du poumon provoqués chez le rat par le radon et ses descendants à diverses concentrations - Biological and environmental effects of Low-level radiation - Vol. II - International Atomic Energy Agency VIENNA, 1976.(223-228).

10. BLONDEAU E., DUPORT P., FRANCOIS Y., MADELAINE G. - Ensemble d'étude expérimentale du cancer pulmonaire chez le rat par irradiation - Comm. Energie At. (France), Rappt. STEPPA B.P. N°1 - 87640 RAZES (1973).

11. CHAMEAUD J., PERRAUD R., LAFUMA J., COLLET A. et DANIEL-MOUSSARD H. - Etude expérimentale chez le rat de l'influence du radon sur le poumon normal et empoussiéré - Archives des Maladies Professionnelles, de Médecine du Travail et de Sécurité Sociale (PARIS) ; 1968, T.29 - N° 1 - 2 . janvier-février (pp. 29-40).

12. PERRAUD R., CHAMEAUD J., MASSE J., LAFUMA J. - Cancers pulmonaires expérimentaux chez le rat après inhalation de radon associé à des poussieres non radioactives - C.R. Acad. Sci. PARIS (1970) t. 270 pp.2594-2595.

13. CHAMEAUD J., PERRAUD R., LAFUMA J. MASSE R. - Cancers du poumon expérimentaux provoqués chez le rat par des inhalations de radon. C.R. Acad. Sci. PARIS, Sér. D 273 (1971) 2388.

14. LAFUMA J., MASSE R., METIVIER H., NOLIBE D., FRITSCH P., NENOT J.C., MORIN M., SKUPINSKI W., CHAMEAUD J., PERRAUD R., et J. CHRETIEN - Etude expérimentale des polluants radioactifs inhalés - Données actuelles Inventaire lésionnel - Validité du modèle animal - Relations dose-effet Coll. INSERM 18-19 Janvier 1974 - Vol. 29 - pp. 307 à 324.

15. CHRETIEN J., MASSE R. - La cancérogène broncho-pulmonaire - Revue des faits expérimentaux - Revue Française des Maladies Respiratoires - N°1 janvier 1976 - Tome 4 - pp. 23-45.

16. LAFUMA J. - Les radioéléments inhalés - Séminaire de Radiobiologie - Contamination radioactive interne - Sté Fse de Radioprotection - PARIS 1974 DUNOD 1974 - Vol. 9 - N° 1 pp 15-25.

17. ARCHER V.E., WAGONER J.K., LUNDIN F.E. - Jr. (1973) - Uranium mining and cigarette smoking effects on man - Journal of Occupational medecine, 15, N°3

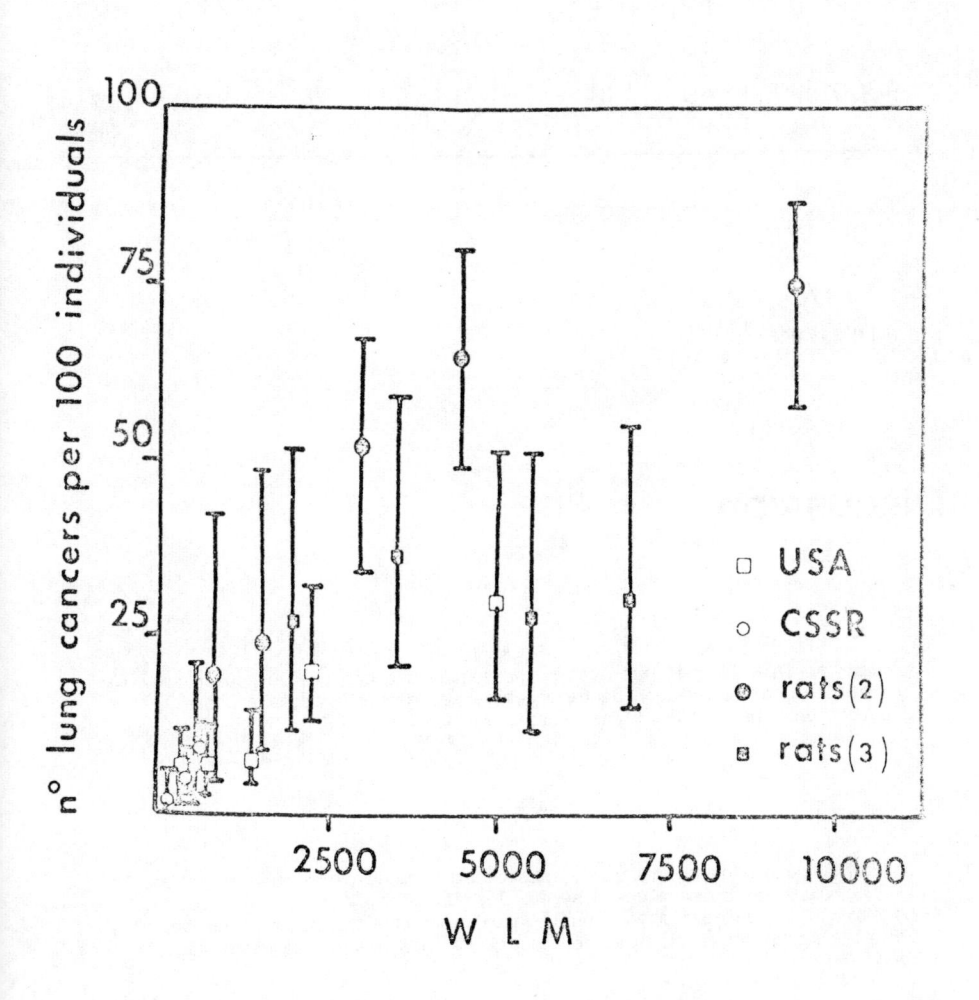

LUNG CANCERS

WLM	n° rats	bronchogenic	bronchiolo alveolar	'mixed'	total	%
750	20	1	3		4	20
1500	20	3	2		5	25
3000	40	3	14		17	52
4500	40	12	14		26	65
9600	40	13	14	3	30	75

2500 WL
5 h/day
5 days/week

Discussion

Question

Comment se fait-il que vous ayez pu provoquer des cancers du poumon chez le rat par des inhalations de radon et de ses descendants, alors que d'autres chercheurs n'ont pas réussi ?.

Réponse

La chance nous a peut-être aidé, mais c'est surtout parce que nous avons fait inhaler aux animaux une dose cumulée de radon et de ses descendants, comparable à celle qui a été inhalée par un mineur d'uranium placé dans les plus mauvaises conditions de travail et ceci pendant la même fraction de vie.

SOCIETAL ASPECTS OF HAZARDS IN URANIUM MINES

Dean James M. Ham, Sc.D.
University of Toronto
Toronto, Canada

In the period September 1974 to June 1976 a Royal Commission on the Health and Safety of Workers in Mines was conducted in the Province of Ontario, Canada. It was my privilege to be the Commissioner for this study and I am pleased to speak briefly about its work as reflected in its report. The general term of reference was to investigate all problems related to the health and safety of workers in mines - a very difficult task indeed. The uranium mines formed a special facet of the study.

I believe that the public, industry, unions, and the Government recognize that more serious attention must be given to the problems of occupational risks at the work place. I think this is so throughout OECD countries. This concern for the work place has risen in significant measure from the general public concern over the past decade about toxic substances in the outer environment of land, air and water. We need to remind ourselves that all toxic substances that do not occur naturally are generated in places of industrial activity, and before they are dispersed as effluents, through tailings, through stacks, through bleed streams, etc., they are created at places where persons work. One of the things that we discovered is that as government regulations have become tougher in limiting the amount of toxic substances that may be released from industrial locations, a result has been to contain more and more of these in internal processes associated with places of work in metallurgical plants. For example, toxic streams which formerly were vented to the air or to tailings, are increasingly recirculated within metallurgical plants and this means that the level of concentration of toxic substances near places of work is rising as a result of the increasing severity of constraint upon the amount of effluent that can leave the place of generation. As a consequence one of the things that we have called for is a formal auditing of the origin, holdup and destination of toxic substances inside metallurgical plants. This is really an extension of conventional metallurgical mass balancing which is carried out for purposes of tracing the flow of valuable product.

Occupational risks ought to be considered in their totality and not separated on the one hand into biological risks and on the other hand into physical risks of traumatic accident. I believe very deeply that the total risks of work which involve accidents and injuries, industrial disease and all other manifestations of loss of human well-being must be looked at in total.

In our province, the tendency has been to separate what we might call health aspects from safety aspects, and I do not think that this is to the good of the working people. It may be suitable to separate scientific knowledge into that of toxicology, epidemiology, and the ergonomics of work, but as far as the good of the worker is concerned I deeply believe that it is necessary to retain a perspective on the total risks. Let me illustrate this point by quoting you a few figures that derive from the study of the uranium nominal roll for our uranium mines. A search of national vital statistics was conducted for some 15,000 persons who have worked one month or more in dust exposure in our uranium mines between 1955 and 1974. This search of the deaths among these 15,000 persons identified 81 confirmed cases of lung cancer and a study of the relationship of these to expected mortality indicated that approximately one half of them were unexpected. To emphasize the issue of considering occupational risks in their totality we calculated the potential life/years lost among the excess cancer cases. That is to say from vital statistics tables we determined the number of years that a person might expect to continue to live, having arrived at the age at which death from cancer occurred. We also examined the number of fatal accidents from industrial sources among these 15,000 people and calculated the number of potential life/years lost. We discovered that 8 times as many life/years have been lost from industrial accidents as from excess lung cancer. So, when we think about the problem of excess lung cancer among uranium miners, we have to see this unhappy phenomenon in some total context of occupational risks. I believe that when decisions are made as to what constitutes a "safe" level of radiation exposure it is not a satisfactory procedure to isolate the lung cancer risks from all the other risks that exist including that of smoking. This poses real difficulties to authorities which, like the Atomic Energy Control Board, bear the responsibility to set the standard for exposure to ionizing radiation in the uranium mines. The uranium mines have also had a silicosis problem. We have two groups of uranium mines, of which the ones that you have been visiting are in one group. There is another set of mines that have similar grades of uranium ore but have different geology and in particular are very different in their silica content. The local host rock and ores here in Elliot Lake have something like 60 to 70% free silica. The host rock and ores in Bancroft area, which is the other area of uranium mining have a silica content of something like 5 to 15%. In the Bancroft mines, there has been essentially no silicosis; in the mines here there has been a very startling outbreak of silicosis. Among the persons on the uranium nominal roll there have been approximately 100 cases of silicosis identified and charged to the uranium mines for purposes of workmen's compensation. It is my personal opinion that the continuing risks of silicosis in uranium mines here is at least as big a problem as the risk of lung cancer.

I won't comment on dust because I gather that's not a subject that is the principal one of concern to you here, but I will comment on the problem of radiation in terms of my own personal puzzlement, because I have no capability to add to your technical understanding of this problem. Let me cite you a few figures. Among the 81 lung cancer deaths that were identified, the imputed cumulative working level month exposures had a maximum value of 375, a minimum value of 0, and an average value of 75. These figures concern the total number of cancer cases identified in Ontario in the period 1955-1974 since the uranium mines opened. The difference between these figures and the ones in U.S.A. is very notable indeed. Among 113 cases that were reported by Archer, Lundin & Wagner from the Colorado Plateau in the U.S.A. the average imputed exposure in Working Level Months was 2000. One of the peculiarities as between Ontario experience and U.S. experience to date is that we have excess lung cancer at radiation levels more than an order of magnitude lower than the corresponding historical levels in the U.S.A. I am aware that there are very eminent people who question whether the kind of excess lung cancer that we have observed is indeed attributable to radiation ionizing radiation at all. I had a recent discussion initiated by the Ontario Mining Association with Dr. Robley Evans in which he reasserted his view based on U.S. data that there is no dose dependent risk of cancer mortality from ionizing alpha radiation that is statistically significant below a cumulative exposure of about 800 WLM. If that is so, then the excess cancer that we have observed might have little to do with ionizing radiation. Were this so there must be some unknown mysterious carcinogenic element in hardrock mines of diverse kinds in all of which radon has been found. Frankly, as a Commissioner who has discussed this problem as a lay person with experts in several countries, I do not accept Evans' position. It is absolutely essential to resolve what I regard to be an absurd situation where there is imputed to be some carcinogenic agent common to hardrock mines unrelated to radiation.

I asked Dr. Evans whether in the U.S.A. anyone had done any well defined epidemiology comparing the occurrence of cancer in hardrock mines that were known historically to have had negligible radiation levels with the uranium mines on the Colorado Plateau. He said that to his satisfaction such epidemiology had not been carried out and I would be interested to know whether any of you around the table, in your countries, have any well defined epidemiological studies which compare the occurrence of lung cancer in non-uranium miners with the occurrence of lung cancer in uranium miners. I am aware of the Swedish data.

I am not an epidemiologist, I have no capability to interpret epidemiological detail, but I find it very mystifying, that there are views as markedly contrasting as I have suggested. In the study that we did on the basis of sampling methods we tested the hypothesis that the radiation exposures experienced in Ontario mines has no effect on cancer mortality and this null hypothesis was rejected. The Commission did not have the time to develop a dose-response relationship for Ontario experience to supplement those that are in the BEIR report, the UNSCEAR report, Swedish and Czechoslovakian sources. The one thing that is different about the Ontario situation and the U.S. situation is that we have over 100,000 man/years at risks, at levels of radiation exposure that are down in the range from 300 WLM to 0. It is absolutely essential that the issue of what radiation exposures are, and have been, be cleared up. Are the historical records of radiation exposures in mines that exist reasonable? In Ontario, we have a very remarkable set of records that go back to 1957. With the exception of the French records, they are possibly the best existing set of records in the sense of their completeness. My sense of the U.S. data on the Colorado Plateau is that it is not founded on nearly as complete a set of historical records. There is also the fundamental question in the face of latency of what in fact is the exposure effective in contributing to excess cancer mortality. I am told for example that a given WL in a well ventilated mine is more toxic than a given WL in a mine that is less well ventilated. The WL is a very indirect indicator of irradiation. I don't for one minute, begin to understand the details of the eminent work of Dr. Jacobi on lung and irradiation models. But as a person called upon publicly to examine some of these problems I perceive there to be need for specialists such as yourselves to provide clarifications for the public good of workers themselves, and obviously for the social good related to understanding the risks of the nuclear fuel cycle. The costs of such work are negligible compared to the costs of nuclear power. The complications of human response in the presence of a combination of cigarette smoking, ionizing radiation, diesel emissions and dust are clearly very great.

Obviously if we could stop uranium miners or any other workers from smoking, all respiratory risks would go down. We examined some recent government data on a study done in one of our metallurgical plants on the frequency of symptomatic chronic bronchitis in relationship to exposure in a copper refinery to sulphur dioxide, metal fume and dust. The comparative epidemiology that was done showed that there was three times as much chronic bronchitis among the group exposed to sulphur dioxide, metal fume and dust, as in a control group that was not so exposed and yet when one looked at the data one discovered that essentially all of the persons that were chronic bronchitics had been smokers. I am aware that the same thing pertains in uranium mining and there is of course the unresolved problem of diesel fumes.

From a public health standpoint, it is essential to assess for the worker and the public the total occupational risk and the components of this. From a scientific point of view, how to break down the total occupational risk into identifiable and attributable components, strikes me as being an extremely difficult task but, nevertheless is one that must be faced through more research based on careful dosimetry.

I believe that we should be using the finest technology we have to verify what the actual exposures of people are, and then of course to carry on and to support toxicological and epidemiological work to find out what the consequences of these things are. In Ontario, we have been suffering seriously from a lack of unified approach to these issues. There simply hasn't been a coherent government policy backed by adequate resources. The epidemiology that has been done in the mining industries has been done on a crisis basis. This is not now a socially acceptable situation. If good epidemiology is going to be done, one has to follow whole populations of suitable people before they get sick, to establish nominal rolls for different worker groups, to discover accurately what their environmental exposure is and then investigate what the consequences of their living in those exposures are. Dr. Chameaud's pioneering work in France is part of this necessary pattern. Radiation dosimetry has a central role to play in this work. Despite the extensive knowledge that exists on the human effects of ionizing radiation, there are disturbing areas of ignorance related particularly to mining. And I reject the extrapolation of U.S. data to average exposure levels lower by an order of magnitude.

Discussion

Dr. W. JACOBI - F.R. of GERMANY

Dr. Ham, with regard to what you said concerning occupational risks that we should not consider only lung cancers, but we should also take into account all types of risk that are involved in mining, I would like to know, have you some idea which methods can be used to derive something like an acceptable risk from particular agents such as dust? Then there's another remark I want to make. You mentioned the opinion of Robley Evans that there is no indication about a significant increase of lung cancer risk among miners below 800 WLM. I think especially the data from Czechoslovakia uranium miners clearly indicate that a significant excess is observed at a level above about 100 WLM, and I think Dr. Snihs will agree that among Swedish miners a significant increase occurs at a lower level and surely we don't know how those risks relationships will be at low levels below 100 WLM. There are some indications that it's not linear in this region, it depends on the latency period from dose rate. We have some examples where it seems so in the case of lung cancer from uranium miners; the statistics are too limited to make a decision.

Dr. Ham:

On the very basic question that you asked about risk acceptance, I don't believe that you believe as scientific people, that it is a scientist's responsibility to make this decision. The ultimate decision as to what constitutes an acceptable risk must be a sociopolitical one. But it is part of our several responsibilities to provide the most truthful set of data as a basis on which that judgment can be made. Now my perception from working with our own community is that socially, we have tended to "duck", I mean to have avoided the question of speaking openly about acceptable risks. We use the word safety without defining what we mean and good data on risks are hard to find. I think one of the reasons why industry is reticent to have the facts presented publicly is in part because of concern for misinterpretation. But I have a very deep conviction that if we are really going to answer sensibly the question of acceptable risks we must discuss these risks openly. Also I think acceptable risks are in part determined internationally because we live in a competitive system and if one country adopts high standards for health and safety and another doesn't there may be unacceptable price differentials for products. But to come back to the question, I think it has to be answered in terms of clear public understanding of what I would call voluntary risks and involuntary risks. We are aware that if we smoke or if we get in our automobile, we voluntarily undertake risks; if we go into uranium mines and smoke, we take larger risks. Now the difference is of course that when we go to work the risks are involuntary and it is obvious that the public is much more alarmed about a number of lung cancer deaths in uranium mines than it is about the deaths that occur in automobile accidents. That doesn't mean that I have any callous view and I am sure none of you do, of human life, but if we are going to have industrial activity, we in fact are going to have deaths attributable thereto and I think the sooner we recognize that as an inescapable characteristic and wrestle with it the more we can intelligently discuss what constitute acceptable risks. It is my understanding that it is not possible to predict who will come down with lung cancer so that each uranium miner is running a risk but no one can say in which one carcinoma will appear. I will be very bold and say this: if 8 times as many life/years of living with one's family and enjoying one's life are lost because of accidents than are lost from all of industrial diseases then that to me is an argument for being deeply concerned about accidental injuries as well as industrial disease. At the same time I recognize the necessity of accepting a certain number of deaths from accidents and disease. Then of course you say, well how many will you accept? Here, I think the answer partly comes from the assurance that risks are in fact being defined through dosimetry, toxicology, epidemiology and technology assessment to determine what can be achieved to minimize loss of well-being. In other words, I think the public has the right to know that uranium mines are using accessible technological and managerial means to control risks, that they are using these methods with care and effectiveness, that exposure conditions are being carefully assessed, that the human implications of exposures are being studied and that the state of health of workers is being reviewed. There is no simple answer to ultimate human questions and few of us want to talk about acceptable deaths. When I talked with the people in the British coal mines, I was told that the current British dust hazards are based on the expectation of a certain number of cases of pneumoconiosis after a certain elapsed time. I quite firmly believe as a person that we have to talk more openly about these things and compare the risks of smoking, of accidental death, of lung cancer death. In the end these decisions will be made on a political basis; in our province, by a combination of Atomic Energy Control Board and provincial governmental decisions. That's a long winded answer to a very difficult question.

If I may make a final comment, I am pleased to hear you say, Dr. Jacobi, that there is evidence to which you attach some significance and to which Dr. Snihs attaches some significance about increased risks of cancer mortality at radiation exposures significantly below 800 WLM because if that's not true, then it is somebody's responsibility to find this mysterious carcinogenic agent that exists in hardrock mines.

Mr. R.L. ROCK - United States of America

Dr. Ham, I think one of the big problems with safety related traumatic accident and health problems, is of course, as well you know, we are not compelled to report our health statistics in the States. There's no way we can do it because we don't have a handle on them. So everything is focussed on lost time manhours from traumatic injury due to safety related injuries. We don't have people actually working on environmental health problems, we don't actually have that much emphasis on these problems, and from what I have been able to see, where we do have actual realistic studies going, I would say that contrary to your statement, that 8 times as many man/days or life/days are lost due to safety related accidents, we're probably kidding ourselves greatly because of this lack of emphasis. I am studying health problems. Now, you see all the time that we're talking here, we are talking deaths, we are not talking injuries. Certainly there is loss of lung capacity and it is my feeling, of course, I can't prove this, but it is my feeling in our uranium mines probably the radiation hazards are probably at least as great to the miners as are the normal safety hazards.

Dr. Ham:

God bless you for saying that because that's one of the things if you do have the time to read some of this (Ham report), I point out! The only information that we have access to in this province, on the impact of toxic materials basically comes from our Workmen's Compensation Board and that means that the evidence is limited to those things that have already been agreed as being attributable and therefore chargeable to the workplace. Now, our Workmen's Compensation Act and our Workmen's Compensation system in this province is, I think, eminently superior to much of the situation that exists in the U.S.A. Nevertheless, our Workmen's Compensation Board adopts the policy position that it's not its responsibility to be seeking out what constitute the various toxic effects that contribute to a loss of well-being. This is one of the jurisdictional problems I have addressed in my report. I would agree with you on this, that in terms of the <u>currently recognized industrial diseases conditions</u>, recognized for compensation, these constitute a small part as you say of the traumatic injury loss.

MR. ROCK - U.S.A.

If a man dies of lung cancer, it's not written up in the newspapers, but if a rockfall kills a man, or if an explosive accident results in a man's death, of course, it's really played up. So this is the real problem and it's one of pure straight education that we got to get used to, a dead man is a dead man, and regardless how you die, I'd say I'd rather die of a rock on the head than I would of 9 months of lung cancer before I finally expired, and of course, without compensation loss, lots of these people in 9 months have exhausted lifetime savings, and there they are, the family is left with absolutely nothing.

Dr. Ham:

I want to make another comment. In my report, there is a study of the number of what are called medical aid events and compensable accident events involving toxic substances. I think what concerns you and concerns me is that people may have one or two incidents of intense inhalation of some toxic substance such as nickel carbonyl, the long range effects of which are not known. What I have recommended is as follows. One is that whenever any worker has a chemical incident, that is to say, an encounter with a chemical that leads to not necessarily a loss of time, but he has to report to a doctor to deal with some symptomatic response that arises from that occurrence, the company is required to maintain a register of the use of that toxic substance; and I have also recommended that a nominal roll be established for a selected toxic substance so that at least one begins to gather the data that is necessary to have confirmed epidemiological data. If we have to use a body count method, to be vulgar, let us make sure that at the time we have to do that, there is a clear basis of record of exposures, and so on, that makes that study then meaningful, so that we don't have to go back into history and start making all kinds of inferences, etc., about what things were like.

Dr. J. CHAMEAUD - FRANCE

Je voulais simplement dire quelques mots au sujet de la silicose. Je crois que M. Ham a tout à fait bien fait de parler de ce problème quand on sait le nombre de morts que peut provoquer la silicose, jamais il n'y aura autant de morts par cancer dans les mines qu'il y a eu de morts par silicose, Dieu soit loué! Mais je suis un petit peu étonné de voir que dans une mine même à proportion de quartz très élevé, on ait constaté autant de silicose récemment, car la protection contre le radon en général convient aussi à la protection contre la silicose, et dans une mine qui est bien aérée, où la foration à l'eau est systématique, il est quand même assez rare de voir des teneurs en poussière qui permettent de provoquer des silicoses rapidement. Sur le plan silicose et cancer, je ne pense pas que la poussière de silice favorise l'apparition du cancer. Je peux vous signaler qu'il y a plusieurs études qui ont été faites

sur le cancer à silicose, en Allemagne en particulier, et il ne
semble pas qu'il y ait davantage de cancer du poumon dans les silicoses que chez les autres personnes exposées. Au contraire même.
Je dois dire aussi comme je vous l'ai dit lundi en ce qui concerne
l'expérimentation animale, nous n'avons pas provoqué des silicoses
graves en faisant inhaler du radon à des rats à qui nous avions
provoquer des silicoses expérimentales. Il en est de même d'ailleurs
avec le minerai d'uranium, nous n'avons pas favorisé l'apparition de
cancer du poumon avec du minerai d'uranium riche, associé au radon.
Je voudrais dire aussi un tout petit mot en ce qui concerne les
enquêtes épidémiologiques. Ces enquêtes épidémiologiques pour les
gens qui ont été peu exposés, c'est-à-dire de 0 à 300 WL, moi ça
fait quand même une exposition qui est très faible. Il faut pour
ces enquêtes-là, pour qu'on ait une bonne réponse que la dosimétrie
ait été très précise. Est-ce que en 1957-1960, vous avez eu une
dosimétrie très précise? D'autre part, vous avez parlé des cancers
en général, il faut savoir à l'heure actuelle que le cancer du
poumon est un cancer qui représente un quart des cancers de l'homme,
puisque c'est un cancer qui n'existait pratiquement pas chez la
femme il y a une quinzaine d'années, et qui à présent est devenu le
4ième cancer de la femme. Est-ce que vous avez des échantillons,
absolument qui vous permettent de faire des comparaisons tout à
fait valables? Parce qu'il y a une évolution du cancer du poumon
dans la population qui fait qu'il faut des échantillons absolument
comparables. Ce sont les deux questions que je voulais vous poser.

Dr. Ham:

If I may respond quickly to the first one. Your sense of
surprise that there is a silicosis problem, and also your sense of
surprise that there is a cancer problem relates to what is reported
in my report, which gives an historical picture that covers the
period of 1955 to 1974. In this report, there are detailed figures
which I think are the best available, giving the actual Working
Level figures for the mines and there has been a vast improvement.
Currently, the average Working Level Month exposure per annum among
the population in the two mines here, is a number something like 1½
WLM per annum. There is of course a distribution here. There are
a few people up at the level of 4 WLM per annum. I remember looking
at some of the French data that was published at the Bordeaux
Conference, and it seems to me that in the French mines your radiation levels have been much more stable in the sense that the
earliest levels were in many ways relatively comparable to the most
recent. But in Ontario mines, the conditions in the early days were
such that in a few of the mines the WLM per annum would be up in a
level of 50 or 60. They are now down in the range of 1 or 2 or 3.
So, on the question of excess lung cancer, if we were to start with
the current uranium population, in other words, if we were able to
take a completely new group of miners and start them ab initio in
the mines and follow them epidemiologically for 20 years, my
expectation is that there would be very very much less excess lung
cancer unless there is this mysterious carcinogenic agent which is
present in all our hardrock mines. The figures in the report here
in some sense give you the worst picture of the Ontario mines. It
covers the period from the beginning. Now you asked a second question which I am simply not competent to make any response to. In
the report, there is a table of the occurrence of lung cancer in a
general population changing with time and in France as I am sure in
Canada, the occurrence of lung cancer is increasing dramatically,
and of course, the epidemiological calculations that were done in
the report, were done taking account of the fact that lung cancer
is occurring increasingly in the whole population. The medical
advice that I got was that if we are really in some sense to grapple
with these things, we have very carefully to control not only dosimetry, we must very carefully control the populations that we follow.

Now, I know, if I understand correctly, you were doing this. I believe we need to be following populations in our non-uranium mines, we need to define a comparable non-smoking population and follow this. The cases of silicosis I referred to are not deaths caused by silicosis but cases of involving the minimum clinical evidence of lung impairment which is accepted for purposes of workmen's compensation. Silicosis is the certified <u>cause of death</u> in a very small percentage of miners.

Dr. SNIHS - SWEDEN

I think that the labour unions must have deep concern about what you have reported here, about the excess of lung cancers for instance, and you have also suggested that the standard which is used now for WLM should be reviewed. I will just ask what you expect from this review and is the intention to have it lower than 4, and if so, how low?

Dr. Ham:

I asked for a review, I did not say whether it was up or down. I don't know what I expect other than, and I look Monsieur Hamel directly in the eye on this, as one of the members of the Atomic Energy Control Board, I would hope that AECB would do a synthesis of available data and conduct further epidemiological work on the basis of the Ontario data and base its judgment on that. I believe that the Ontario data is in many ways a very remarkable set of data because it is very extensive. There's the whole issue of whether the pattern of area monitoring was enough to create a valid picture for the mines and to permit acceptably accurate personal exposures to be inferred. There is much more epidemiological work to be done on the nominal roll of 15,000 people in which there are over 100,000 man/years at risk involved, and I would hope that AECB would conduct further epidemiology and make a contribution to really the international understanding of dose-response relationships at these levels if it's possible using Ontario data. I would like to complicate life for the Board this much further. Should you measure WLM, should you measure radon, should you measure both of them, should you measure some combination of them? I expect as a citizen, that the AECB will publish some summary of the grounds on which it states what the level is to be, because I believe we have come to a time when the public has a right to statements from regulatory bodies as to the grounds on which these levels are set. Let me go on to a further point. In our uranium mines, we have not monitored whole body radiation except on a sample basis. Now perhaps this is something that doesn't need to be done on the ground that the gamma fields are sufficiently stable that one can just do a few samples, and then make inferences as to what people's radiation burdens are. What confounds me as a citizen, is that around our nuclear reactors, we have meticulous badge control and we worry about persons who accumulate a few rems; but the man-rems of gamma radiation that are being delivered in total, in our uranium mines, I believe is much higher than the man-rems that are being delivered around our nuclear reactors; and, you know, it is man-rems that are being delivered that we really want to worry about, and I ask the question: how is it, and partly it's because of the social concern about nuclear accidents that we do not more carefully monitor gamma radiation where uranium miners are if we are so really meticulous about doing it for the people around the nuclear reactors?

Dr. R.M. FRY - AUSTRALIA

Dr. Ham, I will be very interested to read your report and I just had a copy here which I was able to glance at briefly, and what strikes me as being particularly interesting, is that, if I understand from a quick look at the data, that the curve of your excess cancer data vs WLM curve, is again incredibly consistent with what has been found elsewhere among the Colorado miners, the Czech miners, and the Swedish non-uranium miners. They're all in about a factor of 2 each other. Am I correct that your risk assessment would be something like 200 cases per million WLM? The figures that Dr. Jacobi and I came up with is a rough average of what has been found around the world.

Dr. Ham:

I don't believe that it is proper to make such an inference from our data; our data does not provide a dose-response curve. If you have an opportunity to examine it carefully, you will see that it is a relative risk calculation within our particular mines population, over the designated time, and I am advised that epidemiologically you cannot use this sample data to infer any general dose-response function for populations. Professor Hewitt who did this work based it on samples (that are subject to statistical error) and he would place no more substantive value on it than to give some statistical evidence that cancer risks in the population in Elliot Lake and Bancroft over the period 1955-74 are not independent of radiation exposure. Please do not interpret the sample data that is at the back of the report as presenting itself as being dose-response data of any general significance. It is not. But it is an interesting addition to evidence of risk at low levels of radiation exposure.

STATEMENT OF IAEA ACTIVITIES IN THE FIELD OF RADIATION PROTECTION
IN MINING AND MILLING OF NUCLEAR MATERIALS
AND PERSONAL AND AREA MONITORING IN GENERAL

J.U. Ahmed
International Atomic Energy Agency
Vienna, Austria

The International Atomic Energy Agency has, from its inception in 1957, been active in the field of radiation protection in mining and milling of nuclear materials. Several panels and scientific meetings were convened on this subject and a number of publications have been issued. The first of these activities was a symposium on Radiological Health and Safety in Mining and Milling of Nuclear Materials (STI/PUB/78) held in 1963 in collaboration with the International Labour Organization.

Nuclear mining and milling industries have been growing rapidly with the rapid expansion of the nuclear power industry. This is the only industry in the nuclear fuel cycle which has associated with it a significant incidence of occupational illness. The Agency, recognizing the significance of the radiation protection problems in this branch of the nuclear fuel cycle felt the need for the development of a code of practice and therefore convened a panel jointly with ILO in 1965 for this purpose. This effort resulted in a joint IAEA/ILO publication, Safety Series No.26, entitled Radiation Protection in the Mining and Milling of Radioactive Ores. This publication is a Code of Practice in the Agency's Safety Series and is complete with a technical addendum.

A panel was convened in Vienna in July 1973 jointly with ILO, to deal with the detailed methods of monitoring and surveillance in mining and milling of nuclear materials. The report of this panel has been published recently in the Agency's Safety Series, with the title Radiological Surveillance in Mining and Milling of Uranium and Thorium (Safety Series No.43). This manual gives details of the principles and actual methods of monitoring radon and radon daughters and thoron and thoron daughters, personnel monitoring and engineering and medical controls.

Another panel on Radon in Uranium Mining was convened in Washington in September 1973 to study the effects of stricter radiation protection standards on the price of uranium (Proceedings STI/PUB/391). The panel concluded that the impact of stricter radiation protection standards is difficult to assess.

However, there was no major effect on uranium prices which could be attributable to the introduction of stricter radiation protection standards because the price of ore is increasing steadily.

In September 1974, a joint ILO/IAEA/WHO/French CEA Symposium was organized in Bordeaux, France, on Radiation Protection in Mining and Milling of Uranium and Thorium. The Proceedings have come out in May this year.

Following a recommendation of the UN Conference on Human Environment (Stockholm, 1972) the Agency convened a panel of experts in Ottawa in July 1974, to review the past and present waste management practices in the uranium and thorium mining and milling industry and to develop the background reference material required to formulate a code of practice for waste management in this industry. The job was completed by a second panel of experts held in Vienna in May 1975. This manual will be published before the end of 1976 and will consist of two parts, the Code of Practice and the Guide to the Code. The Guide to the Code explains how the requirements of the Code will be met.

The subject of radiological hazards in uranium mining was also covered in general by the Panel on Inhalation Risks from Radioactive Contaminants. The report was published in 1972 in the Technical Reports Series (TRS No.142). It contains one chapter on uranium mining which deals, among other things, with the principles of dose estimation from radon decay products, working limits and monitoring of radon and radon daughters.

In the area of personnel dosimetry, the Agency has published the following manuals and guide books:

1) The Use of Film Badges for Personnel Monitoring (Safety Series No.8, 1962).

2) The Basic Requirements for Personnel Monitoring (Safety Series No.14, 1965).

3) Personnel Dosimetry Systems for External Radiation Exposures (Technical Reports Series No.109).

In addition, the Agency organized a symposium on Personnel Dosimetry for Accidental High-Level Exposure to External and Internal Radiation in Vienna in 1965 (STI/PUB/99).

In 1970 the Agency published a guidebook on the Monitoring of Radioactive Contamination on Surfaces (Technical Reports Series No.120). This guidebook covers, among other things, various methods of monitoring contamination by alpha emitters, calibration of monitoring instruments etc.

The Agency published a guidebook in 1973 on Measurement of Short-Range Radiations (Technical Reports Series No.150) which also covers measurement of alpha radiation and radon.

The near future plans for related activities are the following:

1) Revision of the 1968 joint IAEA/ILO Code of Practice on Radiation Protection in Mining and Milling of Radioactive Ores in 1978.

2) Revision of the Safety Series No.14 on the Basic Requirements for Personnel Monitoring in 1977.

Session II
Personal dosimetry

Chairman - Président
J. PRADEL
(France)

Séance II
Dosimétrie individuelle

MEASUREMENT OF EMPLOYEES' INDIVIDUAL
CUMULATIVE EXPOSURES TO RADON DAUGHTERS
AS PRACTICED IN THE UNITED STATES

Robert L. Rock
Denver Technical Support Center
Mining Enforcement and Safety Administration
U.S. Department of the Interior
Denver, Colorado, U.S.A.

ABSTRACT

The Radiation Branch, Denver Technical Support Center, Mining Enforcement and Safety Administration, has been placing emphasis on solving the problems involved in evaluating underground miners' individual cumulative exposures to radon daughters for many years (1)[1].

MESA is concerned that the current exposure record-keeping practices may be inadequate and, as a consequence, we are testing integrating-type personal exposure recorders (dosimeters) which are designed to be worn by men at all times while they are underground. Several types of active and passive devices have been tested. The passive units have so far not proven accurate enough under mine usage; therefore we have concentrated our test efforts for the last few years on an active thermoluminescent-type dosimeter. These tests have been very involved because of the many parameters which can affect response to radon daughter exposure. We believe we have now looked at the more obvious potential interference factors and that we have proven a dosimeter which is far superior to any of our previous exposure evaluation procedures or exposure integration mechanisms.

INTRODUCTION

During the uranium mining boom in the United States in the early 50's when the U.S. Public Health Service first commenced monitoring general radon-daughter exposures, it soon became obvious that the reliable exposure data needed for epidemiological studies and for the protection of miners could only be acquired through the use of some kind of personal exposure integrating device.

Radon-daughter exposures are so extremely variable with ventilation, miner mobility, and barometric conditions, among other factors, that

---
[1] Numbers in parentheses are guides to appended references.

reasonably accurate cumulative exposures are usually not possible to obtain through the ordinary "spot-sampling" method available to mine operators. Currently applied processes of assessing individual employee's accrued exposures are much better than in the early days because of greater diligence, but results can still be very deceiving unless the mine environments happen to be generally consistent; this is seldom the case, especially in mature mines.

In addition to environmental stability, the accuracy of the "spot-sampling" method of determining employee's exposures depends greatly on the sampler's time, judgment, and experience (1).

Unfortunately the odds are stacked against the sampler having enough time to extend his best efforts in determining probable accrued individual exposures. Even given enough time to sample each active mine area daily, it is doubtful that satisfactory results (exposures within ± 25 percent[2]/) are often achieved.

An average long-term exposure of no more than 0.33 WL is required in order not to exceed our "4 WLM per annum" standard. Exposures 10 to 100 times 0.33 WL can occur when ventilation is interrupted or when the miner moves into or through poorly ventilated areas. There is little chance for a conservative estimate of exposures to occur because of the relatively massive overexposure potential involved.

The level at which MESA or state inspectors issue a notice for improving the environment is at 1.0 WL or more. Therefore, so far as MESA is concerned, men are allowed excursions near but below 1.0 WL so long as their cumulative exposure records do not indicate they have exceeded 4 WLM of exposure in any calendar year.

We also have the problem of miners moving from company to company or even from state to state; in these instances, their accrued exposure records do not usually accompany them and are therefore not included in their total yearly exposure accounting. Just how many men are involved in this problem is difficult to say but we do know the miner population is generally one of the more transient groups of industrial employees.

We, of course, become intimately familiar with the actual conditions which detract from the accuracy of current record-keeping procedures when we perform our detailed dosimeter studies. During these tests, a sampler accompanies one miner throughout the program for 10 to 15 consecutive shifts taking conventional samples and we are thus able to see such subtle exposures as 15 minutes at 50 WL or more occurring. Even such short duration exposures at this high a concentration require a full 40 hours of zero exposure to counteract their effect.

CURRENT METHODS OF SAMPLING FOR EXPOSURE PURPOSES

Normally sampling of each working place is performed weekly, however, some mines do not have sampling equipment at the mine at all times. This problem must obviously be corrected so that the exposure effects of anomalous ventilation deficiencies can be accounted for. MESA is in the process of promulgating regulations requiring sampling equipment and competent samplers to be present at each mine during all production shifts.

Miners ordinarily keep track of their own daily work-place occupancy times to the nearest half hour and report these on the backs of

---

[2]/ This is the accuracy-goal which we have established for our first generation integrating exposure measurement devices.

their time cards. Where time cards are not used, first line supervisors usually report estimated occupancy times for the men. Inaccuracies in the reporting of occupancy times and work locations are probably a larger source of inherent error in this accounting system than are nonrepresentative samples, but of course the two problems are interrelated.

The Kusnetz sampling method, which almost everyone here is familiar with, is the sampling method most often used. Recently so-called "Instant" Working Level meters have been introduced for rapid field sampling. These instruments measure either successive composite alpha counts to arrive at a working level estimation or they measure alpha and beta activity separately without directly discriminating from which daughter they originate. Through a computerized system related to probable equilibrium conditions and hence the pCi's of each daughter theoretically indicated by the respective counts, it is possible to arrive at a fairly close estimate of the working level except where extreme disequilibrium conditions exist.

My colleagues in the U.S. Bureau of Mines, Denver Mining Research Center (2), have worked out a graphic system of depicting inherent inaccuracies using the various measurement techniques. Their computerized error analytical system considers disequilibrium and its potential effects related to the methodology and theoretical factors employed but does not include such things as counting statistics and mechanical errors. Figure 1 shows the Bureau of Mines graphic structure of equilibrium conditions represented by several hundred samples collected over the last 10 years by various investigators. Everything on the chart is normalized to 1 working level; therefore, one does not have to envision the chart in a three-dimensional aspect. The X-axis represents RaB in pCi's and the Y-axis represents RaA in pCi's both increasing from the origin. The RaC baseline replaces the need for a Z-axis and is established in the plan section by making it zero pCi's at a point where a perpendicular passing through the origin would yield 1 WL (264 pCi) of RaC. The scale up from the "zero pCi's" baseline is such that 100 pCi of RaC coincides with 100 pCi of RaA and RaB. Limits of inherent error for each measurement method can be established and read directly from this chart for the various equilibrium conditions, i.e., pCi's of RaA, RaB, and RaC which are present.

The Kusnetz measurement chart (figure 2) shows that this method is quite good throughout most equilibrium conditions which are normally encountered. The MDA Scientific, Inc. (MDA) [3]/ instrument is not quite as good as Kusnetz according to figure 3 and at extreme disequilibrium, you can see how the error is increased.

AREA EXPOSURE MONITORS

Several types of area monitors have been constructed to allow continuous measurements of both radon gas and radon daughters in the mine environment. The radon-daughter continuous measurement devices operate on alpha, beta, or a combination of alpha and beta monitoring.

An alpha counting system developed by the USBM, Pittsburgh Mining Research Center, is nothing more than a diode or semi-conductor detector mounted over a filtration system so that it continuously counts the alpha particles emitted from the surface of the filter paper. Decay events are automatically stored in an electronic storage bin and can be read out at the end of the shift through a

---

[3]/ Reference made to specific companies or brand names in this report is made to facilitate understanding and does not imply disapproval or endorsement by Mining Enforcement and Safety Administration.

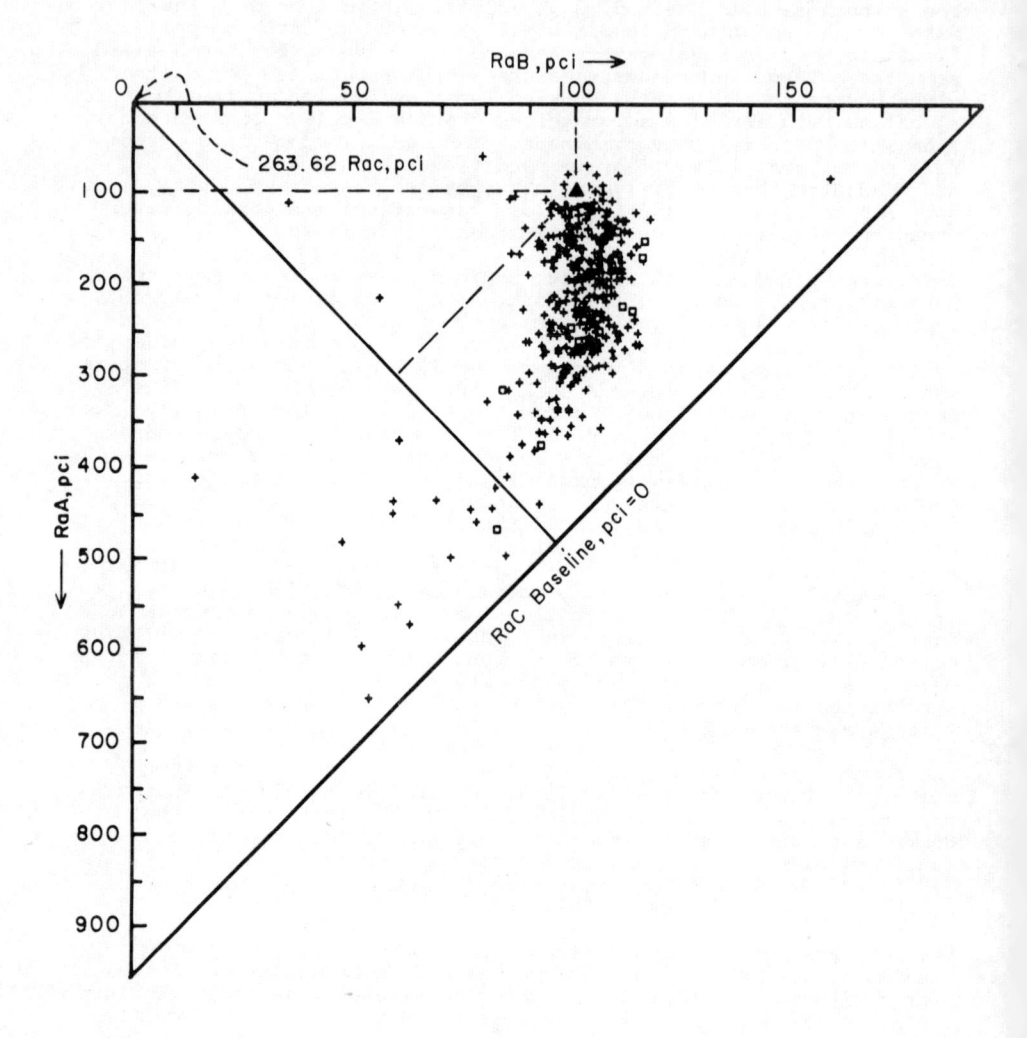

FIGURE 1. - Construction of two-dimensional depiction of 1 working level consisting of different RaA, RaB, and RaC mixtures.

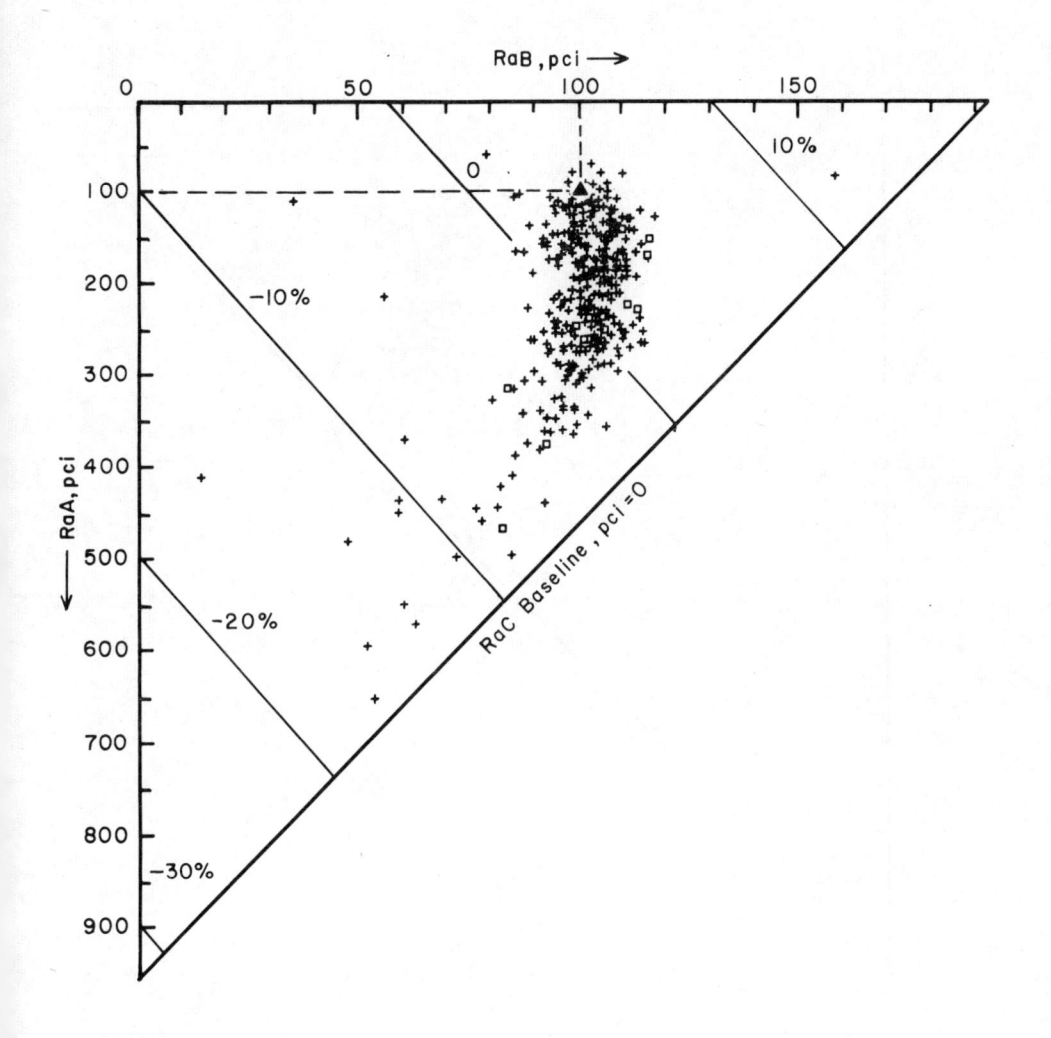

FIGURE 2. - Depiction of inherent error percentages involved in Kusnetz measurements made with variable amounts of RaA, RaB, and RaC present (normalized to 1 WL).

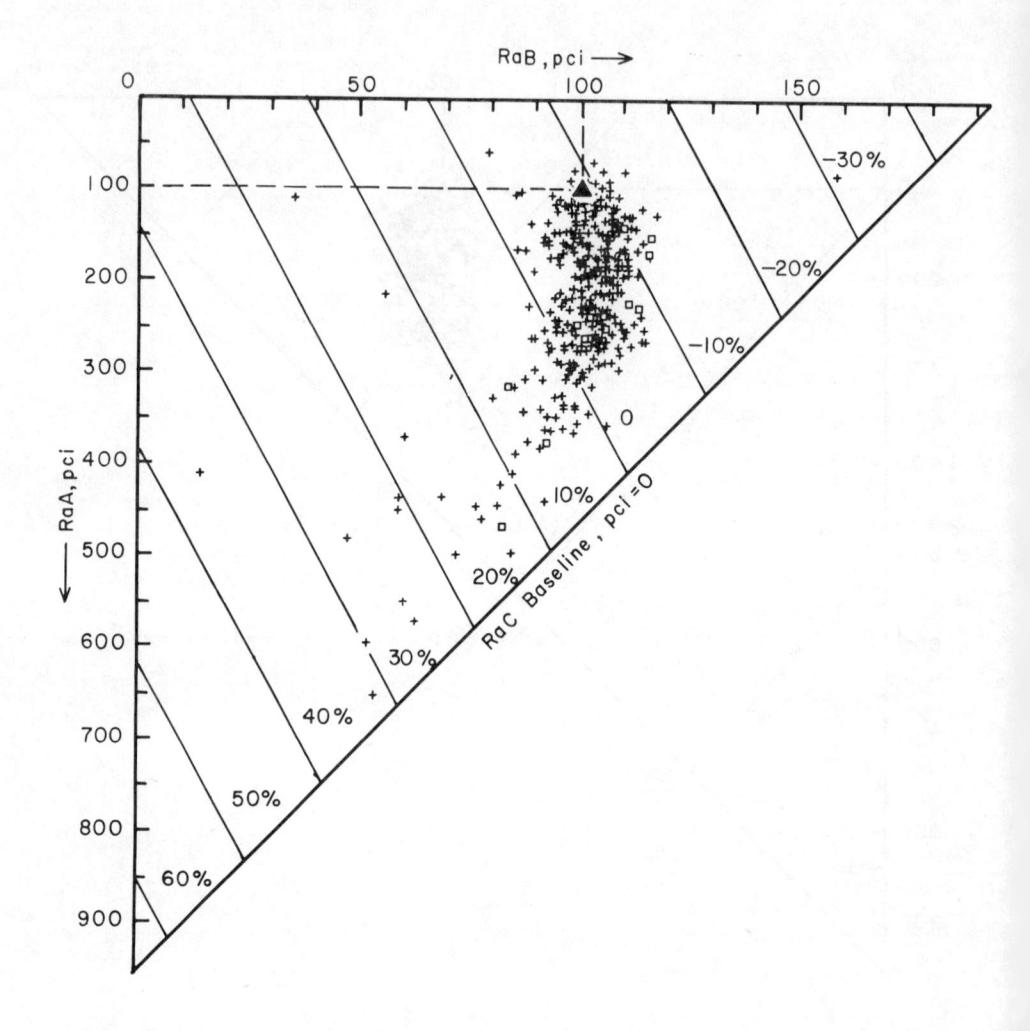

FIGURE 3. - Depiction of inherent error percentages involved in MDA "IWLM" measurements made with variable amounts of RaA, RaB, and RaC present (normalized to 1 WL).

capacitor or an ordinary scaler, depending upon the design. Discrimination between the 6.0 MeV alphas from RaA and 7.7 MeV alphas from RaC' is not required to obtain satisfactory accuracy.

It should be pointed out that the continuous filtration methods of area monitoring require equilibrium conditions to be reached on the filter paper, a minimum of 90 minutes, before short-term working-level measurements can be accurately obtained. Also, there is some hysteresis in response where working level concentrations are quite variable.

Another area monitoring device developed by the USBM, Denver Mining Research Center, utilizes a beta counting system. A pancake geiger tube is mounted underneath the filter paper to count beta and gamma radiation. Gamma background must be subtracted by either measuring it prior to commencing the filtration process or by using an independent detector. This device can be used with a recording strip chart, an electronic storage bin, or can be read out at any time simply by connecting it to a scaler.

Both the "alpha" and "beta" continuous monitors are somewhat sensitive to disequilibrium conditions, however with optimum calibration, the errors for the continuous alpha system seldom exceed $\pm$ 3 percent and the errors for the continuous beta system seldom exceed $\pm$ 8 percent.

A third system of measurement developed by Argonne National Laboratories employs both beta and alpha measurement. Alpha particles in the 6.0 MeV and 7.7 MeV range, representing RaA and RaC' respectively, are counted separately. The RaC' alpha activity represents the RaC present which can then be subtracted from the gross beta activity to obtain the amount of RaB present. Thus an absolute method of measuring working levels is provided which is independent of equilibrium conditions.

INTEGRATED PERSONAL EXPOSURE MEASUREMENT (DOSIMETRY)

General Information

Passive radon-daughter dosimetry methods include track-etch and emulsion film with the latter now being discounted by most investigators because of its high sensitivity to light, temperature, and moisture. The term passive simply means that there is no pump and that the device operates on an infinite-volume principle, recording the registerable radioactive events in the immediate vicinity. Obviously if we could get the necessary accuracy and sensitivity requirements from the passive system it would be far superior to the active system in which a pump and air filtering system are required.

So far track-etch dosimeter experiments have indicated that the method is somewhat lacking in sensitivity and that it also lacks accuracy because of its high degree of variable response with changing radon-daughter to radon gas equilibrium conditions (3,5). There are ways to minimize these deficiencies by measuring the gas separately or by using absorbers to discriminate RaA from RaC', but they become quite complex and have not yet been developed to a practical stage. Another problem with track-etch cellulose nitrate films is that they can register "plated-out" daughters if an intervening layer of dust, moisture, or oil mist forms sufficiently thick on the dosimeter surface to degrade the alpha activity energy. Etching and counting techniques required in track-etch dosimetry techniques also tend to become quite "arty".

For these reasons, we have concentrated our recent dosimetric efforts on the active thermoluminiscent principle. The greatest problem with

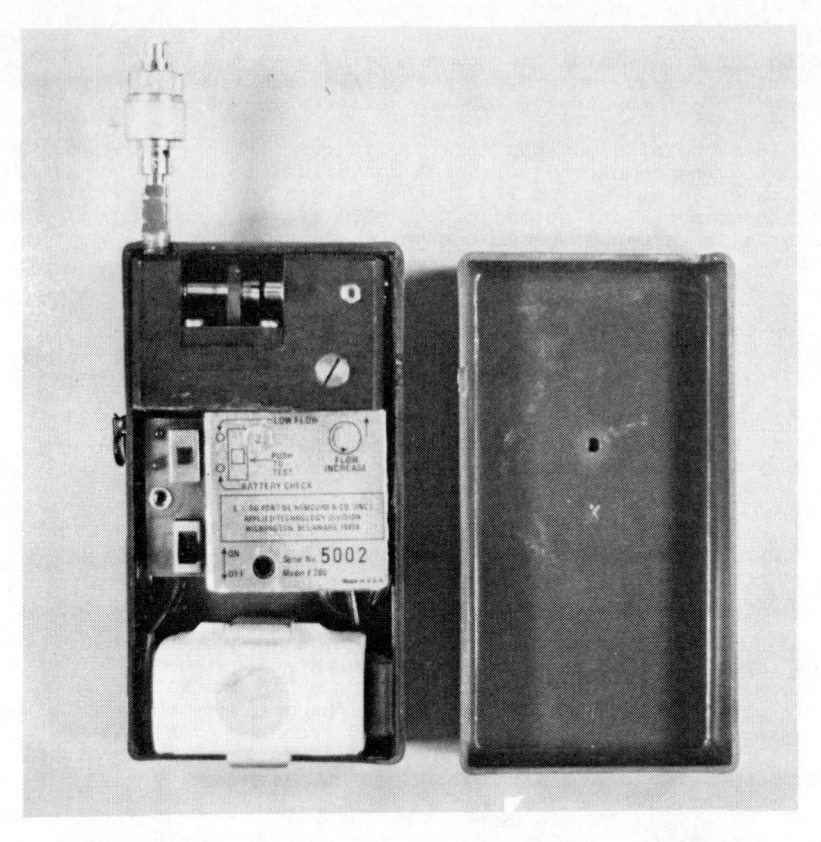

FIGURE 4. - Close up of DuPont Model P-200
            pump with its auxiliary features.

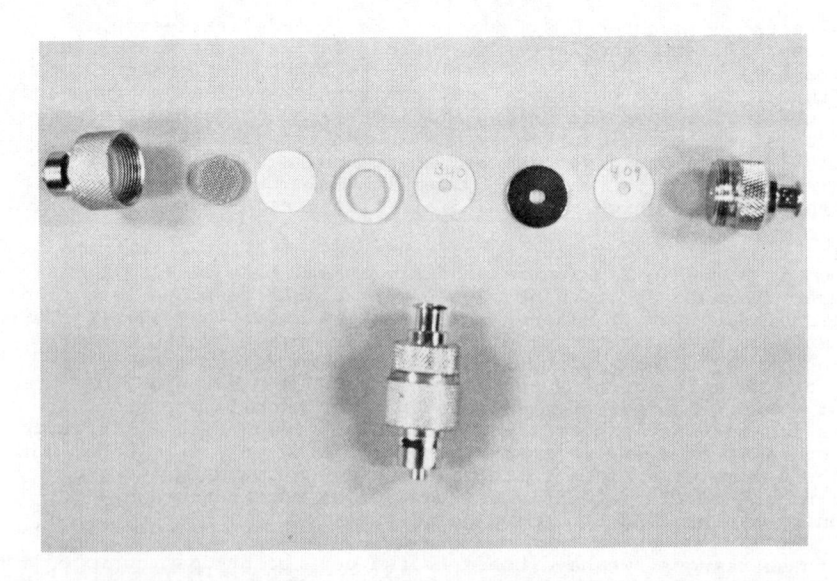

FIGURE 5. - External and internal parts of
            CaF_2:Dy thermoluminescent dosimeter.

this or any other active method is in finding an air-mover which is sufficiently reliable in pumping a constant volume of air through a filter system against highly variable pressures.

We have tested several pumps and are now using the DuPont Model P-200 (see figure 4). This pump has the unique feature of being able to readjust to a uniform flow against varying pressures (0 to 25 inches of water column) through the use of a pressure sensing device located on the exhaust side of the system. The pressure sensing device automatically speeds the diaphram pumping action to maintain an airflow of 100 cc/minute as the pressure drop across the filter paper increases. This is very important because, for satisfactory results utilizing the active system, we must maintain a constant flow rate. The response could be quite biased if we only knew the total volume of air filtered such as is registered by most environmental pump-sampling units.

We are still experimenting extensively with dosimeter chips and filter-holder unitization because this also is highly critical in many respects.

One of our greatest concerns is that the filter and chips not be plugged with water and mud; yet if we make the air inlets too small we may remove some or all of the unattached daughter fraction. Also, if at all possible, we would like to be able to use the same filter for a full month (about 22 working shifts) without changing it. We know from experience that in very wet mines or in mines in which diesel equipment is used extensively that we are going to have to make provision to change filters before the month has expired. Where changing filters becomes necessary, it must be possible to do so without changing the indexing of the active dosimeter chip relative to the filter.

Current dosimeter heads being experimented with are Swinny filter syringe-type holders with Luer-Loc fittings and custom designed plastic heads with radial air inlet openings. The latter head has the advantage of inhibiting the entrance of mud and water into the system but lacks strength at the point where it screws into the pump.

Both types of heads accomodate an Acropor 13-mm filter, an active $CaF_2:Dy$ TLD chip and a gamma radiation background TLD chip of the same material. A steel beta absorber washer prevents the background chip from registering beta activity directly from the nuclides collected on the filter paper. Figure 5 shows the internal parts of the dosimeter head. Figure 6 shows the pump and dosimeter head unit being worn by a miner.

We have decided that it is best to have the dosimeter head mounted on the pump which is to be worn in the area of the miner's right or left breast pocket. Our reason for reaching this decision is that we are certain that there will be work-locations and duties which will require removal of the dosimeter from the miner's person so that it can function as an area monitor. We would like to minimize any need for removing the unit from the man but we know that the dosimeter would become quickly plugged with mud and water if it were worn by a miner drilling overhead holes.

Briefly the way the thermoluminescent material works is that incident ionizing radiation causes electron-hole pairs to be formed in the material. Displaced electrons and holes become fixed in the lattice structure of the material (in this case $CaFl_2:Dy$ in a teflon matrix). When heat is applied to the material, the electrons recombine into the holes and in so doing photons are given off in proportion to the amount of radiation absorbed. A photomultiplier tube is used to

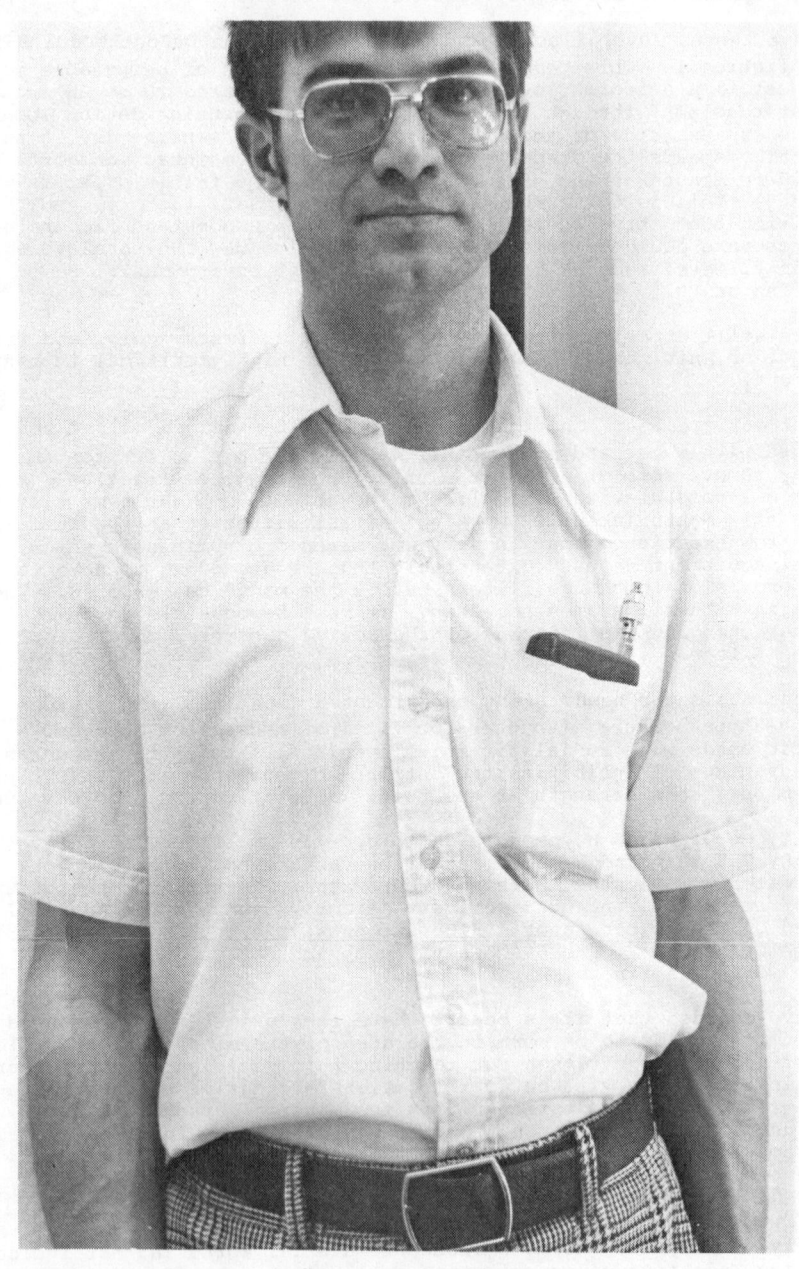

FIGURE 6. - Unitized pump and dosimeter worn by miners.

measure the amount of light given off.

A graph of the light intensity given off by the thermoluminescent material as it is heated is called a glow curve. Glow peaks occur on this curve where temperatures are sufficient to empty a particular energy state. The amplitude or area under a stable portion of the curve can be used as a measure of the radioactivity absorbed.

Low temperature glow peaks are subject to fade at ordinary temperatures and therefore must be removed by preannealing before reading the stable glow peaks. This preannealing process must be carefully controlled to obtain reproducible results.

As we collect the radioactive aerosols on the filter paper at a flow rate of 100 cc/minute, the dosimeter chip closest to the filter is exposed to not only the alpha, beta, and gamma radiation from the decaying daughter nuclides, but also external gamma radiation. Therefore we must have a way of subtracting this variable amount of radiation which has no direct relationship to radon-daughter concentrations. This is accomplished by counting the passive chip which is shielded by a steel washer from alpha and beta activity on the filter. The gamma background chip will also be useful in allowing us to routinely measure miners' exposures to gamma radiation which, although usually of low magnitude, can be significant under abnormal circumstances.

Our original selection of $CaF_2:Dy$ in a teflon chip form (TLD-200) was based on its sensitivity, stability of high-temperature glow peaks, and ease of handling. Although other dosimeter compounds such as $CaSO_4:Dy$, $CaSO_4$-TM, and BeO have found favor with some investigators, the satisfactory results which we have obtained using $CaF_2:Dy$ have led us to believe that we do not need to make extensive tests of other thermoluminescent materials.

The response obtained from a thermoluminescent dosimeter is very dependent on the readout procedure used. Previous investigators have produced glow curves for TL-200 dosimeters exposed to gamma rays and heated at a very slow rate. Depending upon the energy of the ionizing radiation, glow peaks of various amplitudes appear at approximately 115°C, 135°C, 190°C, and 250°C. Faster heating rates, however, will shift the peak maxima toward higher temperatures. Also, TLD's exposed to alpha radiation exhibit glow curves somewhat different from those exposed to gamma radiation. It is possible to obtain the same total light output from a dosimeter annealed for several minutes at a low temperature as is obtained from one annealed for only several seconds at a higher temperature. Both processes will remove the low temperature traps but the shapes of the glow curves will be different. We have studied both low and high temperature glow peak removal with different TLD readers. Either method is adequate, but the short-term higher temperature pre-readout annealing can be accomplished in most TLD readers using their built-in preheat cycle. This is much more convenient than low temperature - longer annealing, which can only be accomplished by a batch process using an oven.

Readout reliability is partially a function of the thermal contact between the heating element and the dosimeter. The best TLD reader from this standpoint is the Harshaw, Model No. 2000A which does not feature a preheat cycle.

We are currently using an Eberline TLR-5 reader set for a 15-second preheat cycle at 167°C and a 30-second readout cycle at 290°C. The ramp function (∿25°C/sec) is quite fast. We achieve about the same results as we did with our previous procedure of annealing the TLD's for 10 minutes in an oven at 120°C prior to readout at 290°C. The readout temperature is not nearly as critical as the preheat temperature.

The Victoreen Model 2800 and Teledyne Model 7300 readers also both have built-in preheat cycles. These readers display very stable heat controls and have a means of holding the TLD chip in positive thermal contact with the heating element. The chip "hold-downs" on these instruments are rather primitive however compared to that of the Harshaw reader and are really unacceptable where fast high volume readout is required. The Victoreen, Teledyne, and Harshaw readers all showed reliable results in preliminary testing, but none has all the features which would be desirable. The Harshaw reader with the Teledyne preheat cycle would seem to be the optimum instrument.

Test Procedures and Results

Our original testing of the TLD-200 dosimeters was performed to confirm the results obtained by Colorado State University investigators (4); to check the reliability of our read-out system, annealing procedures, handling procedures; and to arrive at an empirical equation relating TLD response to total working level hours of exposure.

We found no appreciable fading of the higher temperature glow peaks for the TLD-200 dosimeters after being irradiated with gamma exposures in the range of 300 to 2500 mR. In these fade-tests, as well as earlier tests, we used 10-minute post-irradiation oven annealing at about 120°C and read the dosimeters in an Eberline TLR-5 using a 30-second preheat setting of 150°C. It is now apparent that such a preheat process had little effect on dosimeters which had already been oven-annealed.

Some of our early difficulties involved changes in the sensitivity of the instrument PM tube, due to aging and room temperature changes. This was overcome by warming the instrument up by cycling it several times and adjusting the sensitivity of the PM tube frequently to a known value (the reference light source). Another problem we encountered was the difficulty in maintaining the post-irradiation annealing temperature close to the 120°C value. The temperature in the muffle furnace first used was found to be as much as 15°C in error at times, usually on the low side. When the muffle furnace was replaced with a reliable oven, the results were greatly improved. During these early tests, a muffle furnace was also used to wipe out any portion of the dosimeter glow curve remaining after readout in order to render the chips reusable. This was done by annealing the chips at 300°C for 3 hours. It was found, that when the chips were allowed to cool too rapidly following annealing, warping occurred which adversely affected later exposure and readout reliability.

Even with the difficulties described above, we were able to derive an empirical equation, WLH = 2.048 + 0.006044 (Active-Background) which describes response results quite well (within $\pm$ 15 percent) for TLD-200 dosimeters exposed to known radon-daughter concentrations. The tests which resulted in this equation were performed under very carefully controlled conditions using the sophisticated USBM radon chamber located in Denver. This early derived equation has proven to represent a good average relationship for all laboratory and field data acquired to date.

The original HASL-developed air pumps (later replaced by the DuPont pumps) were unreliable in laboratory tests, but the concurrent laboratory tests of the TLD system itself were encouraging enough to stimulate further efforts. Hence we conducted our first field tests of the TLD system in the Twilight Experimental mine near Uravan,

Colorado. This mine is well equipped with laboratory-type radon and radon-daughter remote monitoring equipment. It also has automatic equipment which records barometric pressure, relative humidity, temperature, condensation nuclei and ventilation. The Denver Mining Research Center, USBM, operates the mine specifically for this type of testing and they have done a remarkable job of instrumenting and equipping both surface and underground facilities.

During the first week of testing, the dosimeter heads were affixed to a manifold system utilizing 10 individually controlled inlets connected to a large vacuum pump. This system provided unreliable flow rates and was replaced with a system using a smaller pump connected to a six-dosimeter inlet manifold with all airflow controlled by a single master valve. The previously mentioned new DuPont "constant flow" pumps were also used during these first field tests. Diesel smoke and drill oil mist were introduced into the mine atmosphere to simulate actual mine environmental conditions; working levels were monitored every 15 to 30 minutes. Even with the relatively unstable flow rates known to be occurring using the first manifold system during the first week of testing, the response vs. exposure results of the first two weeks of field tests were reasonably close (-12 to + 18 percent) to those obtained in the Denver Radon-chamber tests. However, during the third week of testing, exposures calculated using the radon-chamber-developed equation were as much as 40 percent greater than conventionally measured exposures. This prompted additional radon-chamber testing in an attempt to find the source of this abrupt deviation.

The effects of different filter types, MEK (methyl ethyl ketone) and ultrasonic cleaning, white light exposure, and possible intermittent air leakage around the filters were all evaluated. None of these factors were found to account for the sudden divergence of the dosimeter-indicated exposures from the measured exposures. The trouble was found to be the lack of adequate temperature control over the post-irradiation annealing process. Since much of the laboratory testing of the dosimeters following the first field tests also involved improperly controlled post-irradiation annealing, the data obtained is not really meaningful other than that it served to isolate the problem. For those tests in which the dosimeters were known to be properly post-irradiation annealed ($\sim 120°C$), the empirical curve is still very close to that which was initially derived from radon-chamber tests.

Following resolution of the post-irradiation annealing problem, a second field test of the dosimeter system was carried out at the Twilight Experimental Mine. Mine conditions were much the same as during the previous field test. Dosimeters were tested using both the DMRC improved 10-inlet manifold air-mover system and 10 DuPont pumps. The empirical curve derived from this test data was WLH = -3.21 + 0.006556 (Active-Background).

Filter-loading tests were also performed at this time using 13-mm Acropor filters at flow rates of 100 and 200 cc/min. At a flow rate of 200 cc/min the filters plugged after more than 20 hours of continuous operation. However, later tests of the same filters, after allowing 16 hours of drying time, show that most of the pressure loss was due to moisture. Normal humidity does not seem to present a serious filter-loading problem if the flow rate is reduced to 100 cc/min and overnight drying is allowed following each shift. However, longer term use in a very wet environment may cause plugging even at 100-cc/min flow rates. Tests are currently under way to determine the potential severity of this problem.

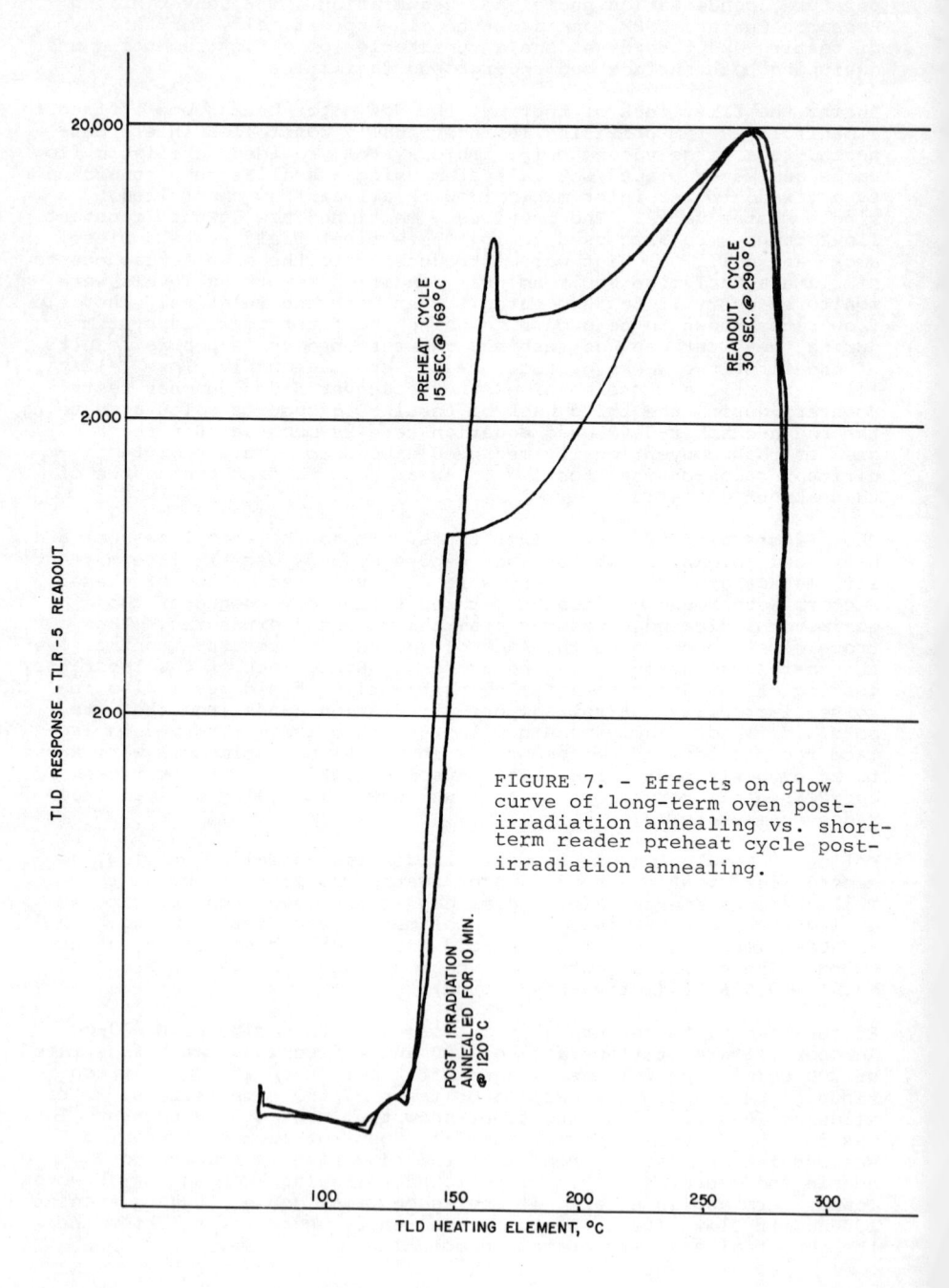

FIGURE 7. - Effects on glow curve of long-term oven post-irradiation annealing vs. short-term reader preheat cycle post-irradiation annealing.

Following the second Twilight mine tests, additional laboratory testing and glow curve analyses were conducted to further define the best post-irradiation annealing and readout temperatures. The values arrived at were the same as those mentioned previously and are still in use.

Potential problems from rock dust loading of the Acropor filters was also studied. This does not seem to pose a problem.

The new DuPont pumps continued to be tested and performed well in the laboratory. We discovered some pump flow rate instability caused by loose needle valve threads which was later corrected by the DuPont Company.

The first field tests in a working mine were conducted during April and May of this year using completely unitized (DuPont pumps with Swinny heads) personal exposure integrating devices. We were primarily interested in introducing the dosimetry system to industry personnel and in testing the ruggedness of the pumps and dosimeter heads under actual usage. An acceptable degree of reliability and ruggedness was demonstrated by the DuPont pumps, but some of the Swinny dosimeter heads became plugged with mud and water during drilling operations. Dosimeter-indicated exposures, however, still appeared quite reasonable (within $\pm$ 20 percent) compared to the more nebulous exposures which we determined by time-weighting conventional samples.

Current testing involves the new SIENCO plastic dosimeter head, DuPont pumps, and Acropor filters.

The preheat cycle in the Eberline TLD reader has been adjusted to completely eliminate the need for the oven post-irradiation annealing process (See figure 7).

Dosimeters are currently being used to measure low-level radon-daughter concentrations in National Park Service caves and in the laboratory. In both situations they appear to be responding well. Another "operating-mine" test was conducted during mid-September but the results have not been compiled.

Figure 8 shows the relative response vs. exposure for laboratory data while figure 9 shows response vs. exposure for mine test data. Between 160 and 170 counts indicates 1WL-hr using our current readout system.

CONCLUSIONS

Our current method of spot-sampling work places at weekly intervals, or less frequently, generally yields unreliable radon-daughter exposure data. The need for more reliable exposure data is apparent. Reliable data is needed both for evaluating compliance with our current "4 WLM per annum" standard and for relating miners' actual exposures to the incidence of radiation-induced lung disease.

We believe our current dosimeter is reasonably reliable and practical to use. Current data would indicate an accuracy better than $\pm$ 20 percent within 95 percent confidence limits. We recognize that component improvements will develop with experience, but we do not believe it prudent to await near perfection before adopting the use of this device which offers such a great advantage in accuracy over current exposure estimation practices.

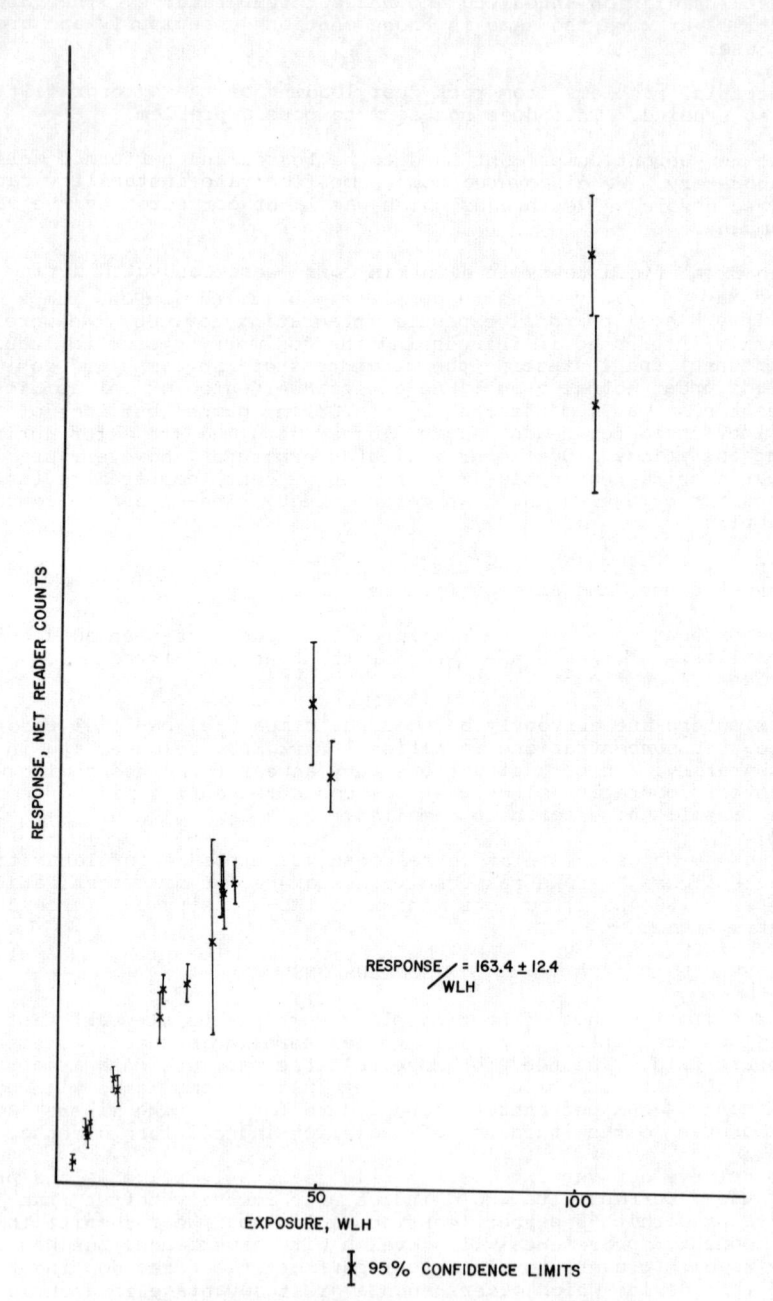

FIGURE 8. - Response vs. exposure correlation of $CaFl_2$:Dy chips exposed in the laboratory.

FIGURE 9. - Response vs. exposure correlation of $CaFl_2:Dy$ chips exposed in mine environments.

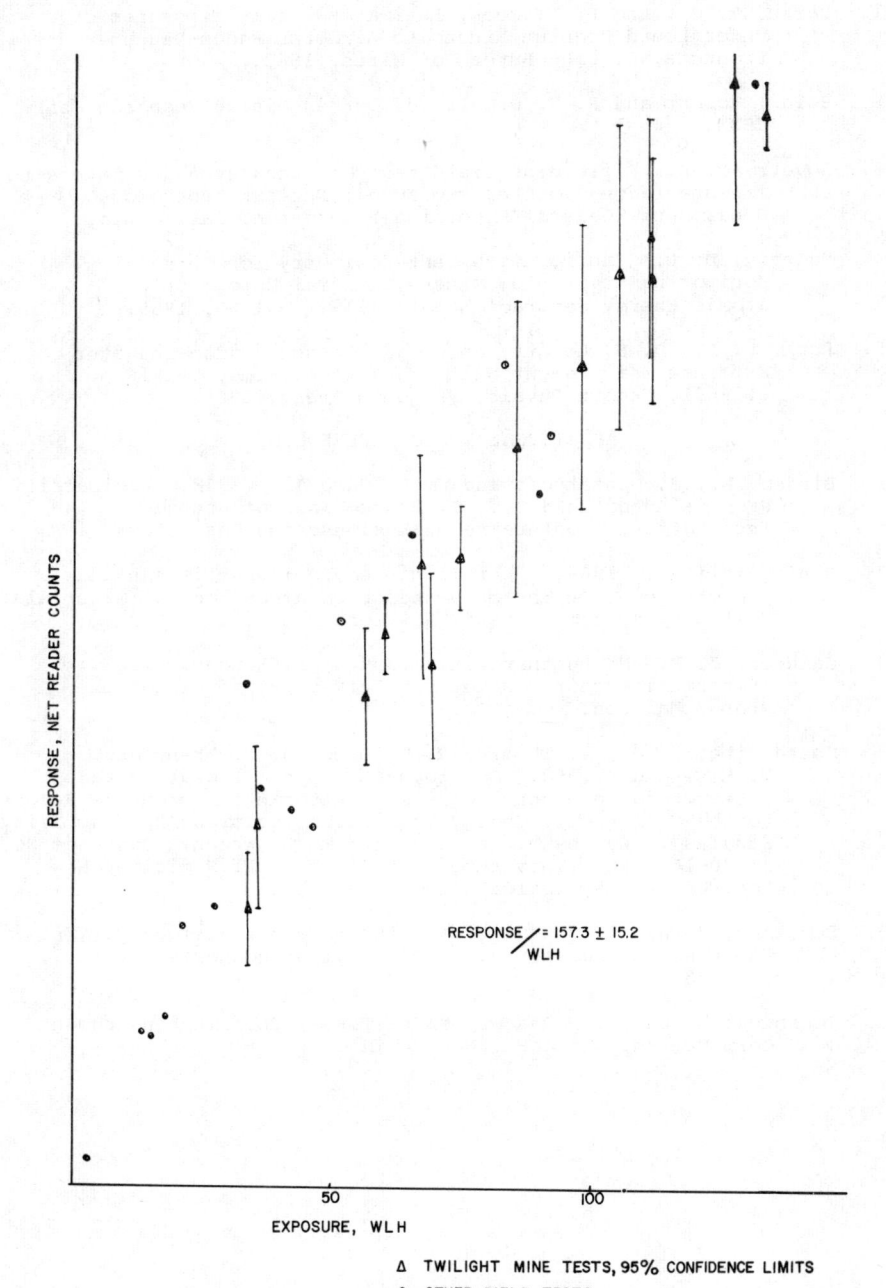

REFERENCES

1. Bates, R. C., and R. L. Rock, Estimating Daily Exposures of Underground Uranium Miners to Airborne Radon-Daughter Products, RI 6106 Bureau of Mines, 1962.

2. Holub, Robert and R. F. Droullard, Denver Mining Research Center, USBM.

3. Lovett, D. B., Rifle mine field test of track etch dosimeters to measure radon-daughter exposure. Nuclear Technology Department, General Electric, Pleasanton, California.

4. McCurdy, D. E., Thermoluminescent Dosimetry for Personal Monitoring of Uranium Miners, Special Report on U.S. Atomic Energy Contract No. AT (11-1) - 1500, 1969.

5. Rock, R. L., D. B. Lovett, and S. C. Nelson, Radon-Daughter Exposure Measurement With Track Etch Films, Vol 16 pp. 617-621, Health Physics Pergamon Press, 1969.

ADDITIONAL TOPIC REFERENCES

1. Binder, W., S. Disterhoft and J. R. Cameron. 1968. Dosimetric Properties of $CaF_2:Dy$. In Proceedings of Second International Conference on Luminescence Dosimetry.

2. Bonfiglioli, G. 1968. Thermoluminescence: What it can and cannot show. In Thermoluminescence of Geological Materials, Edited by Academic Press, New York. p. 15.

3. Cameron, J. R., N. Suntharalingam and G. N. Kenney. 1968. Thermoluminescent Dosimetry. University of Wisconsin Press, Madison.

4. Costa-Ribeiro, C., J. Thomas, R. T. Drew, M. E. Wrenn and M. Eisenbud. 1968. Development of an integrating radon detector for personnel or area monitoring. In Annual Report to the U.S. Atomic Energy Commission entitled "Radioactivity Studies". Ed. by M. Eisenbud and M. E. Wrenn. Contract No. AT(30-1)-3086. Institute of Environmental Medicine, New York University Medical Center. P. VI-1.

5. Schulman, J. H. 1967. Survey of luminescence dosimetry. In Luminescence Dosimetry. A.E.C. Symposium Series, Vol. 8. p. 3-33.

6. Technical Manual, TLD Reader, Model TLR-5, Eberline Instrument Corporation, Santa Fe, New Mexico.

Discussion

1. Q. Is your dosimeter available commercially in the United States and, if not, how soon do you expect it to be made available?

   A. All the components used in the dosimeter are available commercially but a unitized system is not for sale at this time. We expect to make some improvements on the dosimeter - filter holder component which would then become a custom item. I would expect a commercially available dosimeter within the next 2 years, although this could certainly be shortened by the promulgation of regulations requiring their use.

2. Q. How serious is the exposure record-keeping problem? Do you feel that diligent weekly sampling would allow an acceptable level of accuracy?

   A. The seriousness (inaccuracy) of exposure record-keeping practices is of course variable with mine conditions, ventilation consistency, miner mobility, and the effort put forth to assure the adequacy of the record-keeping program. Certainly there must be sampling equipment available at the mine to allow accounting for exposures related to obvious changes in ventilation. In general, I would say the problem is serious. We are not getting reliable exposure accounting figures according to our own independent sampling and exposure studies. Reported exposures may well be underestimated by a factor of 10 in some instances but it is my opinion that the average underestimation would be closer to a factor of 2. Weekly sampling would certainly be an improvement over monthly sampling, but I would not consider this sampling frequency the answer to our exposure accounting problems. We need an integrating personal exposure recording device.

3. Q. Where and how does the current exposure accounting practices most often fail?

   A. Our standards for record-keeping are actually of the "honor system" variety. Where first-line production personnel supervise ventilation and radiation control personnel, there is a conflict of interest. Because of lost production, while radiation control measures are being undertaken, and accompanying pressures to increase production by upper management, the first lime supervisors are caught up in a dilemma. Few supervisors intend to allow workers to be exposed to hazardous radiation but they often must make a compromise. If the pressures for production are great enough, the first line supervisors must often have to give production priority over health and safety matters. Small or financially troubled mines most often give us problems in this regard although this is not always the case.

4. Q. You are probably one of the most well informed persons in the U.S. on practical radon-daughter control measures because of your long and intimate association with these problems in the mines. Could you give us a brief description of which control measures are being most successfully applied? Are rock sealants effective and are respirators widely used? Also is air cleaning utilizing filters or electrostatic precipitators an answer to the problem?

A. The best control methods incorporate mine design relative to ventilation patterns so that as little contamination from mined out areas is allowed to affect persons working in active production areas. In essence this means split ventilation systems and retreat mining toward intake air.

Much experimentation has been done on the possibility of using rock sealants. So far they have not been proven either effective or practical. Respirators can effectively protect miners who do not have to perform sustained arduous work but MESA takes the position that they should be used only until and while environmental control is being established. Exposure credit is not given for wearing respirators unless a formal variance has been given.

Air filtration and electrostatic precipitators remove daughter products very efficiently but do not remove radon. Their main application therefore is near the terminus of the ventilation circuit where only a small amount of ore remains to be mined and the filtered air can be exhausted after use immediately into the return air system. Air cleansing is not a planned control measure but rather is most often applied as a matter of expediency. High maintenance costs are involved in both electrostatic and filtration cleansing systems.

5. Q. Do you foresee any major break-throughs in the development of new effective environmental control measures for controlling exposures to radon daughters?

A. Not really. The ultimate would be either the development of a means of preventing radon gas from entering the mine or the ability to remove radon from the mine air. Both have been investigated with rather discouraging results.

6. Q. What has been your experience with the accuracy of the so called "Instant" working level meters?

A. I showed you the inherent accuracy chart for the MDA Scientific Inc. instrument. Some people have reported favorable results while others have not. Our experience has been that the accuracy of the instrument suffers just as the U.S. Bureau of Mines chart shows, that is at very low equilibrium values. Company representatives believe they can compensate for this. Other IWLM's have either been too bulky for practical field use or mechanically unreliable. We have not tested all of the instruments which are available but will as time permits.

7. Q. We have several countries represented here who are concerned about radiation in mines other than uranium mines. Is this also of concern in the United States?

A. Yes it is. In fact I believe there are probably more non-uranium miners exposed to potentially hazardous amounts of radon daughters than there are uranium miners. Fortunately the non-uranium miners are not subject to massive overexposure as is possible for uranium miners.

MOD WORKING LEVEL DOSIMETER

A. J. Breslin

Health and Safety Laboratory, U.S. Energy Research and Development Administration
New York, New York 10014 (United States)

The Health and Safety Laboratory (HASL) became involved in the evaluation of radon and working level (WL) dosimeters for uranium miners in the late 1960s when a number of prototype instruments were introduced by various research and development facilities. Some of the dosimeters performed satisfactorily in laboratory tests but none were found to be reliable in tests conducted in uranium mines. The most common deficiency was mechanical or electrical vulnerability to the harsh environmental conditions.

Because of the lack of a satisfactory instrument for use in uranium mines, HASL undertook development of a working level dosimeter, adopting some of the technology pioneered by the earlier groups and placing special emphasis on sufficient durability to withstand mine conditions that had caused failures in previous units. The selected mode of detection was thermoluminescent dosimetry that had been shown to be well-suited for this application by the workers at Colorado State University[1] and at Massachusetts Institute of Technology.[2]

The detector itself, consisting of a lithium fluoride chip and a membrane filter enclosed in a small metal cap, was patterned on the basic arrangement devised by those two groups but incorporated features introduced by HASL to simplify handling and improve reliability. A schematic of the detector head, which is intended for mounting on the underside of the hard hat brim, is shown in Figure 1. Outside dimensions are 2 cm diameter and 3 cm long and the weight is 45 gms. Air is drawn from outside through three radial apertures and thence through a 1 cm diameter 0.8 µm pore size membrane filter. A .32 x .32 x .04 cm lithium fluoride chip is supported above the membrane filter at a distance of 1.6 mm. An identical second chip is housed in a separate compartment where it is shielded from alpha and beta radiation from the filter. This chip is used to correct for external gamma radiation to which both chips are subjected.

The pump, shown in Figure 2 with the detector mounted on a hard hat, received special attention, a deliberate decision having been made to strive for reliable operation and durability at the sacrifice, if necessary, of convenience with regard to size and weight. Although not as readily attained as anticipated, those objectives are met in the present model which has undergone several alterations since the prototype. The pump, powered by a self-contained, 12v, 225 mAh rechargeable Ni-Cd battery, draws air through the detector head at 90 ml/min for

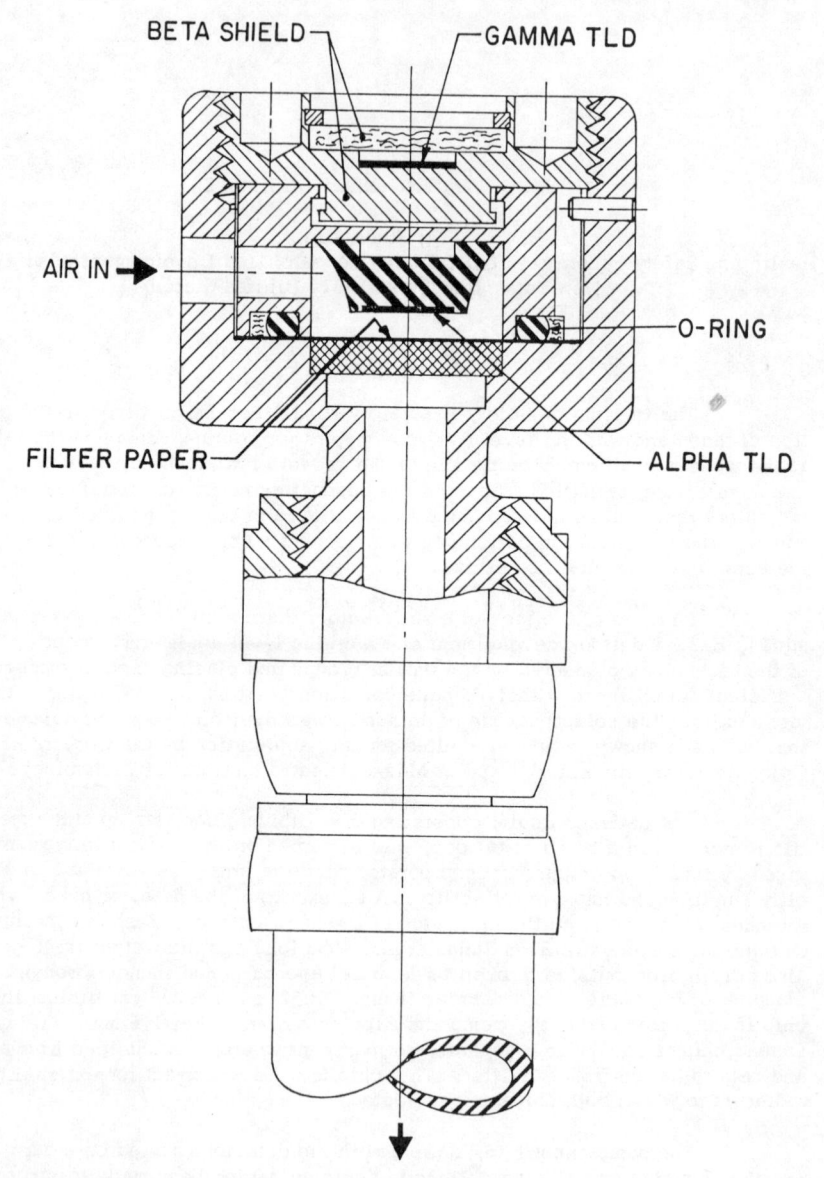

FIGURE 1. CROSS SECTION VIEW OF MOD DETECTOR.

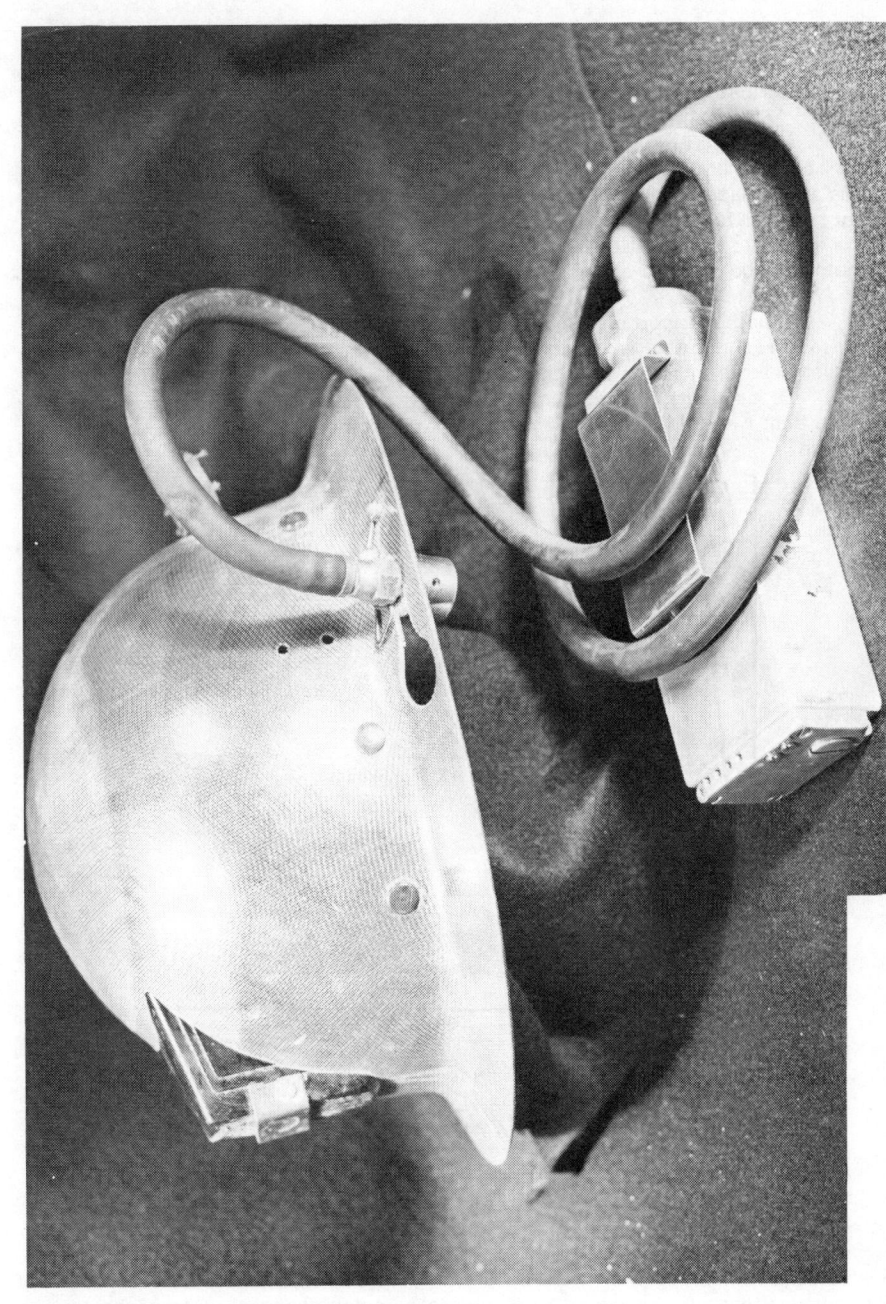

FIGURE 2. MOD WORKING LEVEL DOSIMETER

as long as 10 hours after the battery is fully charged. The pumping element is a single-acting diaphragm with dynamically activated flapper valves. It is driven by a solenoid at a pulse rate determined by an electric timing circuit normally set at 1.5 s^{-1} for the desired flow of 90 ml/min.

Batteries are charged by inserting the pumps in a special rack that accommodates six units. Placement of a pump in the rack automatically establishes electrical contact with charging electrodes.

An important feature of the pump is the extruded aluminum case which is very durable and strongly resistant to mechanical shock. It also incorporates filters and a maze to prevent the intrusion of dirt and moisture at the inlet and outlet ports. The case comprises nearly half of the pump's total weight of 1 kg. This weight has been criticized by some prospective users as too burdensome but it could only be reduced at possible sacrifice in mechanical integrity.

Another important aspect of reliability is the relative insensitivity of flowrate to resistance. From a typical curve shown in Figure 3, it can be seen that even a large increase from the initial resistance of a detector head causes only a modest reduction in flowrate. In actual practice, increases of resistance by more than two-fold are seldom observed after the usual 40 hours of operation in a mine.

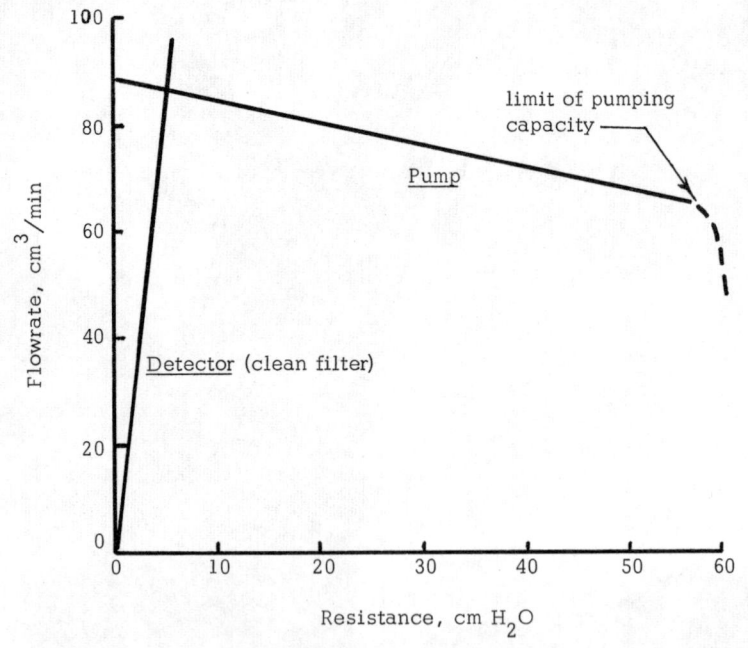

Figure 3. Flow-Resistance Characteristics of MOD Pump and MOD Detector

The successive models of the working level dosimeter have been subjected to a series of laboratory and field tests over a period of five years. Performance in the laboratory has been satisfactory from the beginning but mine tests, in which the dosimeters were worn by working uranium miners, revealed a number of malfunctions that were remedied by appropriate design changes. One of the most persistent problems was the intrusion of dirt into the pumping mechanism. Eventually, this was eliminated by the aforementioned filters and mazes at the inlet and outlet ports.

By arrangement with the Canadian Ministry of Natural Resources, HASL conducted collaborative tests of the working level dosimeter with Rio Algom Limited in a uranium mine at Elliot Lake, Ontario during the summer of 1975 and, in the summer of 1976, additional tests were sponsored by the Atomic Energy Control Board at the same location. In the latter tests, HASL provided the dosimeters but all field work was performed by an Atomic Energy Control Board consultant and TLDs were analyzed by the Radiation Protection Bureau of Health and Welfare Canada. In both tests, the dosimeters were worn by miners engaged in normal mining activities. Pump malfunctions still occurred in the 1975 tests but none were observed in the following summer. Further testing ought to be performed to substantiate the pump's reliability but this latest result indicates that basic flaws have been eliminated.

A useful indication of pump performance in the 1976 tests is the record of flow rates that were obtained as a check at the beginning and end of work shifts.[3] End-of-shift flow rates were within ±4% of initial flow rates.

In addition to documenting pump performance, the mine tests in both 1975 and 1976 provided comparative measurements of miners' exposures by the MOD WL dosimeter and by conventional air sampling methods. The results from both years are shown in Figure 4. There is considerable scatter in the data but the average pair difference was 31% in the 1975 tests and 28% in the 1976 tests. The coefficient of correlation for the combined sets of data is 0.94.

It seems reasonable to conclude that the MOD WL dosimeter, having undergone repeated testing in the field, is now a reliable instrument for use in uranium mines. Its weight and size are reported to be inconvenient by some of the miners who have worn the dosimeter. Both could be reduced by redesigning the case but this might result in a reduction of resistance to mechanical and environmental stresses. The cost for the commercial production of small quantities (50-100) of the working level dosimeters is estimated to be $200-$250 per unit.

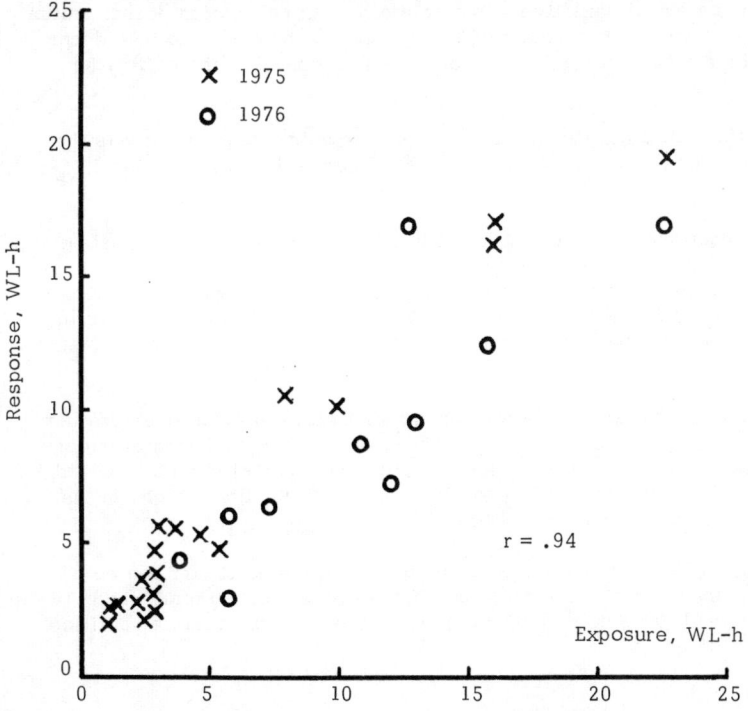

Figure 4. Uranium Mine Tests of MOD WL Dosimeter

References

1. McCurdy, David E.
   Thermoluminescent Dosimeter for Personal Monitoring of Uranium Miners
   Colorado State University Report COO 1500-16, June 1969.

2. Evans, Robley D. and Robert J. Kolenkow
   Massachusetts Institute of Technology Annual Progress Report MIT-952-5,
   Part I, May 1968.

3. Personal Communication from G. R. Yourt.

Discussion

Session II, Discussion following Mr. Breslin's communication.

Several questions were raised concerning the effects of different variables on the response of the working level monitor. The first concerned the possible effect of radon daughter equilibrium ratios. Because of the integrating manner in which the monitor operates, it provides an accurate measurement of working level independent of radon daughter ratio. In effect, the total, potential alpha energy of radon daughters in the sampled air is measured exactly consistent with the definition of working level.

A related question concerned the effect of beta decay on the detector. The detector is somewhat sensitive to beta radiation but calibration of the monitor is empirical and the small component of TLD response attributable to beta decay from the collected radon daughters is automatically included in the calibration factor.

A third atmospheric variable that was brought into question was the effect of humidity on response. None has been observed in a wide range of laboratory and field test conditions.

Another questioner asked why lithium fluoride was selected instead of a more sensitive thermoluminescent material such as calcium fluoride. The basic reasons are that lithium fluoride has adequate sensitivity for most applications (exposures as low as 1 WL-h can be measured satisfactorily), is almost entirely free of fading, and is much less affected by light and temperature than the other materials.

A panel member asked about the upper limit of exposure measurement. According to McCurdy (Thermoluminescent Dosimeter for Personal Monitoring of Uranium Miners, Colorado State University report COO-1500-16, June 1969), the maximum detectable exposure for lithium fluoride is 10^4 WL-liter which, in the HASL dosimeter, is equivalent to 10^5 WL-h.

The final questioner asked if the weight of the dosimeter can be reduced. Since the weight of the pump case is nearly half of the total weight of 1 kg, reduction could be achieved by redesigning the case but at some sacrifice in strength.

MEASUREMENT AND RECORDING OF OCCUPATIONAL

RADIATION EXPOSURES FOR URANIUM MINERS

D. Grogan
Radiation Dosimetry Division, Health Protection Branch
Department National Health and Welfare
Ottawa, Canada.

Introduction

The Radiation Dosimetry Division is currently involved in two areas relating to Personnel Dosimetry of Uranium Miners. The first concerns development of a reliable working level personnel dosimeter which would be suitable, if possible, for automated processing, and the second, the collation of all uranium miner dose records in the computerized National Dose Registry (NDR) maintained by the Radiation Dosimetry Division. The current status of each of these projects is briefly described below. A third involvement, the provision of the established Film and TLD Monitoring Services for external radiation measurement is not discussed.

Personnel Dosimeters

Through the kind assistance of Dr. A. Breslin of the U.S.E.R.D.A., New York City, thermoluminescent dosimeters were exposed to known radon daughter working level (WL) atmospheres. The results appear to give reasonable correlation with WL over a limited exposure range. Lithium Fluoride-100 chips, 1/8" x 1/8" x 0.015", manufactured by Harshaw Chemical Co. Ltd., were used along with the dosimeter housing and pump assembly designed by Breslin. Linear response curves were obtained in the range 0-20 WLH which were used to interpret exposure data during subsequent field tests being conducted at Elliot Lake by R. Yourt, Consulting Engineer. Although it is not proposed to go into detail, it is perhaps interesting to note that the calibration response curve did not go through the origin at zero exposure and this later resulted in some field measurements giving negative results at low exposures. This anomaly is probably caused by the inherent error in the TLD chips estimated to be ±20% or less. Breslin and George have indicated a similar anomaly (1).

Several assemblies have been tested by Yourt, at Elliot Lake over a period of approximately 4 months. The results of these tests are included in his subsequent presentation (2). The Radiation Dosimetry Division is interested in contributing its expertise to the development of an assembly which will be suitable for monitoring large numbers of workers on a reasonably long frequency, say two weeks. If Breslin's dosimeter in its present form proves reliable under field conditions both in relation to the pumping component and the TLD response, modification of the dosimeter head to allow incorporation of a TLD plaque containing permanently mounted chips would appear perfectly feasible. This would then permit speedy readout of WLH exposures at preset intervals on an automated basis using already developed and tested readout equipment which is currently operational. Computer interpretation of exposures would not seem to present a significant problem.

Yourt has suggested (3) that the chips used to measure background external gamma radiation could perhaps remove the necessity for miners wearing a second dosimeter for measurement of external dose. If it is deemed acceptable to ignore the non penetrating radiation component, and hence to accept on occasion a slightly low estimate of external skin dose, this proposal would certainly seem worthy of further investigation.

During a recent visit of a U.S.S.R. delegation to the Department, there was some discussion of an interlaboratory collaboration study on response of TL dosimeters to known radon daughter atmospheres. No decision regarding further action has been taken at this time.

Registry of Exposures

The large majority of radiation workers in Canada subscribe to the Dosimetry Services provided by the Radiation Dosimetry Division and as part of this service their doses are automatically recorded by computer in the National Dose Registry (NDR). Current efforts, under the auspices of the National Health and Welfare Working Group on Dosimetry, are well underway to obtain past, current, and future records for that information which, and those radiation workers whom, are not presently included in the national registry system. Uranium miners represent one of the most significant omissions at this time and the Working Group has prepared a protocol which, subsequent to final review, will be presented to the mines, the provinces, and the A.E.C.B. for discussion. It is hoped that implementation of the recommendations will follow soon after. The original implementation will require two distinct and separate efforts. One will be to obtain, on a cumulative yearly basis if possible, all past exposure records including complete identification of the worker, and the other to develop and implement administrative procedures for routine periodic submission of future exposures, probably on a quarterly basis. Every effort will be made to keep the requirements as simple as possible to alleviate potential administrative problems. It would appear that centralization of this data is desirable, and necessary, if epidemiological studies of Uranium miners are to be conducted efficiently and at reasonable cost. In fact, a specific request to N H and W has been made in this regard (4) and has subsequently been approved (5). Canada is fortunate in that many of the vital statistics of its population are available in machine readable form. Dr. H. Newcombe, of Atomic Energy of Canada Ltd., and also a member of the Working Group, has developed procedures for cross linkage of computerized Social Insurance Number, Cancer, and Death Index files and it is felt that these will in fact allow for efficient economical epidemiological evaluations of uranium miners. As a secondary, but nevertheless very important benefit, this data will of course also be invaluable for regulatory and legal investigations.

Summary

　　　　The above represents very briefly the current involvement of the Radiation Dosimetry Division in the field being discussed at this session. The development of a reliable automated WL dosimeter and the collation of occupational WL exposures for epidemiological studies would seem to be of paramount importance in safeguarding the health of uranium miners. Continuing efforts will therefore be made to attain these objectives.

References

1. Personal Correspondence, A. George to D. Grogan, June, 1976.

2. R. Yourt, Proceedings of this meeting, October, 1976.

3. R. Yourt, Personal Correspondence, July, 1976.

4. A. T. Prince to A. B. Morrison, March 9, 1976.

5. A. B. Morrison to A. T. Prince, April 5, 1976.

Summary of Discussion

Dust Measurements

Concern was expressed by several delegations regarding the omission of dust counts from the occupational dose information to be included in the Canadian National Dose Registry. The current feeling of the Canadian authorities that this information should not be included on an individual's record was explained on the basis of the following:

It was felt that any numbers available are extremely difficult to interpret because 1) the method of detection is extremely sensitive to the varying size of particles 2) the dust count can vary greatly over short periods of time and 3) measurements are area measurements, which because of 2) cannot be realistically assigned to individuals. It was therefore concluded that a meaningful dust count, would be impossible to maintain, and therefore it would be better to investigate the possibilities of keeping such information as is available on an organizational rather than an individual basis. Presumably the same rationale would hold for diesel fumes.

It is perhaps interesting to note that a similar view was not held in regard to cigarette smoking. However problems of a legal and administrative rather than a technical nature would likely cause similarly annoying difficulties. In this regard the possibility of obtaining worker, management and union cooperation in setting up a mechanism to record this data in a separate, strictly confidential file, to be used only for statistical purposes is being considered.

Record Keeping for Non-Uranium Miners

It is not the intention of the authorities to restrict the National Dose Registry to uranium miner exposures. Should it be deemed necessary and should appropriate information become available, non uranium miner dose information will most certainly be included.

FIELD TEST OF MOD WORKING LEVEL DOSIMETERS*

Sponsored by

Atomic Energy Control Board

in Co-operation with

Rio Algom Limited and

Health and Welfare Canada

by

G.R. Yourt, P. Eng.
Consulting Engineer

November 16, 1976

*On loan from ERDA Health and Safety Laboratory (HASL)

FIELD TEST OF MOD WORKING LEVEL DOSIMETERS

G.R. Yourt
Consulting Engineer
Ontario, Canada

Introduction

This paper briefly describes the second field testing of the ERDA MOD Working Level Dosimeter in Rio Algom's New Quirke Mine, Elliot Lake, Ontario. The study was conducted in May to September, 1976 under a Federal Government (Department of Supply and Services) contract with the writer and was sponsored by AECB (Atomic Energy Control Board).

The earlier field testing was conducted in July to September 1975 and is described in HASL Technical Memorandum 76-4 Field Test of the MOD Working Level Dosimeter by Andreas C. George.

The two principal objectives of the 1976 study were:
(1) To develop an arrangement for wearing TLD dosimeters that would be convenient and acceptable to the miners. This is considered by the mining industry to be a fundamental requisite.
(2) To obtain if possible a closer correlation between conventional radon daughter measurements and dosimeter readings using MOD dosimeters on loan through the courtesy of HASL officials.

Summary of Findings

1. Good progress was made in developing a compact pump that is convenient for the miner to wear and operate with the HASL TLD dosimeter head.

2. Notwithstanding frequent conventional radon daughter measurements taken during the exposure of TLD dosimeters a close correlation was not obtained. Several probably reasons are discussed in the report.

3. The HASL pumps performed well but miners indicated strong preference for the lighter units.

4. A mulfunction in the du Pont pump prevented a satisfactory test.

Recommendations

It is recommended:

1. That additional field testing be conducted specifically to obtain a closer correlation between TLD dosimeters and frequent conventional sampling.
2. That the three light weight pumps, purchased by AECB, be operated with TLD dosimeters throughout the entire study.
3(a) That one be worn by a miner and two by a technologist while taking conventional samples as close to the miner as feasible.
3(b) That all three pumps be worn in the same position, for example, on a hip.
4. That brief observations of ventilation conditions and changes be made to the extent feasible during each shift.
5(a) That the two pumps fabricated by H & H Custom Work be modified to incorporate the sampling head wholly or partly within the bottom of the case.
5(b) That an optimum charging schedule be established before actual field testing is initiated.
5(c) That a spare motor and pump assembly be purchased.
6(a) That a spare constant flow control unit be purchased for the du Pont pump.
6(b) That arrangements be made with H & H Custom Work for the fabrication of a sturdy aluminium case incorporating a sampling head.
7. That the effect of lubricant and moisture aerosols on TLD readings be investigated early in future testing.
8. That the feasibility of measuring both internal and external exposure with TLD dosimeters be investigated.

MOD Working Level Dosimeter

For the convenience of readers, the description of the MOD Dosimeters is included as Appendix A and is taken directly from the HASL Technical Memorandum.

Arrangements

Eleven dosimeters and ancillary equipment were obtained on loan from HASL but only up to three were used at any one time.

In the 1975 study TLD chips were mailed to HASL in New York for reading. This involved significant delays and loss of one shipment. In order to eliminate this problem, and also to establish a service in Canada for the future, arrangements were made with the Radiation Protection Bureau (Health and Welfare Canada) in Ottawa to process exposed chips. This was initiated after the second period of testing. Possibly this transition introduced some differences in readings due to limited experience in reading TLD chips exposed to alpha radiation.

In order to minimize this problem a number of TLD chips were exposed to known quantities of alpha radiation at HASL and then read at the Radiation Protection Bureau.

Arrangements for the field testing were made with management and other officials of Rio Algom Limited. These included:

(1) Employment of a technologist, a graduate from the Haileybury Provincial School of Mines who had considerable previous mining experience. The arrangement provided coverage of benefits. The cost was charged to the contract.

(2) Training of the technologist in sampling procedures and familiarizing him with the mine by having him travel with supervisors.

(3) A meeting with Rio Algom officials and union representatives to explain the purpose of the study.

Major purchases included:

(1) 25 TLD - 700 lithium fluoride high sensitivity ribbons (1/8" x 1/8" x 0.015") from The Harshaw Chemical Company, Solon, Ohio. These supplemented chips provided by HASL.

(2) Two diaphragm pumps from H & H Custom Work, 63 Syracuse Cr., West Hill, Ontario. Dimensions: 3-1/2" x 1-7/8" x 1-3/4" Weight: 10 oz. Price: $450.50, including FST and PST.

(3) One constant flow sampler pump (Model P20) from E.I. du Pont de Nemours & Company, Wilmington, Delaware. Dimensions: 5-1/4"x 2-15/16" x 1-1/2". Weight: 14 oz. Price: $562.92, including duty and sales tax.

Other expenses for equipment involved mainly modifications to HASL pumps and the du Pont pump, transportation, taxes, duty and brokerage on equipment borrowed from HASL and also a few tools and supplies.

Procedures

(1) Pump batteries were put on charge at the end of each shift. Flow rates of pumps were measured before and after charging. They were operated at about 90cc per minute.

(2) Pumps were started simultaneously when they were fitted on the miner(s) and technologist. This was done by the latter. They were stopped simultaneously at the end of each shift.

(3) Operating time of pumps was carefully recorded.

(4) TLD chips and filters were changed at the end of each period (up to 5 days). In a few instances filters were changed more frequently.

(5) An attempt was made to get a broad range of radon daughter concentrations. The procedures listed above are similar to those used in the 1975 study. However, the procedures adopted in the 1976 field tests differed significantly as follows:

(1) Only one of at most two miners were involved instead of six in the 1976 tests.

(2) This enabled the technologist to take 25 to 30 conventional samples per shift (about every 15 minutes) near each miner instead of two to four samples.

(3) In an effort to meet the first objective, it was decided not to request a miner to wear the dosimeter attached to his hard hat.

Tha additional restriction of head movement, caused by the hose connecting the sampling head on the hat with the pump attached to the belt, constituted a problem in the 1975 tests. This was especially true for drillers who usually wear ear muffs.

(4) During the early part of the field tests, the miner (and technologist) wore the sampling head on his chest by attaching it to a shoulder strap. The cap lamp belt was passed through a loop at each end of the shoulder strap. An aluminum shield was incorporated with the attachment to the shoulder strap to give similar protection from direct splashing of water or dirt into the orifices of the sampling head as is provided by the brim of the hard hat.

The technologist also wore a sampling head attached to his hard hat in order to provide a comparison of readings in the two positions.

(5) During the middle part of the study the miner and technologist wore the sampling heads and protective shields attached directly to the bottoms of the pumps suspended on their belts at about hip location. The technologist continued to wear a sampling head on his hard hat.

(6) During the latter part of the study, two new compact and light but sturdy Ni-Cd battery driven pumps were tested. They were fabricated to the writer's specifications by H & H Custom Work. They were made so that the threaded end of the HASL dosimeter head could be readily screwed into a protective shield attached to the bottom of the aluminum case.

(7) In view of the strong preference indicated by Denver Mining Research of the Bureau of Mines for the Dupont Model P20 pump, one unit was purchased. Unfortunately due to some malfunction in the airflow control, it operated only one day.

(8) During most of the field testing the slushing operation was chosen because it facilitated sampling close to the miner by the technologist. The technologist followed and took conventional samples when the miner departed from his usual position at the slusher controls. This included adjusting pulleys and cables, eating lunch, and travelling to and from his work place at the beginning and end of the shift.

Findings

Significant progress was made in achieving the principal objective, namely, to develop an arrangement for wearing TLD dosimeters that would be convenient and acceptable to the miners. The two light compact and sturdy pumps incorporating the sampling head operated satisfactorily for several weeks.

The larger and heavier ERDA pumps also performed well as indicated later herein.

Despite the frequent conventional samples (about every 15 minutes) taken during the field testing, a good correlation with dosimeter readings was not consistently achieved.

The readings are summarized in Table 1. The principal operations are indicated in the column under "Remarks". In most instances some slushing was also done during the drilling cycle. Drifting includes both drilling and loading cycles.

Correlation is better during most slushing cycles mainly because the miner spends most of his time in one position, usually where the air enters a stope and also because the technologist was able to take samples closer to one slusher operator than he could in the case of drillers. Where two drillers or loaders wore dosimeters in periods 8 to 11 the technologist had difficulty in getting good coverage because they moved around considerably. For this reason in future testing the technologist should follow only one miner closely.

There is some suspicion that lubrication and moisture aerosols affected the readings. This warrants investigation early in future testing. Possibly a different filter should be tried, for example, Acropor which is used by MESA.

Table 2 indicates the day to day variations in average radon daughter measurements made with conventional sampling instruments.

The right hand column indicates percentages of mean deviations from the weekly averages. This varies from 9 to 52 and averages 20 percent.

The substantial changes from shift to shift indicated in period 7 may be due to temporary short circuiting, for example, when a box hole is inadvertently pulled empty.

Table 3 indicates variations in WL readings throughout a typical slushing shift in a stope. Note somewhat higher readings when samples were taken in the main part of the stope, that is, away from the slusher control area where air usually enters. Radon daughter concentrations normally increase as air passes over broken ore.

Table 4 shows some variations in readings throughout a drilling shift in a stope.

Table 5 is a record of readings taken near miners in a typical drift with good ventilation while they are mucking and drilling.

Pump Performance and Proposals

Table 6 summarizes the airflows of the various pumps tested at the start and end of the shifts during which they were operated. The shift readings are recorded in Appendix B. The accuracy of the airflows are subject to the limitations of reading the flowmeter. This was facilitated by the use of a large pulsation damper provided by ERDA.

Contrary to experience indicated in the HASL report, the HASL pumps performed well in the recent study. There was no significant drop in flow rates encountered between the beginning and end of a shift. A temporary problem was experienced with one charger, resulting in a number of blown fuses. This was the only reason that pump Nos. 5M, 11M and 15M were temporarily taken out of service. Only one of the spare pumps failed because of defective wiring.

In the opinion of the writer it would be difficult to persuade miners to wear this pump if lighter and satisfactory pumps were available. However, the arrangement with the dosimeter head attached directly to the bottom of the pump was found considerably more acceptable than with the standard arrangement involving the hose. The two light weight prototype pumps fabricated by H & H Custom Work also performed well. The only interruption in operation of one unit (No. 001) was due to a faulty resistor in the charging circuit.

As these pumps will operate for more than one shift following an over-night charge, an optimum frequency and period for charging should be determined when preparations are made for subsequent operation.

A further improvement should be made in future units by incorporating a sampling head (filter and TLD chip holder) into the bottom of the case. This would eliminate existing projection of the head below the case.

The unit can be worn: suspended from the miner's cap lamp belt, back or front, by a separate shoulder strap or attached to the lapel area.

Unfortunately a malfunction in the electronic flow control of the du Pont light weight pump prevented a satisfactory test of this instrument. In view of the favourable experience with this pump reported by MESA, it should be included in any future field testing. If this is done, essential spare parts should be purchased. It is suspected that experience in further testing will indicate the need for a more durable case.

External (Gamma) Radiation

The MOD dosimeter head contains two lithium fluoride chips. The one facing the filter absorbs energy from gamma as well as aplha radiation. The second chip, shielded from beta radiation absorbes only gamma radiation. The difference provides the alpha reading.

It would be useful if this arrangement could be used routinely to measure both external and internal radiation.

Conclusion

A light compact dosimeter pump has been developed which is quite convenient for miners to wear. It could also be used for continuous area monitoring. Several weeks of testing has indicated satisfactory performance.

The HASL working level dosimeters pumps were found to perform satisfactorily with or without the hose attachment. They are not as light and compact as the prototype unit manufactured by H & H Custom Work.

The duPont (P-20) pump is also compact. It requires field testing to determine its durability and servicing requirements during extended use in a mine. It is suspected that a stronger case would be required.

Further field testing is indicated in order to establish a closer correlation between conventional measurements and dosimeter readings.

Since the use of TLD chips for measuring radon daughter working levels involves both alpha and gamma readings it appears advisable to combine the instrumentation to provide both types of exposures.

Acknowledgements

The writer appreciates the co-operation provided by officials of ERDA, AECB, Radiation Protection Bureau, Rio Algom Ltd., USWA and the two technologists wo persevered in taking numerous measurements.

TABLE 1 - KUSNETZ Vs. DOSIMETER WORKING LEVELS

Periods	Kusnetz (Period Avg.)	TH	TC	TB	MC	MB	Remarks
1	.13	.15	.12		.16		Slushing
2	.36	.29	.21		.26		Slushing
3	.24	.21	.13		.21		Slushing
4	.19	.09	.10		.09		Slushing
5	.75	.56	.63		.67		Slushing
6	.45	.32	.34		.45		Slushing
7	.4	.24				.49	Drilling
8	.52	.41		.19	.55	.99	Drifting
9	.42	.56		.59	.67	.84	Drilling
10	.19	.20		.13	.05	.09	Drilling
11	*.35	.17		.2	.10	.25	Drilling
12	.21			.21			Drifting
13	.024						Drifting
14	.19			.1			Drifting

Note: TH, TC, TB - Technologist's hat, chest, belt
MC, MB   - Miner's chest, belt

*Error suspected

TABLE 2 - Kusnetz Daily Working Level Averages

Period	1	2	3	4	5	Average	% Deviation (mean)
1	.09	.17	.13	.13		.13	15
2	.39	.34	.30	.32	.45	.36	14
3	.25	.26	.24	.19	.28	.24	9
4	.20	.23	.15	.19	.18	.19	11
5	1.05	.68	.72	.54		.75	20
6	.41	.51	.52	.39	.42	.45	11
7	.74	.13	.56	.16	.39	.40	.52
8	.49	.58	.55	.61	.35	.52	15
9	.44	.41	.50	.37	.40	.42	9
10	.17	.19	.23	.16		.19	12
11	*.74	.27	.36	.17	.19	.35	41
12	.15	.28	(two separate drifts)			.21	
13	.02	.03	.02	.02	.03	.024	20
14	.28	.22	.20	.16	.11	.19	25
							20

*Error suspected

TABLE 3 - SHIFT VARIATIONS WHILE SLUSHING

Date: June 3   Location: A Stope   Type of Work: Slushing

TIME OF SAMPLING	W.L.	REMARKS
8:20	.13	SILLDRIFT
8:30	.05	SLUSHER
8:45	.06	SLUSHER
9:00	.40	STOPE
9:15	.38	STOPE
9:30	.44	STOPE
9:45	.04	SLUSHER
10:00	.03	SLUSHER
10:15	.07	SLUSHER
10:30	.05	SLUSHER
10:45	.05	SLUSHER
11:00	.05	SLUSHER
11:15	.15	SILL DRIFT
11:30	.04	LUNCH ROOM
11:45	.04	LUNCH ROOM
12:00	.45	STOPE
12:15	.07	SLUSHER
12:30	.05	SLUSHER
12:45	.05	SLUSHER
1:00	.06	SLUSHER
1:15	.04	SLUSHER
1:30	.06	SLUSHER
1:45	.23	STOPE
2:00	.07	SLUSHER
2:15	.32	STOPE
2:30	.05	SLUSHER
2:45	.01	MAIN DRIFT
3:00	.14	STATION
3:15	.12	STATION
3:30	.12	STATION

Average   .13

TABLE 4 - SHIFT VARIATIONS WHILE DRILLING

Date: July 15   Location: G Stope   Type of Work: Drilling

TIME OF SAMPLING	W.L.	REMARKS
8:02	.07	MAIN DRIFT
8:16	.38	SILL DRIFT
8:40	.17	BOTTOM STOPE
8:50	.14	TOP STOPE
9:00	.20	TOP STOPE
9:15	.20	TOP STOPE
9:30	.17	TOP STOPE
9:45	.18	TOP STOPE
10:00	.13	TOP STOPE
10:15	.19	TOP STOPE
10:31	.15	TOP STOPE
10:46	.18	TOP STOPE
11:02	.12	TOP STOPE
11:15	.13	TOP STOPE
11:47	.11	SILL DRIFT
12:05	.14	TOP STOPE
12:16	.13	TOP STOPE
12:31	.16	TOP STOPE
12:48	.13	TOP STOPE
1:00	.16	TOP STOPE
1:15	.15	TOP STOPE
1:31	.19	TOP STOPE
1:47	.17	TOP STOPE
2:00	.19	TOP STOPE
2:15	.18	TOP STOPE
2:30	.06	SILL DRIFT
2:45	.03	MAIN DRIFT
2:58	.24	STATION
Average	.16	

TABLE 5 - SHIFT VARIATIONS WHILE DRIFTING

Date: Aug. 30   Location: E Drift   Type of **Work:** Mucking & Drilling

TIME OF SAMPLING	W.L.	REMARKS
9:30	0.01	MUCKING
9:45	0.02	MUCKING
10:00	0.01	MUCKING
10:15	0.01	MUCKING
10:30	0.02	MUCKING
10:45	0.02	MUCKING
11:00	0.02	FINISHED MUCKING
11:15	0.02	PREPARING TO DRILL
11:30	0.01	PREPARING TO DRILL
11:45	0.03	ROCK BOLT
12:00	0.03	ROCK BOLT
12:15	0.02	DRILLING
12:30	0.02	DRILLING
12:45	0.01	DRILLING
1:00	0.02	DRILLING
1:15	0.03	DRILLING
1:30	0.02	FINISHED DRILLING
1:45	0.02	
2:00	0.02	LOADING
2:15	0.03	LOADING
2:30	0.03	FACE WIRED
Average	0.02	

Vent Duct 20' From Face
F.P.M. = 75' @ Face
C.F.M. = 4000

TABLE 6 - SUMMARY OF PUMP PERFORMANCE

Instrument	Completed Shifts	Average Start	Airflow (cc) End	Start/End
ERDA - 5M	44	99.2	95.3	1.04
" - 11M	40	95.9	93.5	1.03
" - 15M	38	89.5	88.9	1.01
" - 6M	5	86.2	87	.99
" - 8M	3	82.6	83	.99
" - 19M	2	88.5	91	.97
H & H Custom Work - 001	33	96.3	94.6	1.02
" - 002	11	105.1	103.5	1.02
duPont	1	74	72	1.03

Appendix A

Description of the MOD Working Level Dosimeter

The MOD, shown in Figure 2*, consists of a hat-mounted detector connected by flexible tubing to a pump-battery pack that attaches to the miner's belt. The pack is 6.3 cm x 3.8 cm x 15.3 cm and weighs 0.82 kg. The air pump (HASL TF-6), powered by a 12 v, 225 mAh rechargeable Ni-Cd battery, draws air through the detector at 90 cc/min for as long as ten hours after the battery is fully charged. The pumping element is a single acting diaphragm with dynamically activated flapper valves. It is driven by a solenoid at a pulse rate determined by an electronic timing circuit nominally set a $1.50s^{-1}$ for a desired flow of 90 cc/min.

The detector head shown in Figure 1** is 2.0 cm in diameter and 3.0 cm long. Air is drawn from outside through three apertures and thence through a 1 cm, 0.8 µM pore size membrane filter. A .32 cm x .32 cm x .04 cm lithium fluoride chip is supported above the membrane filter a distance of 1.6 mm. The chip is held in a specially molded RTV rubber pedestal with a dove-tail annulus that overlaps the four corners, only, of the chip. A second chip of identical size is housed in a compartment above where it is shielded from alpha and beta radiation from the filter. This chip is used to correct for external gamma radiation to which both chips may be subjected when the dosimeter is in use.

---

* See page 93.
** See page 92.

Appendix B

DAILY INDIVIDUAL PUMP PERFORMANCE
(cc at start and end of shift)

Day	5M	15M	11M	001	8M	6M
1	92-92	99-92				
2	92-99	92-92				
3	92-99	92-92				
4	96-99	92-78				
5	99-99	78-76				
6	99-90	76-76				
7	101-99	90-90				
8	106-106	92-90	96-94			
9	96-101	90-85	92-92			
10	101-99	90-87	94-92			
11	101-101	90-90	90-90			
12	106-104	92-90	96-92			
13	101-101	90-90	94-87			
14	101-101	90-92	94-94			
15	101-101	92-90	94-94			
16	94-94	101-99	90-90			
17	106-104	94-92	101-99			
18	101-101	94-92	99-96			
19	101-104	92-92	96-99			
20	101-104	92-92	96-96			
22	104-106	94-94	96-99			
23	106-106	94-96	101-96			
24	104-106	94-90	99-99			
25	104-106	94-94	99-101			
26	104-104	96-92	101-99			
27	104-104	81-90	92-92	114-114		
28	(104-Nil)	(87-Nil)	(90-Nil)	104-106		
29	104-101		87-90	109-114	83-83	
30	101-104		90-90	114-101	83-83	
31	104-104	19M	90-90	104-111	85-87	
32	104-104	90-90	92-92	94-92		85-87
33		87-92	90-92	96-92		87-87
34		(90-Nil)	90-92	87-92		87-87
35	(101-Nil)	13M	92-92	94-90		85-87
36		87-87	(92-Nil)	90-94		87-87
37	101-104		92-90	96-92	81-81	
38	104-104	15M	92-90	96-92	81-81	
39	94-92	87-90	104-104	96-92		

- 115 -

Appendix B

	5M	11M	15M	001
40	104-104	90-90	85-87	96-87
41	104-104	92-92	87-87	96-92
42	104-104	90-92	83-74	94-101
43	104-106	92-92	76-85	106-92
44	104-104	90-90	83-72	96-92
45	106-106	92-92	72-87	99-96
46	106-106	96-94	92-94	120-132
47	104-104	94-94	90-94	111-111
48	104-104	92-94	(90-117)	101-114
49	104-106	94-96	92-94	106-101
50	104-104	92-94	92-94	92-92

	001	002	duPont
51	(139-Nil)	(210-72)	74-72
52		104-117	(104-Nil)
53		104-101	
54	101-92	104-101	
55	92-85	104-90	
56	92-78	101-114	
57	65-74	104-96	
58	85-78	104-90	
59	81-85	104-114	
60	81-74	114-109	
61	81-70	104-96	
62	90-87	109-111	

Note: Readings in brackets not included in Table 6.

APPAREILS DE MESURE EN CONTINU DE LA CONCENTRATION DES
EMETTEURS α DE LA CHAINE DE ^{238}U (POUSSIERES A VIE LONGUE ET
DESCENDANTS DU RADON) EN SUSPENSION DANS L'AIR

- J. Pradel - Ph. Duport - P. Zettwoog -

C.E.A. - PARIS (FRANCE)

Résumé

On présente d'abord le principe de ces appareils, destinés à mesurer dans l'air, d'une part la concentration en aérosols de ^{218}Po (radium A) et de ^{214}Po (radium C'), et par conséquent l'énergie potentielle volumique, et d'autre part l'activité volumique des émetteurs α à vie longue existant dans les poussières de minerai en suspension.

On décrit ensuite une première version, destinée à la mesure des doses individuelles inhalées par le personnel employé dans les mines d'uranium et au traitement du minerai.

Une autre version permet de surveiller la concentration des émetteurs α de l'air sur les lieux de travail, ou dans l'atmosphère libre et les habitations.

Une version applicable à la prospection des minerais d'uranium est enfin proposée.

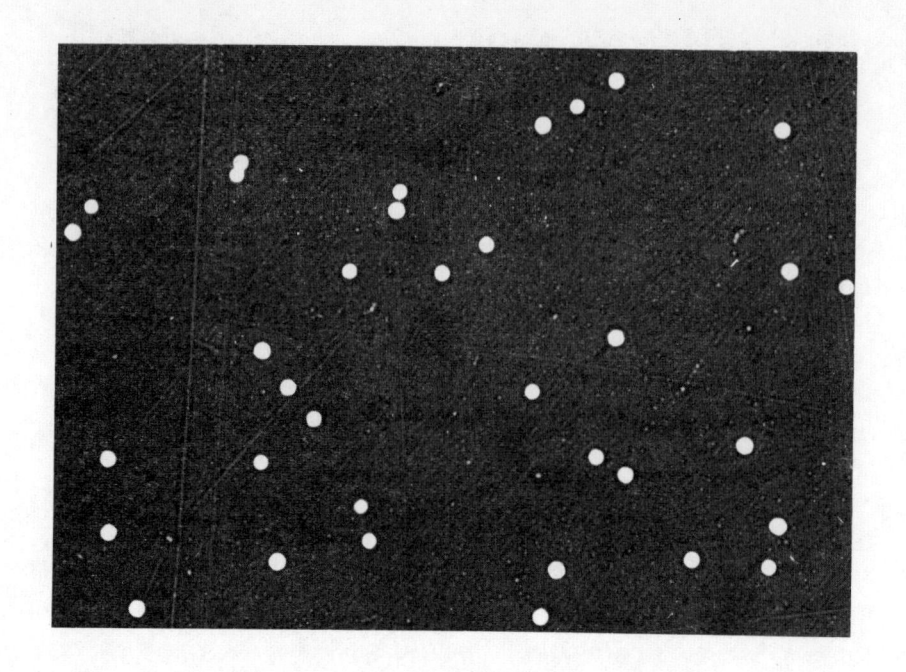

Traces de particules α dans le film détecteur

FIG. 1

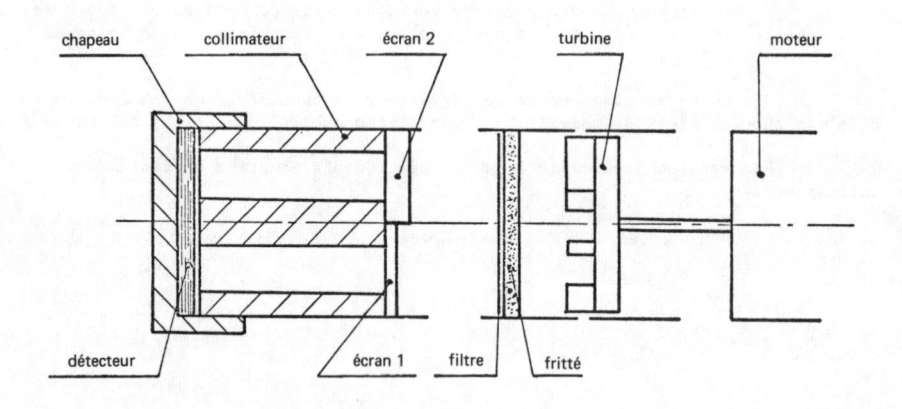

Schéma de la tête de mesure

FIG. 2

Introduction

Les doses de rayonnement α reçues par les travailleurs (ou le public) ayant inhalé des aérosols radioactifs, soit descendants à vie courte du ^{222}Rn, soit poussières de minerai d'uranium à vie longue, n'ont jamais en pratique été comptabilisées que de façon indirecte, car la dosimétrie individuelle des rayonnements α, contrairement à celle des rayonnements β, γ, ou X, n'était pas au point.

Pour ce qui concerne les poussières à vie longue on aurait pu envisager depuis longtemps des appareils de prélèvements individuels, munis d'un filtre à poussières dont la radioactivité aurait pu être mesurée en fin de mois au laboratoire par une électronique de comptage classique. Malheureusement cette technique ne convient pas pour les descendants à vie courte du ^{222}Rn, qui représentent le risque majeur. Ces descendants ayant une vie courte même par rapport à la durée du poste, il fallait que le dosimètre garde en mémoire le nombre des désintégrations α effectuées au cours du prélèvement par les particules de ^{218}Po et ^{214}Po retenues sur le filtre. Il a donc été proposé d'inclure dans le dosimètre des détecteurs solides de particules α [1], [2], [3]. Les détecteurs ionographiques nous ont paru à cet égard préférables aux détecteurs thermoluminescents, car, permettant une meilleure discrimination des énergies des particules α, ils se prêtent à la réalisation d'un spectromètre rudimentaire, distinguant les désintégrations de ^{218}Po de celles de ^{214}Po, et donc aboutissant à la connaissance de l'énergie α potentielle de l'air inhalé [4].

Après plusieurs années de mise au point technologique, nous disposons actuellement d'appareils construits en série et destinés, soit aux travailleurs des mines d'uranium, [5], soit au contrôle des irradiations subies par le public, soit à la prospection des gisements d'uranium.

I - LA TETE DE MESURE

I.A. - <u>Principe de la détection des descendants à vie courte du ^{222}Rn</u>

On utilise pour la détection des particules α le film de nitrate de cellulose LR 115 de KODAK-PATHE. La couche sensible, rouge, épaisse de 13 μm, est déposée sur un film de mylar incolore.

Le passage d'une particule α dans le nitrate affecte la structure de la matière de telle sorte qu'un bain d'attaque basique fait apparaître un trou de quelques μm à l'emplacement de l'impact.

On obtient une trace visible dans certaines conditions de révélation, lorsque l'énergie de la particule est comprise entre quelques centaines de keV et 5 MeV. Cependant, si l'énergie est trop faible, la trace observée ne traverse pas la couche sensible, et si l'énergie est trop élevée, la trace, de faible dimension, est longue à apparaître. En outre, les traces obliques ne traversent pas toujours le nitrate, et sont de toute manière difficiles à dénombrer.

Pour obtenir des traces de taille homogène, bien visibles par contraste sur le fond rouge non détruit, on doit sélectionner les α d'incidence normale au détecteur et réduire leur énergie dans les limites de 1 à 3,5 MeV. Dans des conditions d'attaque bien précises, on obtient alors des traces circulaires très régulières, les traces dues à des α d'énergies suffisamment éloignées des limites n'apparaissent pas à la lecture (fig. 1).

Ces conditions sont réalisées par un dispositif breveté comprenant un filtre millipore sur lequel sont retenus les aérosols, un collimateur à deux canons, terminés chacun par un écran et un chapeau contenant le film détecteur (fig. 2). La distance franchie par les α dans l'air et l'épaisseur des écrans sont telles qu'à la sortie d'un canal, les particules α émises par le RaA n'émergent pas de l'écran, tandis que celles émises par le RaC' émergent avec une énergie résiduelle de 3 MeV et sont donc enregistrées. A la sortie du second canal, les particules α émises par le RaA émergent de l'écran avec une énergie de 2,8 MeV, sont enregistrées, tandis que les α émis par le RaC' ont encore une énergie de 5,6 MeV, et ne laissent pas de traces visibles dans nos conditions de développe-

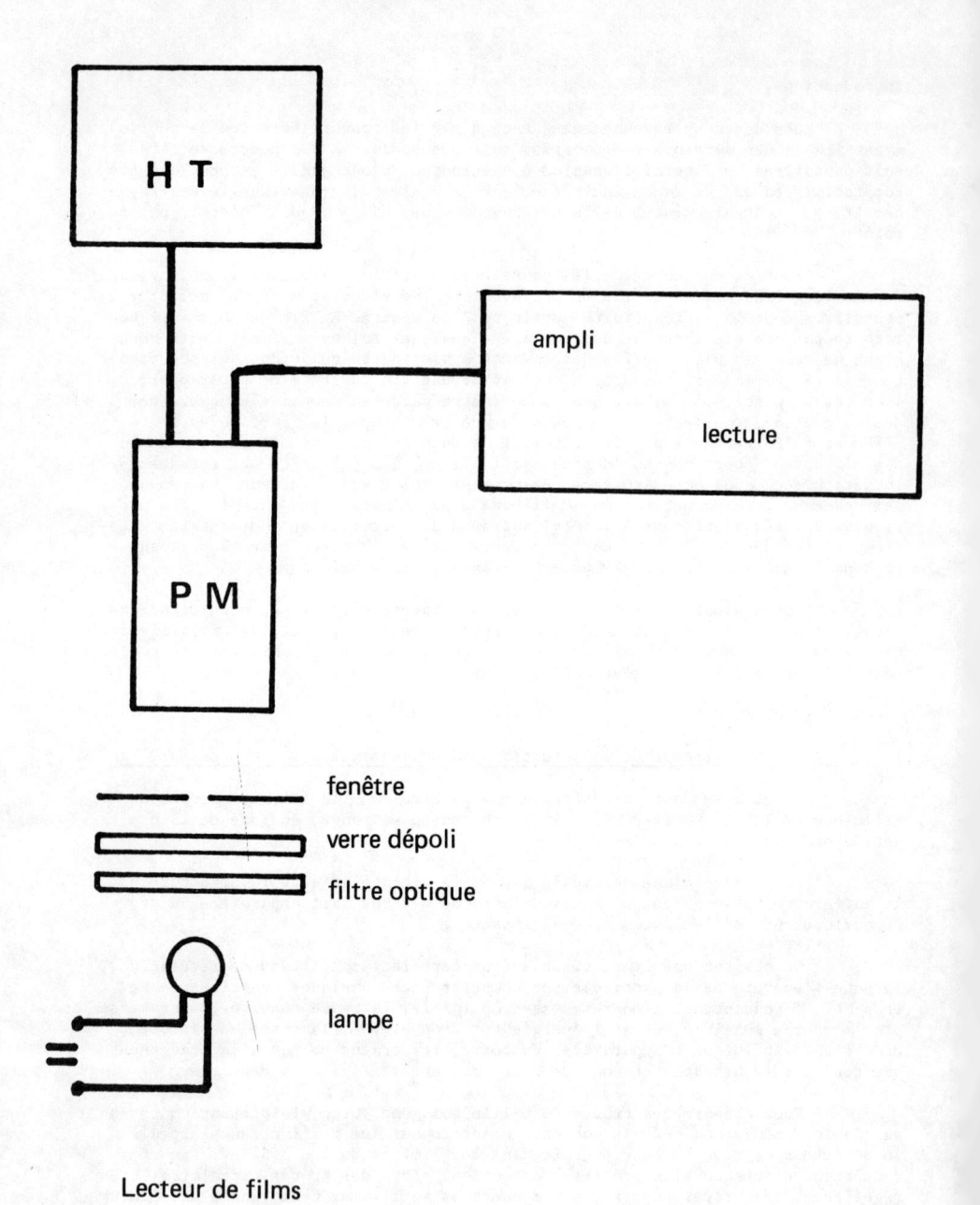

Fig. 3

ment. Nous avons donc un discriminateur d'énergie qui permet au film détecteur d'intégrer sur la période choisie la quantité de particules α potentielles contenues dans un volume d'air connu, donc de connaître la quantité d'aérosols radioactifs descendants du radon que les mineurs ont effectivement rencontré dans leur environnement.

I.B. - Principe de la mesure du nombre de traces

On ne compte que les traces qui traversent entièrement la couche de nitrate de cellulose, apparaissant au microscope comme des tâches lumineuses blanches sur fond rouge.

On peut dénombrer les traces au microscope optique ou encore, pour éviter ce procédé fastidieux, évaluer la quantité de lumière transmise par ces traces.

On utilise pour l'éclairage du film une lumière de longueur d'onde complémentaire à celle du nitrate rouge, de manière à obtenir la transmission minimum à travers la couche colorée.

Dans ces conditions d'attaque chimique déterminée, on peut relier directement la quantité de lumière transmise au nombre de traces.

En effet, grâce au collimateur, seules les particules dont la trajectoire est perpendiculaire au film sont comptées, les traces étant toutes circulaires et de même diamètre, et il y a proportionnalité entre le nombre de traces et la lumière transmise. L'étalonnage de la réponse du film à l'aide d'une source connue est possible.

On mesure alors l'intensité lumineuse transmise à un photomultiplicateur. Une fenêtre de section variable permet de choisir la forme et la surface de film à mesurer (fig. 3).

I.C. - Mesure des poussières à vie longue

L'activité des poussières à vie longue retenues sur le filtre millipore, environ 10^{-5} fois plus faible que celle des descendants à vie courte, est mesurée au laboratoire par une électronique de comptage classique.

II - ETALONNAGE DE LA MESURE DES DESCENDANTS DU RADON

Etalonnage au laboratoire

L'appareillage utilisé pour l'étalonnage du système de détection (fig. 2) comprend une source de radon, une pompe, un volume de 200 dm^3 montés en série (fig. 4). La concentration du radon dans le volume est de l'ordre de 4.10^{-8} Ci/l, l'état d'équilibre varie avec le débit de la circulation permanente. Une injection permanente contrôlée de fumée permet de ne pas avoir de fraction libre dans le volume d'étalonnage. Un petit ventilateur lent assure un mixage correct du mélange air - radon - fumée.

On mesure par des procédés classiques la concentration du radon. L'état d'équilibre entre le radon et ses descendants, ainsi que les concentrations des RaA, RaB, RaC sont déterminés par le temps de séjour du mélange dans le bidon, le radon étant filtré à l'entrée.

L'action sur le taux de recyclage permet de moduler la valeur de l'état d'équilibre.

La figure 5 montre la relation entre la densité des traces lues sur le film et le nombre de particules α émises par les aérosols radioactifs collectés. La proportionnalité des facteurs est satisfaisante.

Boucle d'étalonnage
FIG. 4

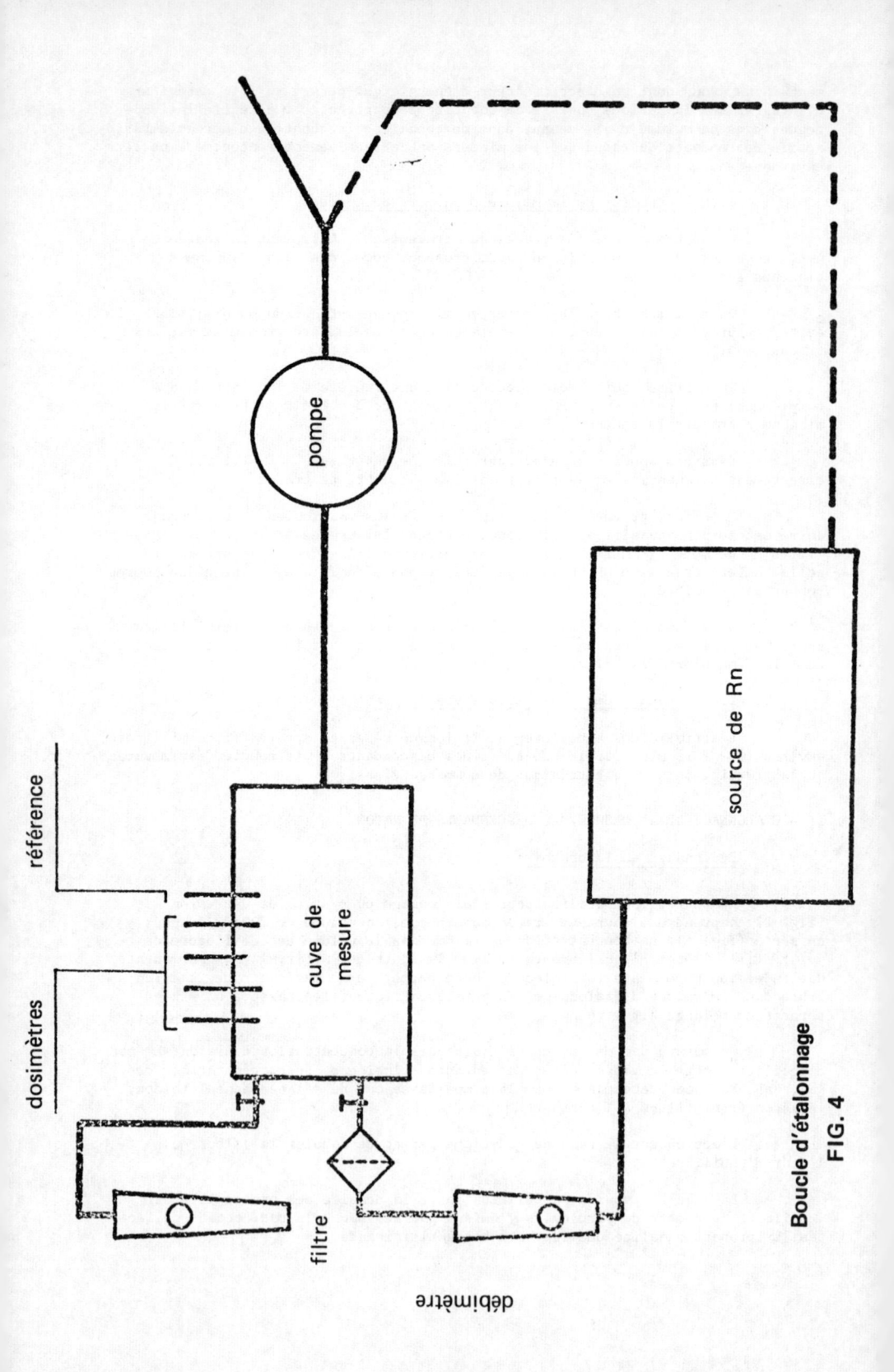

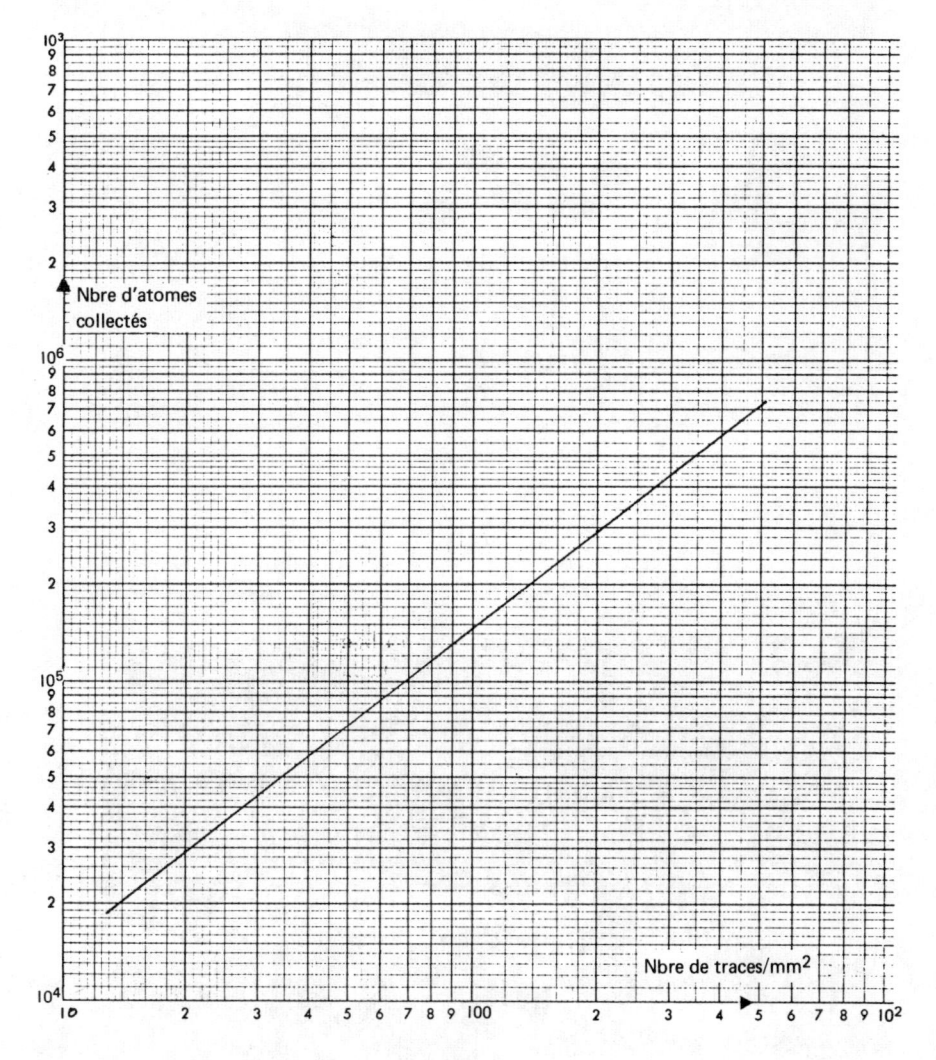

FIG. 5

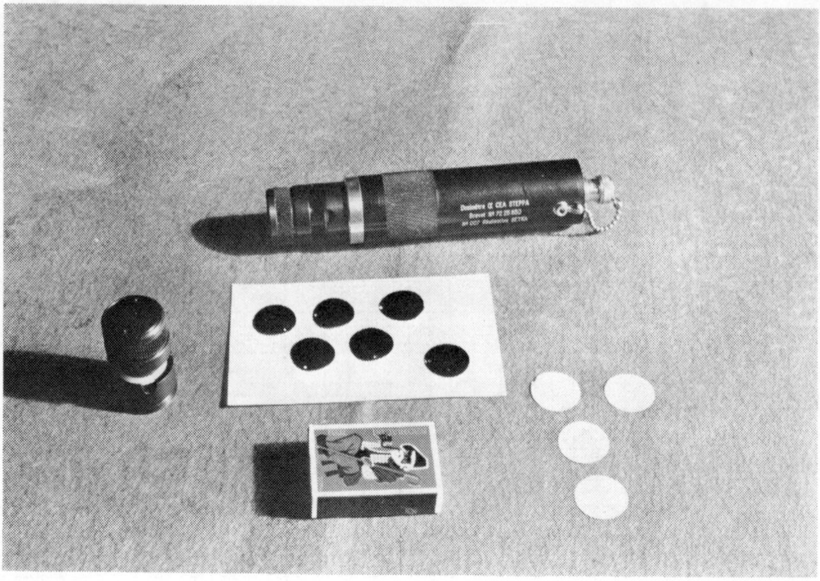
CEA STEPPA

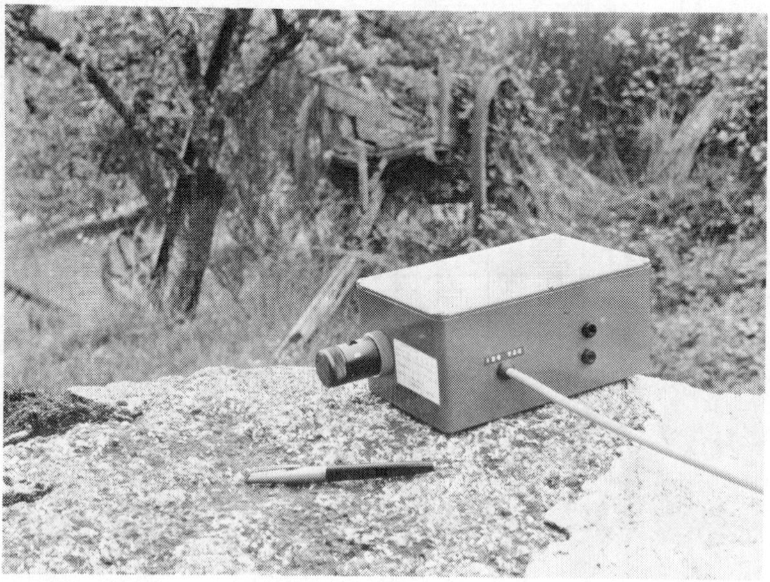

Une série de 33 mesures d'étalonnage, conduites avec 4 dosimètres en fonctionnement simultané, a donné les résultats suivants :

$$\text{Rapport observé } \frac{\text{nombre de traces RaA}}{\text{nombre de traces RaC'}} = 0,287$$

$$\text{Rapport théorique } \frac{\text{concentration RaA (en atomes/litre)}}{\text{concentration (RaA + RaB + RaC) (atomes/litre)}} = 0,294$$

Rendement géométrique théorique : $6,9.10^{-4}$

Rendement expérimental (d'après la mesure des concentrations en radon et d'équilibre radioactif théorique) : $6,7.10^{-4}$.

Pour une énergie α potentielle de 0,1 WLM, on obtient de l'ordre de 10 traces/mm^2 pour ^{218}Po (Radium A) et 35 traces/mm^2 pour ^{214}Po (Radium C'). La surface irradiée est de 28,3 mm^2.

Contrôle en mine de la reproductibilité des mesures

Après la série de mesures d'étalonnage en laboratoire une série de mesures conduite avec 4 dosimètres en fonctionnement simultané pendant un mois dans un chantier en exploitation sur le siège minier de Margnac a donné des résultats cohérents à 10 % près.

III - DESCRIPTION DU DOSIMETRE INDIVIDUEL DESTINE AUX MINEURS

Le dosimètre individuel est constitué (photo n° 1) d'un corps cylindrique en duralumin contenant la batterie d'accumulateur, l'interrupteur charge-marche, le moteur et la turbine.

Le moteur entraîne la turbine centrifuge à 15 000 t/mn sous 1,2 volt. L'autonomie est d'environ 15 heures. La durée de vie des moteurs est estimée à environ 4 000 h.

La tête de mesure est constituée du porte-filtre, des collimateurs et de leurs écrans, du film de nitrate et d'un chapeau assurant le maintien mécanique de l'ensemble.

Le dosimètre complet dans son étui de cuir pèse 400 grammes.

La turbine peut créer une dépression de 20 mm d'eau sous 1,2 volt, à débit nul. Le débit à travers un filtre Millipore RAWP de surface utile 1,75 cm^2 est de 4 à 5 l/h.

Un essai en mine, dans une galerie fortement polluée (retour d'air général) a montré que le colmatage est faible. Pour un débit initial Q_o de 4,7 l/h on a après 200 heures de prélèvement $Q_{200} = 4,2$ l/h.

IV - DESCRIPTION DU DOSIMETRE DE SITE DESTINE AUX CONTROLES D'AMBIANCE

Un appareil destiné à la mesure de l'énergie α potentielle de l'air, soit dans les chantiers miniers ou d'autres lieux à forte exposition, soit dans l'atmosphère naturelle ou dans les habitations a été réalisé (photo n° 2).

La tête de mesure est la même que celle décrite précédemment. La pompe et son débit sont adaptés à la concentration de l'ambiance à contrôler. L'alimentation électrique peut être au choix le secteur, ou des accumulateurs de quelques dizaines d'ampères/h.

V - DESCRIPTION D'UNE SONDE POUR LA PROSPECTION DES GISEMENTS URANIFERES

L'établissement des plans "radon" se fait par prélèvement en ampoule scintillante de l'air existant au fond de séries de trous de reconnaissance creusés dans le sol à quelques décimètres de profondeur. Les concentrations en radon dans ces trous varient fortement avec les paramètres atmosphériques et l'interprétation des résultats de prélèvements instantanés en est perturbée. L'emploi du dosimètre α permet un prélèvement continu sur plusieurs jours et permet d'éliminer ces variations à courtes périodes et assurent la bonne simultanéité des prélèvements.

La tête de mesure est cette fois montée sur la partie supérieure d'un tube à enfoncer dans le sol, qui, à sa partie inférieure, possède une grille de protection et un filtre, et qui constitue, par son volume interne, une chambre de désintégration pour le radon. L'alimentation par accumulateur permet de laisser l'appareil plusieurs jours sur le terrain.

REFERENCES

[1] R.L. ROCK, D.B. LOVETT, S.C. NELSON "Radon - daughters exposure measurement with track etch films"
Health Phys. Pergamon - Press - 1969 - Vol. 16 (617 - 621).

[2] J. ANNO
Thèse de spécialité.
Faculté des Sciences de Toulouse - 1970 -

[3] J.A. AUXIER, K. BECKER, E.M. ROBINSON, D.R. JOHNSON, R.H. BOYETT, C.H. ABNER
"A new Radon Progeny Personnel Dosimeter"
Health Phys. Pergamon - Press - 1971 - Vol. 21 - p. 126 - 128 (1971) -

[4] A.M. CHAPUIS, D. DAJLEVIC, Ph. DUPORT, G. SOUDAIN
Communication au Congrès de Bucarest Juillet 1972

[5] Ph. DUPORT, Guy MADELAINE, J. PRADEL
Un appareil personnel pour la dosimétrie des descendants du radon.
Rapport CEA - STEPAM - 1976 -

Résumé de la Discussion

- Précisions techniques sur l'appareil et ses périphériques
- Problèmes du prix de revient du dosimètre
- Comparaison des prix de revient de la surveillance selon la technique traditionnelle et une dosimétrie personnelle.

A COMBINED PERSONAL SAMPLER FOR DUST AND
RADON-DAUGHTER EXPOSURE IN MINES

G. Knight[1], R.A. Washington[1] and W.M. Gray[2]
Mining Research Laboratories
[1] Elliot Lake, Ontario
[2] Ottawa, Ontario

ABSTRACT

A personal dust sampler has been developed at the Elliot Lake Laboratory for use in hard rock mines. It features respirable (alveolar) dust size selection, single full-shift operation, gravimetric assessment for total dust and x-ray assessment for the quartz component.

A thermoluminescent chip for cumulative measurement of alpha radiation has been mounted adjacent to the filter in the dust sampler to provide for the assessment of exposure to radon daughters.

Possible sampling strategies are briefly discussed.

INTRODUCTION

This paper reports on progress in the development of a personal sampling instrument for the combined measurement of dust and radon-daughter exposure in mines. The new instrument is based on the Canadian Mining Personal Dust Sampler (CAMPEDS), which was developed at the Elliot Lake Laboratory for use in hard-rock mines.

The CAMPEDS was designed to allow the measurement of quartz* by x-ray diffraction as well as total dust. Experience in its use has shown that particulates from diesel exhaust and oil mists from compressed air operated equipment can also be measured successfully. Since radon daughters are akin to the lattter pollutants in being collectable on a filter it is logical to consider ways of measuring them on the same filter.

The CAMPEDS and its modification to allow the measurement of radon daughters will be described. Possible sampling strategies will be discussed briefly.

---

* Quartz, a form of free silica, is a very common mineral and is particularly hazardous to health.

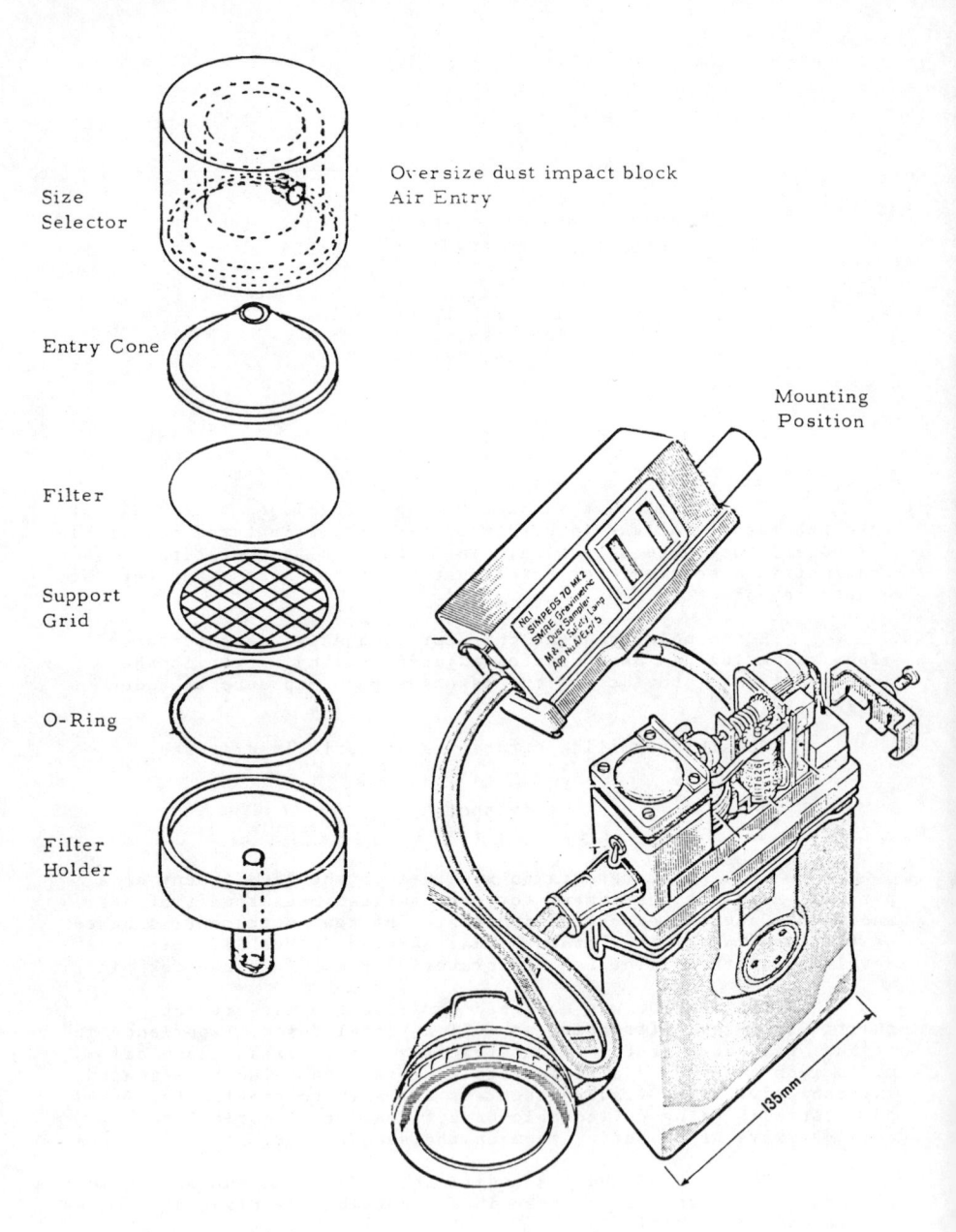

FIGURE 1: The Modified Filter Holder and Size Selector for the SIMPEDS Pump.

THE CANADIAN MINING PERSONAL DUST SAMPLER (CAMPEDS)

Between 1965 and 1969 the Elliot Lake Laboratory carried out a comparison of dust sampling instruments (1). It was found that there was a very wide variation in indications of relative hazard by different instruments in dust clouds of differing characteristics. Attention was then turned to the development of sampling and assessment techniques to give a better indication of the health hazard.

Preliminary studies (2, 3, 4) showed that full-shift portable self contained samplers could collect enough dust for assessment of total dust by weighing, and of quartz by x-ray diffraction. It was also found that a personal sampler indicated, on the average, a 25 per cent higher dust concentration than a fixed sampler at the nearest safe location in the working place. Thus the use of personal samplers would provide samples that were larger and therefore somewhat easier to assess and would simplify the logistical problem of taking full-shift samples for men scattered in small groups.

To fill the need for a suitable personal sampler, the CAMPEDS, sketched in Figure 1, was developed (4, 5). Its particular features are:

1. respirable size selection to simulate the deposition of dust in the alveolar region of the lungs where it offers the major health hazard,

2. full-shift period sampling,

3. combination with the cap-lamp battery to minimize the logistical problems of issuing it to underground miners,

4. a filter cassette to eliminate handling of the delicate filters outside the laboratory,

5. an impaction size selector, which is more compact and more readily machined reproducibly than other designs,

6. dust collection on a silver membrane filter for subsequent assessment by

    a) weighing (tare and gross) to give total respirable dust,
    b) ashing and reweighing to give respirable combustible dust or its complement, respirable mineral dust,
    c) x-ray diffraction to give respirable quartz dust.
    d) development of elemental (e.g. lead) analysis techniques is in progress.

The CAMPEDS has been used for investigations in a number of mines. Many of the problems associated with the use of such instruments in mines (6) have been overcome and CAMPEDS has been developed into a rugged instrument with satisfactory performance.

MEASUREMENT OF RADON DAUGHTERS BY MEANS OF THE CAMPEDS

Thermoluminescent dosimetry has been studied extensively by other investigators (7, 8, 9). It is based on the fact that certain materials, e.g., CaF_2 and LiF, are able to absorb energy from ionizing radiation and to release it again in the form of light when they are heated. The intensity of the light output is proportional to the cumulative total energy of the ionizing radiation. Thus, if radon daughters are collected on a filter to which a chip of such a thermoluminescent material is exposed, a measurement of the light output on heating of the chip can be made. After suitable calibration this measurement will yield the cumulative exposure to radon daughter in WLH* represented by the sample on the filter.

In the most recent studies of this process (10, 11) chips of thermoluminescent dosimeter (TLD) material (embedded in teflon) were exposed to airborne dust during its collection on a filter at a low flow rate (100 cc/min) for a 40-hour work week. Thus the total volume sampled was 240 l. In one of these studies (10) the following conclusions were reached:

"1. The air pump should be an integral part of a miner's lamp battery rather than a separate unit from the lamp battery. The air pump should be wired in such a manner that when a miner turns his lamp on the air pump is automatically turned on. Such an arrangement would obligate each miner to transport the monitoring system throughout his working area in a mine because a miner's light is usually the only source of illumination underground.

"2. The air pump must be completely sealed to prevent water and particulate matter from corroding and binding the moving parts of the pump.

"3. The filter-head unit should be designed as an integral part of the miner's lamp. The air inlet of the filter-head should be shielded from direct impaction of mud or other particulate matter.

"4. The entire unit should be as compact and light as possible.

"5. The final testing procedure should involve miner usage under actual mining conditions. Laboratory testing usually underestimates the severe and abusive treatment encountered by equipment in the mining operations."

It is intended to attempt to meet these requirements by incorporating a TLD chip in the entry cone of the CAMPEDS filter cassette (Figure 2). Because the CAMPEDS samples for an 8-hour shift at a flow rate of 2 l/min, however, the total volume sampled is 960 l. In order to keep the exposure of the TLD chip at approximately the same level as in the previous studies the entry cone has been modified to accommodate it in such a way that the chip is expected to "see" about one quarter of the dust sample on the filter.

The modified CAMPEDS, incorporating a TLD chip, is about to be calibrated and field tested. It is expected that a negligible fraction of the radon daughters will be associated with the coarse dust sizes rejected by the size selector, but this will be checked. Field tests will include the periodic measurement of WL by standard methods in parallel with the modified CAMPEDS, and correlation of the estimated WLH exposure with the integrated light output of the TLD chip indicated by the TLD reader.

One other possible technique for exposure extimation should be mentioned. During and after collection of a respirable dust sample, the associated radon daughters decay to Pb^{210}

* WLH = Working Level-Hours, i.e. the product of the radon daughter concentration in working levels and the exposure time in hours.

FIGURE 2: Section Through Modified Campeds Entry Cone (Not to Scale)

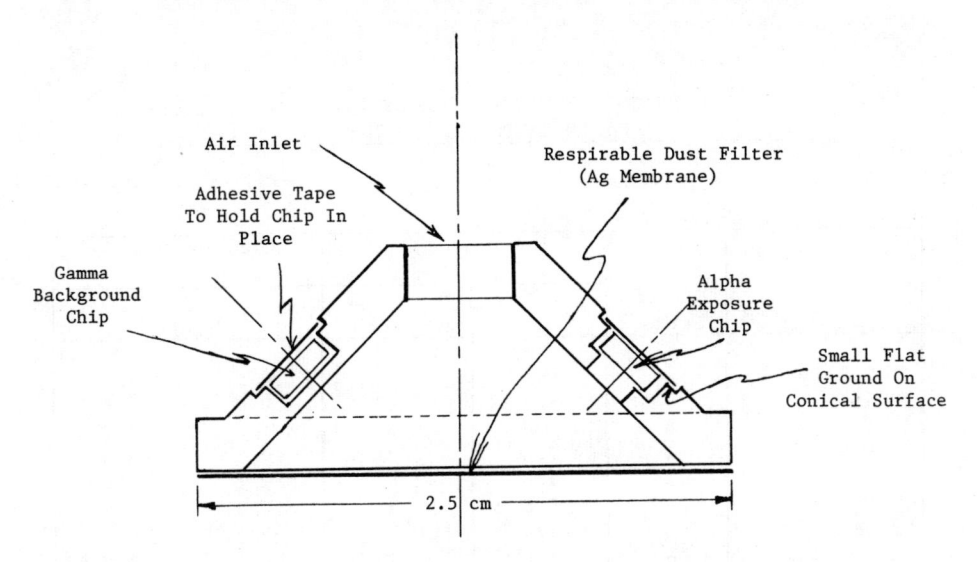

(RaD, $t\frac{1}{2}$ = 22 years). This in turn decays to Bi^{210} (RaE, $t\frac{1}{2}$ = 5 days) and Po^{210} (RaF, $t\frac{1}{2}$ = 138 days). The latter is an alpha emitter (E = 5.3 MeV) which may be measured as it grows into equilibrium with Pb^{210}. Preliminary tests on old (> 6 months) respirable dust samples have indicated that the technique may be practical, but much additional work needs to be done.

SAMPLING PROGRAMS

Routine sampling programs for dust and radon daughter measurement may have various objectives, such as engineering control, estimation of exposure for groups or estimation of exposure for individuals. The following three basic approaches have been suggested for meeting these objectives:

1. Short-period sampling, with periods of the order of minutes.

2. Full-shift sampling, either personal or by area.

3. Continuous monitoring, either personal or by area.

Short-period sampling, frequently repeated if necessary, is used for engineering control and the same type of sampling, repeated at intervals of the order of months, is the basis of the historical system for estimating personal exposure. The CAMPEDS is intended for exposure estimation using full-shift sampling. Holaday (6) has discussed the advantages of continuous personal dosimetry for radon daughters, but, unlike the film badge, the pump- and-filter samplers are expensive. Continuous monitoring is unlikely to give a sufficient improvement in exposure estimates over random full-shift sampling to justify the greater cost of the former.

Full-shift area sampling for dust is used in coal mines to monitor the exposure of the relatively large groups of men who work on long wall faces. This type of sampling cannot be equally satisfactory for uranium mines, where the men are scattered in small groups. In a practical monitoring system it may be necessary to calculate the individual's exposure from results obtained by

TABLE 1

VARIABILITY OF SHIFT MEAN DUST CONCENTRATIONS FOR INDIVIDUAL MINERS

Occupation	No. of Shifts	Mean	Standard Deviation	Range
Shift Boss	6	0.8	0.4	0.5 - 1.7
Driller	5	1.1	0.4	0.6 - 1.5
"	6	1.6	0.3	1.0 - 2.7
"	3	2.5	2.2	0.3 - 5.3
"	3	1.4	1.6	0.1 - 3.7
"	4	0.6	0.1	0.4 - 0.8
Slusher	3	1.4	0.6	0.9 - 2.4
"	6	1.6	0.7	1.1 - 3.0
"	3	1.2	0.4	0.9 - 1.7
Muck & Drill	4	0.7	0.05	0.6 - 0.7
"	6	0.7	0.2	0.4 - 1.0
"	5	0.6	0.2	0.4 - 1.0
"	3	1.5	1.2	0.4 - 2.8
"	4	0.4	0.15	0.2 - 0.6
Pipefitter	6	1.7	0.7	0.6 - 2.5
"	6	1.2	0.5	0.6 - 1.7
Tracklayer	5	1.4	0.4	0.7 - 1.8
"	6	1.2	0.6	0.6 - 2.3
Haulage	5	0.6	0.2	0.1 - 0.8
"	4	0.7	0.2	0.5 - 1.0
"	5	0.5	0.15	0.3 - 0.7
"	4	0.6	0.05	0.5 - 0.6
Cage Tender	5	0.7	0.3	0.2 - 1.2
Mean Values		1.00	.5	-- --

Respirable Quartz Dust Concentration (normalized to mine average)*

* Expressed as a ratio to the average dust concentration found in this mine.

personal full-shift dosimetry on randomly selected shifts, combined with the results of similar measurements on other individuals doing similar work. Full-shift monitoring of the working places on randomly selected shifts may also be needed to minimize the amount of personal sampling required and to check the validity of the results.

It is well known that the details of a sampling program depend critically on the variability of the measurements in space and time, and on the accuracy with which the estimate of exposure is required. The latter factor must ideally be determined from epidemiological studies linking the susceptibility of workers, with its own inherent variability, to the particular type of exposure in question. The variability of the measurements must be established at the outset of the sampling program under the prevailing conditions of mine operation. All important parameters, such as those relating to ventilation, should be under control.

At the Elliot Lake Laboratory work is under way to obtain adequate data on the variability of dust and radiation exposure in the uranium mines. As part of this work the performance of the modified CAMPEDS will be throughly field tested. Through the design and testing of sampling programs to meet various objectives it is proposed to develop a basis for choosing practical programs for routine sampling.

Table 1 shows the high variability of full-shift quartz measurements observed for uranium miners sampled on 3 to 6 shifts doing nominally the same work in the same place.

REFERENCES

1. Cochrane, T.S., G. Knight, L.C. Richards and W. Stefanich, "Comparison of dust sampling instruments", Research Report R250, CANMET, Department of Energy, Mines and Resources, Ottawa, Canada, 1971.

2. Knight, G., R. Kowalchuk and G.R. Yourt, "Development of a dust sampling system for hard rock mines based on gravimetric and quartz assessment", American Industrial Hygiene Association Journal, Vol. 35, pp 671 -681, Nov. 1974.

3. Knight, G., T.E. Newkirk and G.R. Yourt, "Full-shift assessment of respirable dust exposure", Canadian Institute of Mining and Metallurgy Bulletin, Vol. 67, pp 61-72, April 1974.

4. Knight, G., "Detailed studies on the assessment of quartz by x-ray diffraction in airborne dust samples and their collection in hard rock mines", Internal Report MRL 75-17.

5. Knight, G., "Guide to gravimetric sampling with quartz analysis in mines", Internal Report MRL 74/96.

6. Holaday, D.A., "Evaluation and control of radon daughter hazards in uranium mines", HEW Publication No. (NIOSH) 75-117, U.S. Department of Health, Education and Welfare, 1974.

7. McCurdy, D.E., "Thermoluminescent dosimetry for monitoring of uranium miners", USAEC Report No. COO-1500-16, Colorado State University, Fort Collins, Colo., June 1969.

8. McCurdy, D.E., K.J. Schiager, and E.D. Flack, "Thermoluminescent dosimetry for personal monitoring of uranium miners", Health Phys. $\underline{17}$, 414 (1969).

9. White, O., Jr., "An evaluation of six radon dosimeters", USAEC Health and Safety Laboratory Technical Memorandum 69-23, Oct. 15/69; "Evaluation of MIT and O NL radon daughter dosimeters", USAEC Health and Safety Laboratory Technical Memorandum 70-3, July 28/70; "Evaluation of MOD and ORNL radon daughter dosimeters", USAEC Health and Safety Laboratory Technical Memorandum 71-17, Oct. 8/71.

10. Schiager, K.J., and N.F. Savignac, "Radiation monitoring of uranium miners: A comparison of bioassay, TLD, and Kusnetz determination of current exposure", USAEC Report No. COO-1500-21, Colorado State University, Fort Collins, Colo., May 1972.

11. Breslin, A.J., personal communications.

Discussion

It was suggested that the quantity of ^{210}Po on a dust sample would be too small to measure, the authors replied that it had been measured with a 16 hour counting period using α spectrometry in a preliminary test of the method and this view was supported by a further speaker

CODE FOR RADIATION EXPOSURE IN ONTARIO MINES

W. A. Bardswich, P. Eng.,
Ontario Ministry of Natural Resources
Canada

The principal radiation hazard is the irradiation of lung tissue caused by inhalation of air containing radionuclides. The most serious result is the development of lung cancer, which generally does not appear for 10 to 20 years after initial exposure.

In most uranium mines, the measurement of radon daughter concentration is determined by using a specific field method. The results are then converted to the "Working Level" (WL) a unit defined as any combination of radon daughters in a litre of air that will result in the ultimate emission of 1.3×10^5 MeV (Million Electron Volts) of potential alpha energy or is equivalent to 100 picocuries/litre of each of the radon daughters RaA, RaB and RaC. Exposure to radon daughters over a period of time may be expressed in terms of cumulative "Working Level Months" (WLM). Inhalation of air containing concentration of 1 WL for 170 working hours results in an exposure of 1 WLM.

The most recent information regarding cancer incidence outside of Ontario indicates a substantial increase in the risk of lung cancer in miners with over 800 WLM of exposure and a slight increase among those having a lower exposure. Cases have been very rare (one or two) among non smokers. Consequently the following practice must be adhered to in Ontario mining:

1. Effective January 1, 1973, no person shall be permitted to smoke underground in a uranium mine.

2. Effective August 1, 1974, the occupational exposure to radon daughters shall be controlled so that no person will receive an exposure of more than 4 WLM per year.

3. Areas in underground uranium mines, whether normally or occasionally occupied, shall be monitored for the concentration of radon daughters in the air. The location and frequency of taking samples shall be determined in relation to compliance with Item 2. Such frequency shall be at least quarterly and recorded on a suitable form which shall be made available to the District Mining Engineer and forwarded to the Chief Engineer of Mines at the completion of each quarterly survey.

4. A record of exposure to radon daughters shall be kept for each person working underground in a uranium mine. Such records shall be made available to the District Mining Engineer and shall be forwarded in duplicate to the Chief Engineer of Mines at the end of each calendar year.

5. The Instruments used to measure the radon daughters shall be calibrated and certified every six months to the satisfaction of the District Mining Engineer.

6. Mine employees working in areas where their exposure cannot be limited to the requirements stated in Item 2 shall wear an approved type respirator, acceptable to the Chief Engineer of Mines, for the removal of radon daughters from the mine atmosphere. The control of exposure as required in Item 2 and the record of exposure as required in Item 4 must be carefully calculated and accumulated. For this purpose each worker will be required to submit a signed statement at the end of each shift stating the working area and the duration of time (in hours and quarter parts thereof) that he wore his respirator. In calculating exposure, an efficiency figure of not more than 90% may be used for powered type respirators and not more than 80% for non-powered type.

7. Requirements in Items 1 to 6, inclusive, shall apply also to all mines in areas where radon daughter concentrations exceed 0.3 WL.

PERSONAL DOSIMETRY IN THE NINGYO-TOGE MINE

Yoshiyasu Kurokawa*
Kenji Nakashima*
Yoshihisa Kitahara*
Ryuhei Kurosawa**
* Power Reactor and Nuclear Fuel Development
  Corporation (Japan)
** Waseda University (Japan)

1. Measurement of Individual Exposure Dose

The individual exposure dose is the total of the external exposure dose measured by the film badge method and the internal exposure dose assessed pursuant to the regulations of the applicable law.

(1) Measurement and Results of External Exposure Dose

All the mining and milling workers are ordered to wear film badge for the measurement of external exposure dose. This film is sealed with vinyl-film against moisture and is placed in a badge case. As humidity in the undergrounds is high, to prevent the latent image fading, the film badge wearing period is limited to one month. But in the plant, the measurement is done every three-month period. Fig. 1 and 2 represent the exposure dose distribution of the mining and milling workers.

According to Fig. 1, the averaged exposure dose showed an increase in 1973. This is due to the mining of high grade ores in the Nakatsugo District. Likewise, as seen from Fig. 2, the exposure dose per year of the plant workers is in the range of 100 ~ 200 mrem. However, some of the workers in the vicinity of the ore bin have been exposed to fairly high dose of 1.5 rem/yr.

(2) Assessment of Internal Exposure Dose

For the assessment of internal exposure dose, the estimation is performed in accordance with the methods prescribed by the Japanese law established in 1963. That is, in respect of ^{222}Rn and natural uranium contained in the air as radioactive materials, when one breathes for one week (48 hrs) the air with the ^{222}Rn concentration of 2.5×10^{-8} µCi/cm^3, or natural uranium concentration of 5×10^{-11} µCi/cm^3, his internal exposure dose is assumed at 100 mrem respectively. For this radioactive material concentration, the radon concentration is adopted as it is regularly measured every week. Thus estimation of internal exposure dose of each worker is made from the

Table 1. Radon-Daughters Removal Factor of
Protective Respirators

Respirator	Removal Factor at Flow Velocity 25 ℓ/min (%)
Sakai No. 117	88
Sakai No. 117 - washed filter	73

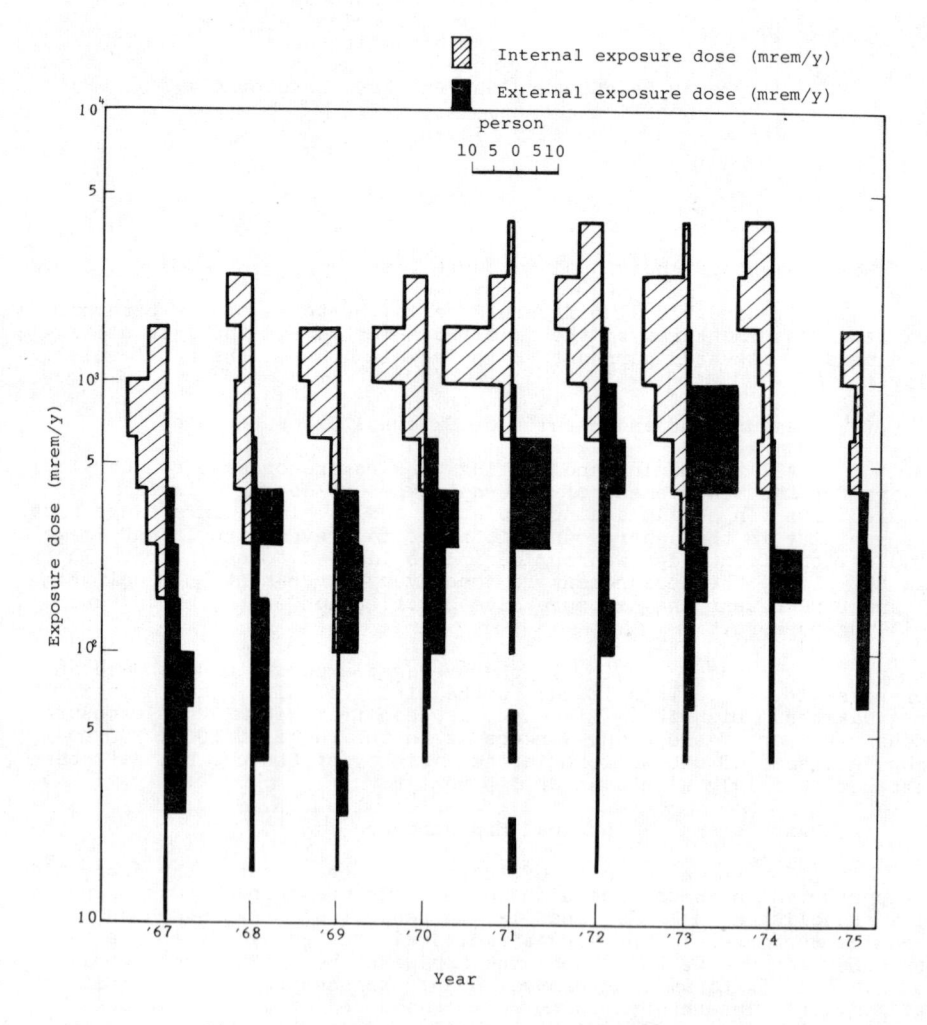

Fig. 1. Exposure Dose Distribution of Mining Workers

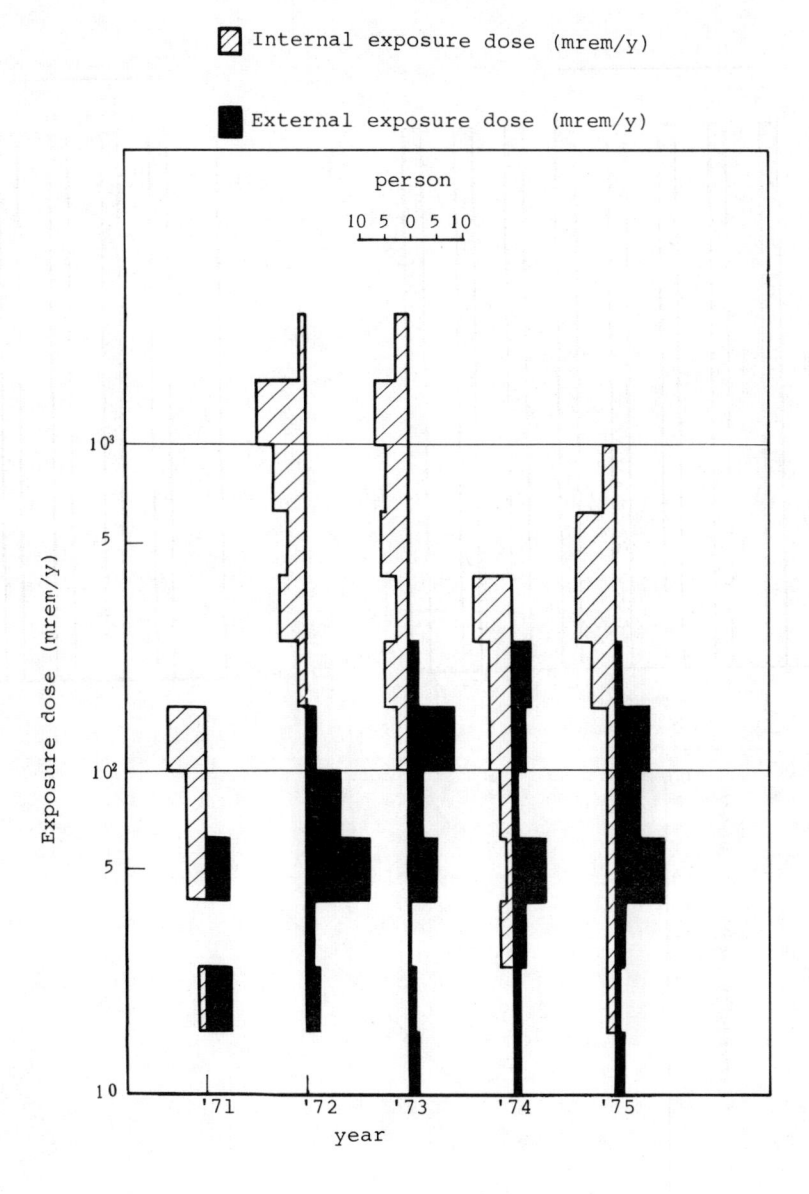

Fig. 2. Exposure Dose Distribution of Milling Workers

Fig. 3. Exposure Dose Distribution of Mining and Milling Workers

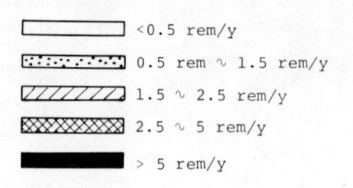

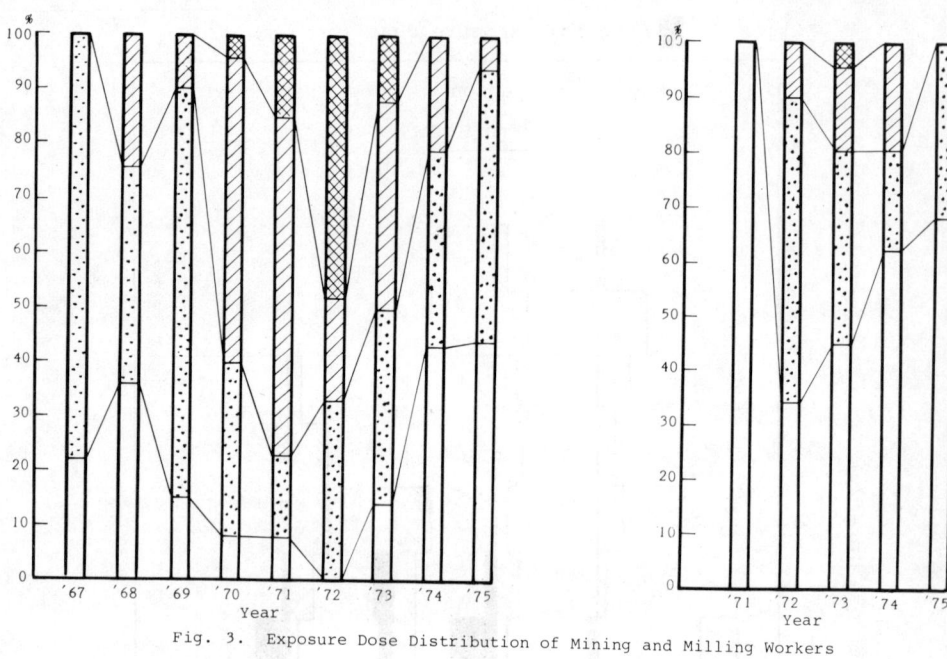

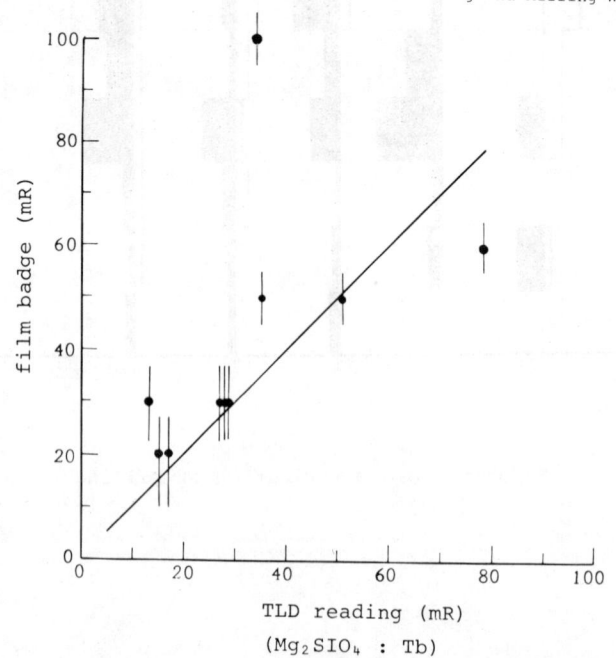

Fig. 4. Comparison between Film Badge and TLD
(Mg_2SIO_4 : Tb)

individual working records. The calculation method is as follows:

Internal exposure dose by ^{222}Rn in the working area (A_1):

$$P_1 = 100 \times \frac{p_1}{2.5} \times \frac{t}{48} \text{ (mrem)}$$

Internal exposure dose by uranium dust in the working area:

$$Q_1 = 100 \times \frac{q_1}{5} \times \frac{t}{48} \text{ (mrem)}$$

where, p_1 = the mean concentration of ^{222}Rn ($10^{-8} \mu Ci/cm^3$)

q_1 = the mean concentration of uranium dust ($10^{-11} \mu Ci/cm^3$)

t = worker's residing hours in the working area.

Provided however the above mentioned radon concentration of $2.5 \times 10^{-8} \mu Ci/cm^3$ as defined by the Japanese law is prescribed to be changed to $2.5 \times 10^{-7} \mu Ci/cm^3$ for the assessment of internal exposure dose only in the undergrounds. Consequently, Fig. 1 represents the figures calculated based on this method.

(3) Individual Exposure Control

Result of individual exposure dose is legisted to in the individual record, and each individual worker is received of his own accumulated dose record in every three-month period. The record is now being made into a centralized computer controlled system. The dose distribution in the mining and milling are shown in Fig. 3.

(4) Results of Medical Examination

There are many workers who have experience of working in other mines and some have been found to be suffering from light silicosis. But no marked symptom of progress of this disease has been noticed with respect to these workers during they worked in our uranium mine.

Following the provisions of the law, workers receive periodical blood tests twice a year to examine their counting of white and red blood corpuscles. Upto the present, there have been reported no special medical comments in this respect.

It is planned to distribute copies of health questionnaire to the workers who left the mine to check on their health condition. The health follow-up check made five years ago revealed no abnormal health condition.

2. Development of Personal Dosimetry

(1) Measurement of External Exposure Dose

Conventionally, the film badge method has been in use for the measurement of individual exposure dose. Currently the use of TLD ($CaSO_4$:Tm) which has high sensitivity and moisture stability is under study as personal dosimetry.

In Japan, there are marketed two kinds of TLD, namely $CaSO_4$:Tm and Mg_2SiO_4:Tb. Fig. 4 shows the results of the comparative use of the film badge and TLD (Mg_2SiO_4:Tb), both of which were worn simultaneously by the mining workers for one month.

(2) Measurement of Internal Exposure Dose of Mining Workers

i) Integrated Values of Radon Concentration

As previously described, according to the Japanese regulations, an attempt was made to obtain the integrated values of Rn concentration in the working area. It is known that when α particles are incidented onto a cellulose nitrate film (celluloid), an etchable latent image can be created. When Daicel-made celluloid is subjected to a 35-minute etching by NaOH of 6M at 45°C, almost all the α-ray incidented onto the film at the angle of 0° ~ 40° with energy of 0.5 ~ 3.9 MeV empirically proved to be detectable. Consequently, when the existence ratio of radon-daughters is assumed, the etchpit density (number of etchpits per 430 μmϕ in one visible field of a 400X optical microscope) when the film is worn on the body surface of a worker during one month work period (assuming 160 hours) can be calculated. According to this calculation, it was 0.57 etchpits (^{222}Rn : ^{218}Po : ^{214}Bi = 1 : 1 : 1) ~ 0.46 etchpits (^{222}Rn : ^{218}Po : ^{214}Bi = 1 : 1 : 0.4) per $3 \times 10^{-8} \mu Ci/cm^3 \times 160$ hrs. This has also been empirically confirmed. But according to the actual case of mining workers wearing the film, attaching of uranium ore dust onto the film surface caused etchpits by α-ray of uranium and its daughters, and thus etchpits have been caused also by other elements than radon-daughters itself. Because of this, there will be a possibility of making overestimation on exposure dose. And, in order to avoid such effect, it becomes necessary to prepare considerably large film cases, and moreover, the etchpit density can not be counted automatically. For these reasons, this is not considered quite practicable.

ii) WL Integrated Values

In order to obtain the integrated values of radon-daughters concentration in the working area, the radon-daughters was collected on the filter (millpore AA) by the personal dust sampler at the flow rate of 2 ℓ/min, while in doing so, the α-ray from the radon-daughters was detected by TLD placed at the position of 5mm from the filter. The used TLD was possible to measure upto 4.3 WLH. Since no regulations have been issued for WL in Japan, no plan is made to apply the results of this research work to the individual control in the actual working sites.

iii) Protective Respirator

According to the safety regulations, wearing of filter-type respirator is compulsory to all mining workers who are exposed to dust so that they are protected from the effect of the mineral dust and uranium dust. With respirator, it is possible to remove not only dust but also most of radon-daughters. Because of this it is officially recommended to wear of respirator for all mining works. The type of respirator worn by mining workers is of Sakai No. 117 which can be used repeatedly by washing the respirator's filtering section when it gets dirty. The removal rate of respirator is as given in Table 1.

Besides, in the actual use of respirators, there are problems of fitness of respirator mask to the user's face, air leak due to defective respirator valves, etc. which are the causes of hindering effective removal factor.

3. Radon-Daughters Removal Device

In order to reduce WL in the working area where ventilation is difficult, research and development efforts have been dedicated for an effective device to remove radon-daughters. This type of device is designed to remove radon-daughters and at the same time to reduce both WL and the primary particle concentration by recharging condensation nuclei.

The sketch diagrams of removal devices are shown in Fig. 5-1 and Fig. 5-2 respectively.

A test gallary was made by sealing a portion of the Kannokura District with vinyl-film sheets to form a wall. The cross section of the test gallary was of 2.2m (floor width) × 1.7m (ceiling width) × 1.9m (height), and the ^{222}Rn concentration went up to 1.9 ∿ 3.3 × 10^{-6}μCi/cm^3 as the test gallary had been sealed for a prolonged time. ^{218}Po concentration was in an equilibrium state with ^{222}Rn, and the primary particle concentration of ^{218}Po had been 1.0 ∿ 1.8 × 10^{-9}μCi/cm^3 before the removal device was brought in. But by the operation of the device, the primary particle concentration of ^{218}Po could be reduced down to 1/10 and 1/4 of the equilibrium state by the operation of high and low volume type devices respectively, and likewise, WL could be reduced down to 1/100 ∿ 1/20 and 1/40 ∿ 1/10 respectively.

As the device becomes comparatively large in size and the existing regulations have not prescribed WL, it is not actually used.

Discussion

Q. Why are condensation nuclei injected after the high efficiency filter, as shown in Fig. 5-1 and 5-2 of radon-daughters removal device? The U.S. reference was said to be to keep the condensation nuclei level at a minimum to allow maximum plating out of the radon-daughters on the pump and walls.

A. It is necessary to reduce a primary ion of RaA-daughter in the air and remembering that the blower connects to a drift. In the situation, miners or operators are present within a few meters of the air stream outlet. Under these conditions, the radon-daughters do not appear to plate out. Adding condensation nuclei as shown, a reduction of measured WL by a factor 3 has been observed experimentally at the working area. It is agreed that much further down the drift, the WL may rise with the addition of condensation nuclei, but there is no one present there.

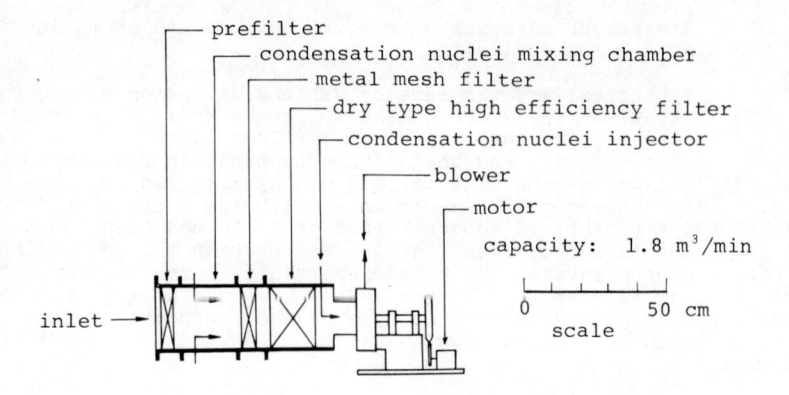

Fig. 5-1   Radon-Daughters Removal Device
(Low Volume Type)

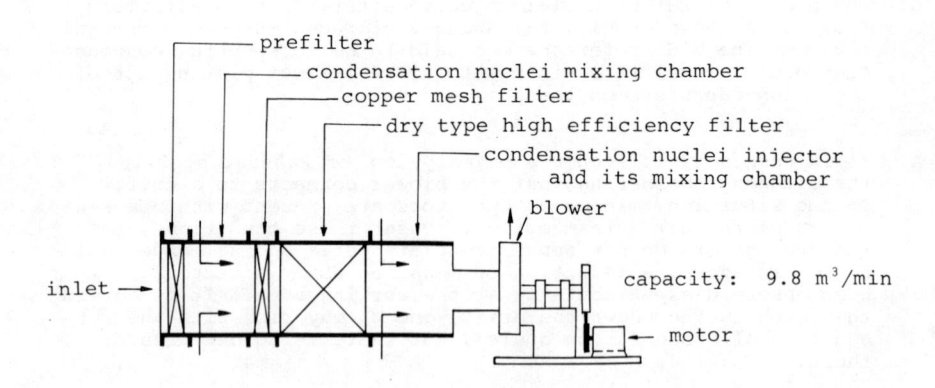

Fig. 5-2   Radon-Daughters Removal Device
(High Volume Type)

HAZARDS TO WORKERS FROM RADON AND RADON DAUGHTER PRODUCTS
IN TUNNELS AND DRIFTS OF SWITZERLAND

E. Kaufmann
Swiss National Accident Insurance Fund
Lucerne (Switzerland)

1. Introduction

Before dealing with the special features of hazards from radon and radon-daughter products, I think it necessary to give a short explanation of the functions of the Swiss National Accident Insurance Fund (SUVA) concerning the protection of workers. Contrary to most other countries, the SUVA is not only the holder of the obligatory insurance against accidents and occupational diseases of the workers in the companies under its control. Authorized by law, it is also responsible for the prevention of accidents and occupational diseases. In this sense and according to the decree of the Bundesrat (Federal Council) from April 13, 1963, revised on June 30, 1976, which concerns the protection against radiation, the SUVA is the official control organ responsible for the protection against radiation in the companies under its control, i.e. in all industrial and trade companies of Switzerland. Particulary in the field of protection against radiation, the SUVA is also responsible for the control of the environment.

2. Mining and Milling of Uranium

Up to now, there has been no mining of uranium ores and no milling of uranium in Switzerland. For the purpose of a survey, however, exploration tunnels have been excavated at several places in the Alps. In 1967 the SUVA was for the first time confronted with the hazards to workers in such an exploration tunnel at Mürtschenalp. Hundred years ago, this tunnel has already been used for the mining of ores containing copper. The badly damaged tunnel was provided with a new entry and was partly enlarged. The total length of the exploration tunnel is 75 meters. At the walls of the tunnel exposure rates of a maximum of 1,0 mR h^{-1} could be measured. Some rocks were of 3 - 8 % grade UO_2. On the average, the excavated material was estimated at 0,2 % grade UO_2.

At that time already, it was clear that we had to develop a measuring system in order to determine the hazards from radon and radon-daughter products. The fact that in these exploratory tunnels no electrical power was available, proved to be a major problem.

In 1964 at the Heidelberg symposium of the IAEA, Jacobi (Jacobi 1964) described the so-called double-filter device method. Our first self-made measuring system to determine radon-daughter products was based on this method.

As early as 1968 the "Studiengesellschaft für die Nutzbarmachung schweizerischer Lagerstätten mineralischer Rohstoffe" (study group for the development of Swiss mineral deposits) selected a construction company in Truns to excavate an exploration tunnel near the Alp of Dalisch, above Truns, where radioactive rock formations were suspected. The geologists of the study group measured surface exposure rates of up to 7 mR h^{-1}.

From the very beginning the construction company was under obligation, to provide mechanical ventilation in the drift as is common practice in tunnel construction. Fresh air had to be delivered as close as 5 meters to the working face using an air change-rate which would provide an air velocity toward the exit of at least 0,2 m s^{-1}.

At that time our measuring set-up was not yet operational. Only after the drift had been closed down we were able to take some measurements without interference. All the equipment necessary to measure radon-daughter products, including a gasoline power unit for the generation of electric power, had to be transported through pathless territory to a altitude of approximately 1400 meters above sea-level.

We have succeeded to establish the concentration of radon-daughter products in the exploration tunnel not beeing mechanically ventilated any more. However this field work in a mountainous area showed very clearly, that the measuring technique applied proves rather awkward in practical use.

3. Portable apparatus for the measurement of Radon-daughter products using the Kusnetz-method

Kusnetz himself described his method for the measurement of radon-daughter products in mines as early as 1956 (Kusnetz 1956) and Groër published in 1972 his investigations as to the accuracy of this method (Groër 1972). As could be expected, the Kusnetz-method could be applied with a substantially decreased expenditure on measuring equipment. In view of the fact, that electrical power was not available, air-intake was realized by the principle of the Venturi-tube. Thereby compressed air is beeing pressed through a nozzle in which a vacuum is generated just like in a water-jet pump. This will draw a well controlled amount of air through a filter. These measurements were taken in accordance with the regulation Nr. 7 of ANSI from 1969 (ANSI 1969).

4. Measuring Results

A good opportunity to test the new measuring equipment in practice presented itself in 1974 in a store room for uranium ores.

This store room contained approximately 10 tons of uranium ore having a grade of approximately 1 % UO$_2$. On our request the store room was equipped with a ventilation system, providing an air change rate of 3 to 4 per hour, even bevore the ore was brought in. Therefore, measurement could be taken with the mechanical ventilation system turned-off.

The concentration of radon-daughter products amounted to:

with mechanical ventilation   0,25 ± 15 % WL

without mechanical ventilation
after 2 1/2 hours   1,5 ± 15 % WL

During 1974 and 1975 measurements of radon-daughter products could be taken in several tunnels and drifts under construction. For other reasons too, these construction sites are always very well ventilated, as long as they are beeing worked on. For one part these are tunnels for the Swiss national highway system for the other they are pressure tubes for electrical power stations. Special attention has always been paid to the geological strata beeing traversed.

In the highway tunnels of the Gotthard and the Seelisberg which present the main obstacles for a north-south connection only minor concentrations of radon-daughter products could be measured (< 0,1 ± 15 % WL in the Gotthard-tunnel; < 0,01 ± 20 % WL in the Seelisberg-tunnel). Similar measuring results could be obtained in the pressure tubes.

One exception was the Rotlaui-tunnel in the Grimsel area. There, the mechanical ventilation system was removed after the break-through. When the natural ventilation of the air was very low, the concentration of radon-daughter products reached values of up to 1,12 ± 0,048 WL.

Literature:

ANSI   American National Standard Supplement to Radiation Protection in Uranium Mines and Mills
American National Standards Institute, Inc. ANSI Nr. 7, 1 a - 1969

Großr   The accuracy and precision of the Kusnetz method for the determination of the working level in Uranium Mines
Health Physics 1972, Vol 23 (July) pp 106-109

Jacobi W.   A Double-Filter Device to measure Radon and Thoron in the Breath
Symposium on the Assessment of radioactive Body Burdens in Man
Heidelberg 11.-16.6.1964, pp 275-289

Kusnetz   Radon daughters in mine atmospheres
Am. Ind. Hyg. Assoc. Quart. 17, pp 85-88 (1956)

Kaufmann E.   Radon und Radonfolgeprodukte in Tunnels und Stollen
Schweizerische Unfallversicherungsanstalt, Luzern
Mitteilungen der Sektion Physik Nr. 4, April 1976

PROBLEMES PRATIQUES RENCONTRES DANS LA DETERMINATION DES DOSES α
INHALEES PAR LE PERSONNEL DES MINES D'URANIUM

J. Pradel - Y. François - P. Zettwoog -

C.E.A. - PARIS - (FRANCE)

Les particules radioactives en suspension dans l'air des chantiers miniers comprennent en particulier des émetteurs α, qui, après inhalation, se déposent dans le système bronchopulmonaire, devenant une source potentielle de destruction des tissus biologiques. Parmi ces émetteurs α on distingue d'une part les descendants à vie courte du radon, avec lequel ils sont plus ou moins en équilibre, et les poussières de minerais, contenant l'uranium et ses descendants à vie longue.

La correspondance entre la quantité inhalée de ces aérosols porteurs de radioactivité α et la dose qui est reçue par les tissus critiques est mal connue. Les normes de radioprotection basées sur la valeur de l'intégrale du produit du temps passé dans chaque type d'activité minière par la valeur de la concentration en émetteurs α, définissent une certaine Quantité Maximum Admissible (Q.M.A.) par année (ou trimestre). On se base en France sur un temps de travail égal à 264 postes de 8 heures chacun soit 2112 heures (11 mois de travail).

Dans le cas des poussières à vie longue, la Concentration Maximale Admissible (CMA_p) est de 35.10^{-15} Ci/l, la QMA_p étant de

$$35.10^{-15} \times 11 \ /(Ci/l) \times Mois/$$

Dans le cas des descendants du radon, l'I.C.R.P. recommande le contrôle de ^{222}Rn, dont la C.M.A. est de 3.10^{-11} Ci/l, la QMA_R correspondante est $3.10^{-11} \times 11 \ /(Ci/l) \times Mois/$. La FRANCE applique la norme de l'IAEA qui est de 3.10^{-10} Ci/l. Aux U.S.A. la règle est de se baser sur l'énergie α potentielle des descendants du radon existant dans l'unité de volume, exprimée en Working Level (W.L). Un W.L. = $1,3.10^5$ MeVα/l. La QMA_R est aux U.S.A. de 4 W.L x Mois.

Les services chargés de la radioprotection souhaitent depuis longtemps que les QMA_R et les QMA_p soient mesurées directement et individuellement, comme cela est fait pour l'irradiation externe. Or si pour ce dernier cas la technologie de la dosimétrie individuelle est au point depuis longtemps, il n'en est pas de même pour la dosimétrie α. On a donc procédé jusqu'alors par voie indirecte, mesurant ponctuellement les concentrations dans les lieux de travail, et tenant compte du temps passé par chaque travailleur. La qualité des estimations des doses individuelles ainsi obtenues repose sur trois présupposés : la représentativité des mesures ponctuelles, à faire dans chaque lieu de travail et pour chaque activité (foration, tir, raclage, boisage, roulage, accès, etc...), la véracité des fiches de postes qui précisent les lieux de travail de chacun, et la neutralité du découpage arbitraire du temps de travail en activités unitaires, qui est effectué une fois pour toutes. On se propose d'examiner dans quelle mesure ces trois présupposés sont acceptables : les erreurs inévitables ont-elles tendance à se compenser statistiquement, ou y a t-il des risques d'écarts systématiques dans un sens ou dans un autre ? Il sera montré que les doses déterminées par cette méthode indirecte ont de bonnes chances d'être systématiquement sous-estimées. Les premiers résultats que nous obtenons depuis Février 1976 avec une série de 6 dosimètres α individuels confiés à des mineurs de fond confirment ce point de vue.

I - Prise en compte des difficultés pratiques -

1/ Sources d'erreur dans la connaissance des concentrations sur les lieux de travail.

Les sources possibles d'erreur au niveau de la connaissance des concentrations sur les lieux de travail sont répertoriées dans le tableau I, qui indique quelles en sont les conséquences sur l'écart type et la valeur moyenne des valeurs mesurées.

Bien que nous ne soyons pas certains que la baisse de la qualité de l'aérage et les transitoires d'exploitations soient toujours une cause de sous-estimation, toutes les autres causes identifiées aboutissent, à la sous-estimation, qui doit être considérée comme probable.

Tableau I

Sources d'Erreurs	Conséquences		
	Elargissement de l'écart type	Déplacement de la Moyenne	
		Sous-Estimation	Surestimation
Conditions de travail -			
Variabilité des chantiers	x		
Variabilité des opérations	x		
Inhomogénéité des concentrations dans le chantier	x		
Négligences insoupçonnées -			
Retour prématuré après-tir		x	
Baisse graduelle de la qualité de l'aérage		x	?
Vieux travaux mal isolés (fuites locales)		x	
Transitoires d'exploitation Chantiers et Mines nouvelles		x	?
Halls de stockage Musées Labo de minéralogie		x	
Locaux construits en ou sur stériles radifères		x	
Passage dans des zones interdites		x	
Incidents non ou mal comptabilisables -			
Arrêt de l'aérage		x	
Poches de radon		x	
Conditions du contrôle -			
Fréquence trop faible des prélèvements	x		
Ecart entre front de tir et point de prélèvement		x	
Annonce de la venue du contrôleur		x	
Pressions exercées sur le contrôleur au cours de la mesure		x	
Vérification systématique des valeurs jugées trop fortes		x	

2/ Sources d'erreurs dans la détermination des temps passés sur les différents chantiers.

On peut considérer que les erreurs dûes au manque de véracité de la fiche de poste se compensent statistiquement sur le temps de vie du travailleur.

3/ Manque de neutralité du découpage du temps de travail en activités unitaires.

La non neutralité du découpage, soit voulue, dans un sens ou dans l'autre, soit dûe à un changement progressif des méthodes d'exploitation (mécanisation plus poussée par exemple) ou de la configuration de la mine (allongement de la longueur des galeries) conduit à surestimer ou sous-extimer les doses moyennes reçues.

4/ Conclusion - Si la possibilité d'une surestimation existe au niveau du découpage du temps de travail, presque toutes les causes identifiées tendent à la sous-estimation.

II - Premiers résultats obtenus à l'aide de dosimètres α individuels -

Depuis le mois de mai 1976 nous avons commencé une campagne de comparaison des valeurs de doses α inhalées par le personnel obtenues soit par le calcul, soit par dosimétrie α individuelle. Un certain nombre de mineurs reçoivent en début de mois un dosimètre personnel qui permet de déterminer la dose inhalée en ^{218}Po et ^{214}Po pendant le mois et donc l'énergie α potentielle inhalée. Parallèlement, dans le cadre normal des opérations de contrôle radiologique, on a inscrit, sur la fiche individuelle des mineurs concernés, une quantité de radon inhalée, qui, au cours du même mois, par utilisation d'un facteur d'équilibre moyen de 0,17, déterminé par des mesures préliminaires, peut être traduite en énergie α inhalée.

La comparaison des valeurs de doses calculées et mesurées est présentée globalement dans le tableau ci-dessous, aussi bien pour les descendants du radon que pour les poussières à vie longue :

Pour 16 mesures nous relevons :

	Radon	Poussières à Vie longue
Surestimation de la dose calculée	0 cas	0 cas
Egalité entre la dose calculée et la dose mesurée	1 cas	0 cas
Sous-estimation de la dose calculée	15 cas	16 cas

Bien que le nombre de résultats soit trop faible pour que l'exploitation statistique soit possible, la sous-estimation soupçonnée dans le chapitre précédent semble bien être confirmée.

III - Commentaires -

1/ Pour les poussières à vie longue le facteur moyen de sous-estimation est de 5,5, il est de 2,5 pour l'énergie α potentielle ; la différence peut correspondre au fait que les poussières de minerai sédimentent plus vite que les descendants du radon et que donc l'inhomogénéité spatiale entre front de taille et point de prélèvement est plus grande pour les poussières.

2/ Le calcul de l'énergie α potentielle à partir des doses de radon gaz fait intervenir un facteur d'équilibre de 0,17 déterminé préalablement par plusieurs centaines de mesures. Vu la faiblesse de l'échantillon, la probabilité que le facteur d'équilibre ait été sous-estimée pour ces 16 mesures n'est pas nulle. Cependant nous ne pensons pas qu'une erreur d'un facteur 2,5 ait été possible ; nous aurions en effet un facteur d'équilibre de 0,17 x 2,5 = 0,42 alors que nous n'avons jamais dépassé 0,3. La sous-estimation nous paraît donc probable.

3/ Bien que nous effectuions de l'ordre de 40.000 prélèvements chaque année pour assurer la surveillance de nos 500 mineurs de fond, il semble que la représentativité des prélèvements ne soit toujours pas assurée. Ceci semble militer en faveur de l'emploi de la dosimétrie individuelle.

IV - Conclusions -

Les valeurs rapportées dans la littérature, utilisées aussi bien pour la pratique de la Radioprotection que pour les enquêtes épidémiologiques devraient faire l'objet d'un examen critique approfondi.

Par ailleurs le risque associé aux poussières à vie longue, très souvent non pris en compte, pourrait se révéler, non négligeable.

Ces valeurs calculées ne nous paraissent pas pouvoir être garanties au niveau de l'individu avec les précisions annoncées par les auteurs de ces enquêtes et le risque d'en déduire des normes mal adaptées doit être envisagé.

Résumé de la Discussion

Examen des causes possibles des écarts constatés entre la mesure traditionnelle et la mesure par dosimètre personnel.

Hypothèse de l'existence de différences importantes en ce qui concerne la distribution des concentrations de descendants du radon suivant le type de ventilation, la taille des chantiers ou le type de mines (sédimentaire ou primaire).

IN VIVO MEASUREMENTS AND BIOASSAY FOR ^{210}Pb, AS AN INDICATOR
OF CUMULATIVE EXPOSURE OF URANIUM MINERS TO RADON DAUGHTERS.

C. Pomroy
Radiation Protection Bureau, Health Protection Branch
Department National Health and Welfare
Ottawa, Canada.

Introduction

The epidemiological data on uranium miners is based on estimates of cumulative exposure made by combining area monitoring measurements with estimated occupancy times in different parts of the mine. Such exposure estimates involve considerable uncertainties.

An alternative method obviously being desirable, the Radiation Protection Bureau undertook to evaluate the in-vivo approach, which appeared to offer some promise.

In vivo measurements.

In 1973-74 a pilot study was done to investigate the feasibility of using in vivo skull counting for 210Pb as an indicator of cumulative exposure in Uranium miners. A report on this survey was presented at the Health Physics Society Mid Year Symposium, Denver, February '76 (1). The method was based on work done by Eisenbud's group at New York University, who had surveyed miners in Utah (2). The method made use of special scintillation detectors known as phoswich detectors, which have high sensitivity at low energy and inherent background suppression. Three of these detectors were used, together with specially designed lead shielding in a mobile laboratory. A total of 14 uranium miners, 15 non-uranium miners and 14 controls were measured. The three detectors were arranged around the miner's head and a one hour count was made. Subject background was estimated by a formula derived from measurements on the normal subjects. The measurements failed to detect any 210Pb and the following shortcomings emerged:

1.- Ambient background in Elliot Lake was higher than in our Ottawa Laboratory, which reduced the sensitivity of the method.

2.- The miners were contaminated to varying degrees with short lived daughter products which distorted the spectrum in the 210Pb region.

The minimum detectable burden was 1.9 nCi, whereas the predicted burdens in the miners would be of the order of 1 nCi or less. Background measurements in the steel room in our Ottawa Laboratory indicate that burdens significantly less than 1 nCi would be detectable there. It was therefore concluded that this method is not suitable as a monitoring method for active miners. It is being pursued however as a laboratory method by which former miners who have either retired or moved to surface work would be measured and the results incorporated in any epidemiological studies that are undertaken.

There is some evidence that, despite the fact that the analytical relation of skull ^{210}Pb burden with cumulative exposure to radon daughters is not firmly established, there is a relation between ^{210}Pb burden and lung cancer incidence.

Bio-assay

Methods are currently being developed for measuring ^{210}Pb in urine and blood, as this approach holds some promise of being of value in monitoring active miners. Blanchard (3) has shown a strong correlation between blood ^{210}Pb levels and cumulative exposure in miners at least six months out of the mine, and a lesser correlation for active miners. It is hoped that such correlations can be demonstrated for Canadian miners.

References

1.- Pomroy, C. "Field Surveys of Uranium Miners for the In vivo Detection of 210Pb."

Proceedings of Ninth Mid-Year Topical Symposium of the Health Physics Society Denver February 1976.

2.- Eisenbud M., et al "In vivo measurement of ^{210}Pb as an indicator of cumulative radon daughter exposure in uranium miners".

Health Physics 16: 637-646, 1969

3.- Blanchard R.L., et al "^{210}Pb blood concentration as a measure of Uranium Miner exposure".

Health Physics 25: 129-133, 1973

Session III

Panel discussion on personal dosimetry

Chairman - Président
R.M. FRY
(Australia)

Séance III

Table ronde sur la dosimétrie individuelle

Summary of Discussion
R.M. FRY (Australia)

This session was devoted to a discussion of personal monitoring for radon daughters, and its rationale. Impetus for the development of personal monitors came from the well-known inadequacies, in many situations, of area monitoring in determining the quantity of an airborne contaminant actually inhaled by an individual in a working environment. However the question arises whether personal monitoring of uranium miners is necessary to ensure safe working conditions, and indeed, whether it is practicable. Is not area monitoring of a mine atmosphere sufficient; if not, why not? There is a difference between ensuring a safe working environment and determining accurately the cumulative exposure to radon daughters received by a miner. Since control of a mine atmosphere will be based largely on area monitoring, why does one need an accurate assessment of individual integrated exposures? These questions and related matters were discussed; a variety of views was apparent among participants.

Before summarising the discussion on these points a more basic question needs to be answered. From the work of Jacobi and others it is known that the working level month is not in itself, necessarily a good physical correlative of the dose to the bronchial epithelium. How important is it in assessing the hazard of a mine atmosphere to measure, as well as the total potential alpha energy, the unattached daughter fraction, the aerosol concentration and activity distribution, and mine ventilation rate? For routine monitoring it would seem sufficient to measure the total potential alpha energy of the radon daughters though some attention should be given to developing dosimeters, whether personal or area, that discriminate for particle size in accordance with appropriate lung models. Work should continue, to get more information about the relationship between aerosol concentration and free ion fraction under various conditions of mining, but in most situations encountered in mines the unattached fraction will be quite low.

There was a spectrum of views on the relative importance of personal monitoring and area monitoring in the routine monitoring situation. Some considered that individual miners were so mobile, and the working levels in a mine so transient with ventilation conditions, that an individual's cumulative exposure is virtually unknowable unless measured with a personal monitor. Even under controlled conditions it was reported at this meeting (Zettwoog) that exposures assessed by area monitoring and occupation time may underestimate the "true" exposure as measured by a personal monitor by a factor of two or three. (On the other hand O'Riordan found reasonable agreement, in some measurements made in U.K. mines, between exposures assessed by personal and area monitoring). Others were not convinced that the accuracy in exposure assessment achievable by personal dosimetry was necessary for routine monitoring; it should be sufficient in this case to ensure that exposures are below a certain level which need not be measured to a high degree of accuracy. All however were agreed that personal dosimetry was highly desirable for the accurate individual exposure assessments necessary for reliable epidemiological studies.

An interesting point was raised in this connection. A number of countries and international organizations appear to be converging on an occupational radon daughter exposure limit of 4 WLM/y. This figure is derived from epidemiological studies on lung cancer incidence in underground miners as a function of cumulative exposure estimated from area monitoring (often carried out retrospectively) and work records. Reasonably consistent correlations between cumulative exposure and excess lung cancer have been obtained in this way though it is recognized that the estimates of individual exposures must be very much in error. Part of the error will be due to having to rely on retrospective estimates of what the radon daughter concentrations might have been to which the miners were exposed; part, due to the inadequacies of area monitoring in determining cumulative exposure discussed above. If now, personal monitoring becomes commonplace, and it turns out that it leads to cumulative exposure assessments consistently some two or three times higher than those based on area monitoring, should not the exposure limit, in cases where exposures are measured by personal monitors, be some two or three times 4WLM/y? It was generally agreed that this was a matter to be borne in mind. However it appeared unlikely that a consistent factor could be applied to area monitoring based exposure assessments that would allow actual exposures to be determined. It had to be accepted that there was a high degree of uncertainty in the assessed exposures used in past epidemiological studies and that this was unredeemable. One of the major reasons for wishing to see personal monitoring techniques developed was to provide a more sure basis for future epidemiological studies, though with the requirements on mine ventilation now being instituted around the world, it was unlikely that any harmful effects of exposure to radon daughters would be observable amongst future uranium miners.

There was universal support amongst participants for continuing development of personal monitoring techniques and for assessment and intercomparisons in the mine environment of their performance against other monitoring methods. At the present time there appeared to be nothing to choose between the approaches of the French, using track-etch detectors, and of the USA, using TLD detection. The French monitor is now commercially available but it is considered by some to be too expensive for wide scale use though it was pointed out that it should not be necessary to equip all miners at all times with personal dosimeters. After further intercomparisons of techniques and dosimeters it may be sufficient to assign one dosimeter to a group of miners doing similar work, and then perhaps only on a one week per month basis. It may turn out that making a measurement, say, of one miner one week in a month would still give a better estimate of annual exposure than is now obtained by existing methods.

Nevertheless the present concept of personal dosimeters which require pumps with their attendant problems of reliability, weight, battery charging and worker acceptability, are inherently rather inconvenient instruments and a plea was made for some thought to be given to the possibility of developing simpler, cheaper, passive devices.

A warning note was sounded that increasing sophistication in measuring techniques and dosimetric concepts did not necessarily lead to increased protection of the miner. The concept of the working level was an advance in scientific thinking but good protection can be achieved by simply measuring radon if it forms part of a coherent set of standards and defined monitoring procedures. Perhaps more comprehensive monitoring, not of dust levels or of radioactivity concentrations, but of ventilation is the most effective way to achieve a good working environment. One problem with personal monitoring, if it is to be used to improve work habits, is the delay between the measurement and the results; a miner is unlikely to be able to remember exactly what he was doing a week or so ago that may have led to a high exposure reading. A second consequence to be guarded against in increasing elaboration of monitoring techniques is the disproportionate amount of effort that can be devoted just to measuring conditions rather than controlling them. On the other hand it was pointed out that a basic tenet of radiation protection philosophy is that exposures must be kept to as low a level as is reasonably achievable. This seemed to imply that individual exposures, not only average exposures, should be kept as low as practicable which could only be achieved through the better estimates of individual exposures made possible by personal monitoring.

It is difficult to summarize adequately a discussion that was wide ranging and reflected a number of sometimes disparate views. However there would be a consensus that personal monitoring techniques have a most important role to play in ensuring the continuing protection of uranium miners and that their further development and systematic intercomparison with more established monitoring methods must continue. Because personal monitors are still under development and their full potential subject to further evaluation, it is not possible at this time either to recommend for routine use one technique or approach rather than another, or to say with confidence just what contribution they should be required to make in the routine monitoring of worker exposures.

Session IV
Area monitoring

Chairman - Président
D. ROSS
(United States)

Séance IV
Surveillance de l'atmosphère

TWO AREA MONITORS WITH POTENTIAL APPLICATION IN URANIUM MINES

A. J. Breslin
Health and Safety Laboratory, U.S. Energy Research and Development Administration
New York, New York 10014 (United States)

Two area monitors designed at the Health and Safety Laboratory (HASL), one for radon and one for working level, are being used primarily for studies of background radioactivity but might also have application in uraniun mines if there is need for measuring average concentrations over periods of several days or more. Both instruments are portable, operate unattended for extended time periods on either line or battery power, and can withstand the environmental conditions of uranium mines. Detection in each case is by means of an alpha-sensitive, thermoluminescent lithium fluoride chip positioned above the collection surface of a membrane filter.

The working level (WL) monitor is a direct descendent of the MOD WL dosimeter described elsewhere in these proceedings.[1] With reference to Figure 1, it consists of a metal box, 15 x 25 x 25 cm weighing 5.5 kg, which contains the MOD pump, a power supply, and a running-time meter. The MOD detector head, containing a 1 cm diameter membrane filter and lithium fluoride chip, is mounted on the front of the box. The monitor is operated with the lid closed. Raising the lid gives access to a fuse, a flowrate adjustment screw, and to the face of the running-time meter. Two power receptacles, one for 110 vac and one for 12 vdc, are on the back of the box.

The monitor is not equipped with an air flow meter because the pump maintains nearly constant flow at a nominal rate of 100 ml/min against a considerable range of inlet resistances. With rare exceptions, the flowrate remains constant within 2% during a 168-hour sampling period. However, to guard against the occasional substantial reduction that may result from unusually heavy dust concentrations or instrument malfunction, flowrate is checked at the end of the sampling period by means of a pulse-damped rotameter which is attached briefly to the detector head.

In our background studies, we normally monitor continuously for one week and in that time can measure concentrations as low as 0.0005 WL. The instrument can be operated for longer or shorter periods, at consequent increases or reductions in sensitivity. Sensitivity in a uranium mine would not be as good because of the much higher levels of gamma radiation compared to normal background but 0.05 WL could be measured reliably in 40 hours.

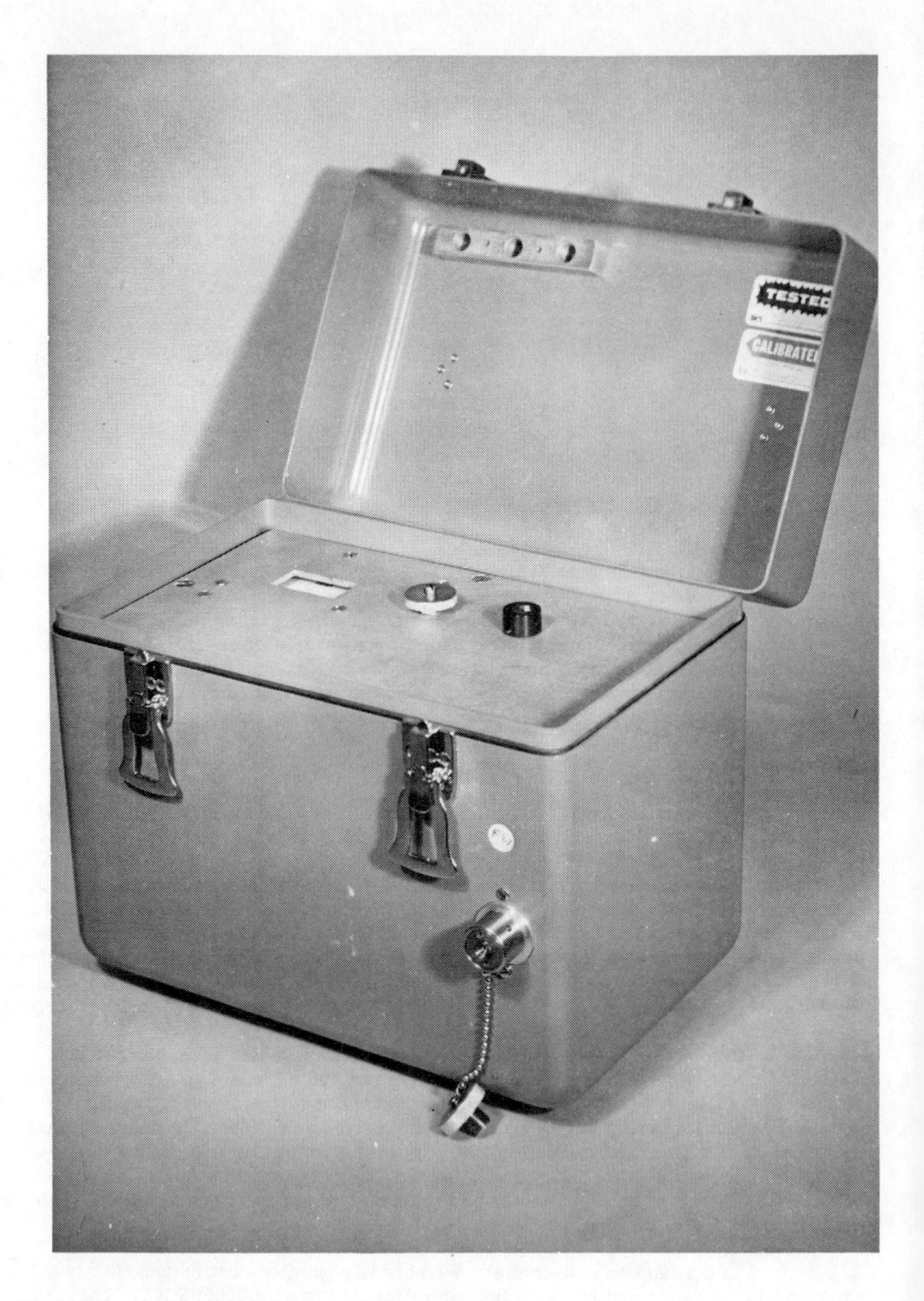

FIGURE 1. HASL AREA WORKING LEVEL MONITOR.

Performance of these monitors during two years of use has been excellent. By the time the first of these instruments was constructed, most of the causes of MOD pump malfunctions (described in reference 1) had been corrected and only minor, temporary difficulties arose. Typically, individual monitors operate for several thousand hours before servicing is required.

The integrating radon monitor[2] shown in Figure 2, operates on the principal of the two-filter method[3] but the downstream membrane filter is small, 1 cm in diameter, and has a lithium fluoride chip positioned just above its collection surface. Therefore, the radon daughters from the decay of radon in flight through the tube are collected on the filter and continuously detected by the chip during sampling periods which, in our applications, are usually one week or longer in duration. The thermoluminescence of the chip, corrected for gamma background by means of a second, control chip, is directly proportional to the time-integrated radon concentration.

The aluminum tube is 15 cm in diameter and 47 cm long. The inlet filter of glass fiber is equal in diameter to the tube. Air is drawn through the tube with a personal air sampler pump at 3 ℓ/min. The total weight including the pump and a battery charger, which can also maintain continuous operation on line power, is 5.5 kg. The unit can operate on its own battery without the charger for about ten hours or it can be powered by a storage battery for longer monitoring periods. In ambient atmospheres, the sensitivity is about 0.1 pCi/ℓ in 168 hours. Shorter or longer periods of operation are feasible, with a corresponding alteration in sensitivity. Sensitivity in a mine would be less because of the higher levels of external gamma radiation.

Sampling rate is virtually invariant with time, an important operational convenience. Predominant resistance to flowrate is the small membrane filter at the tube exit that is exposed only to clean air. Air dust is removed by the entrance filter which, because of its large area, does not develop significant resistance, even after prolonged, repeated use.

Operating experience with this instrument has not been as long as with the working level monitor but the only component subject to malfunction is the personal sampler pump. Reliability of the pump, a commercial product, depends on the quality of the particular model that is selected.

References

1. Breslin, A.J.
   MOD Working Level Dosimeter
   Proceedings of the NEA Elliot Lake Conference on Personal Dosimetry and Area
     Monitoring Suitable for Radon and Radon Products, October 1976.

2. George, Andreas C.
   Scintillation Flasks for the Determination of Low Level Concentrations of Radon
   Proceedings of Ninth Midyear Health Physics Symposium, Denver, Colorado,
     February 1976.

3. Thomas, Jess W. and Philip C. LeClare
   A Study of the Two-Filter Method for Radon-222
   Health Physics 18, 113-122, 1970.

FIGURE 2. HASL AREA RADON MONITOR.

Discussion

The commercial availability and costs of the two types of area monitors were questioned. Neither unit is available commercially but arrangements have been made with a machine shop in the New York area to fabricate a small number of the working level monitors from HASL drawings at an approximate cost of $350 each. Presumably, the radon monitor also could be fabricated by a competent shop from HASL drawings. The total cost would consist of the price of the personal sampler pump, a commercial product which markets for about $300, and the cost of fabricating the monitor proper, which is estimated to be $500.

There was some discussion of the radon monitor designed by M.E. Wrenn and H. Spitz at New York University (NYU), one panel member asking if the HASL radon monitor and the NYU radon monitor had been intercalibrated. The two instruments have not been compared directly but HASL and NYU have engaged in radon intercalibrations using other measurement techniques and it is reasonable to assume that the two monitors in question are in agreement.

Someone else mentioned that G. Cowper at AECL in Chalk River, Ontario, is working on a radon monitor that is similar in some respects to the NYU monitor but utilizes a passive thermoluminescent detector instead of a scintillation-photo-tube detector.

Because Wrenn was not at the meeting to describe the NYU radon monitor, the session chairman asked for a brief description by one of the panel members. The description was presented but subsequently Dr. Wrenn provided his own description as follows:

> The radon monitor designed and constructed at the New York University Medical Center in Sterling Forest is entirely passive. It operates on the basis of electrostatically collecting the decay products of ^{222}Rn after the gas has diffused through a porous foam barrier into a sensitive volume. Radon, but not its daughters, diffuses through a foam filter placed over a hemispherical wire mesh screen which acts as an outer positive electrode. A central electrode is coated with a ZnS phosphor and covered with aluminized mylar maintained at a negative potential. As the radon with the sensitive volume decays, the daughter product, ^{218}Po (RaA), formed mainly as a positive ion, is collected on the central electrode. Alpha particle emission from ^{218}Po and subsequently from ^{214}Po (RaC') is observed by a photomultiplier tube and recorded by the associated electronics and printer. Counts of alpha activity are accumulated and printed every 10 minutes, thus giving a total of 36 values for a 24-hour period.

A panel member suggested that air flow through the radon monitor would vary with time because of filter loading. As explained in the paper, this does not occur because the small exit filter, which is the dominant resistance, is protected from dust loading by the large entrance filter.

The effect of humidity on the performance of the radon monitor was questioned and one panel member stated that humidity was an important cause of error; another member stated that this is an expected result because water molecules attach to radium-A atoms in the two-filter tube, changing the diffusion coefficient of the atoms. Actually, in the original study of the two-filter method[3], response was found to be independent of humidity in the range from 20-90%. This range covers all but the most extreme environments in which the monitor might be applied.

OPERATIONAL RADIATION PROTECTION IN UK MINING

M C O'Riordan

National Radiological Protection Board

Harwell, Didcot, Oxfordshire, England.

The Industry

The mining industry in the UK is described in Table 1. There is no uranium mining, but there are indications of low-grade ore[1,2] in Scotland and South-west England.

Table 1

UK Mining Industry

Type	Mines	Miners
Coal - National	260	220 000
Coal - Private	160	1 600
Non-coal	100	2 600

Nationalized coal mines, operated by the National Coal Board (NCB), are well ventilated because of methane and dust controls and attention to environmental quality. Radon-222 concentrations are consequently low. A median radon activity concentration of 2 pCi/l was found in a countrywide selection of NCB mines[3], and later work supports this value[4,5]. Epidemiological studies[6,7] indicated that coal miners suffer a lower incidence of lung cancer than British men in general. There is no cause for concern therefore about radon exposure in nationalized coal mining.

Nothing was known about airborne radioactivity concentrations in private coal mines until recently. These small collieries, typically employing ten miners, were not nationalized in 1947, but are licensed by the NCB. Many do not have a fire damp problem and are not subject to dust controls; as a result, they are not so well ventilated. Recent work by the Board and the Health and Safety Executive (HSE) mine inspectors shows that the radon daughter concentration is typically less than 0.02 WL with a maximum so far of 0.12 WL. These results are based on a 25% sample and indicate that the position is likely to be satisfactory throughout the private sector. The survey will be completed, however, to make sure that there are no anomalies.

The radiological condition of coal mines is summarized in Table 2.

Table 2

Radon Daughter Concentrations in Coal Mines, WL

Owner	Typical	Maximum
National	0.01	0.02
Private	0.02	0.12

The non-coal category includes mines for fourteen principal minerals. There are no dust regulations, some workings are old and very extensive, and ventilation can be bad. It is not surprising therefore that excessive radon daughter concentrations were found in tin and haematite mines[8]. A linked mortality study of the haematite miners[9] indicated a lung cancer incidence higher than normal. Radon and iron oxide were identified as possible carcinogens. It became clear that a concerted effort was required to evaluate fully the radon daughter exposure of non-coal miners.

Procedures and Instrumentation

This effort was mounted in two phases by NRPB staff and HSE inspectors. In the first phase[10], comprehensive surveys of radon daughter concentrations were made in 40% of the mines by NRPB staff. Daughter measurements were accompanied by radon gas measurements at some of the sampling locations. In the second phase, which is still continuing, mine inspectors take gas samples during routine visits made for other purposes. The samples are mailed to the NRPB laboratory at Harwell, where the radon is measured and the daughter concentration inferred.

This approach has cut costs, but it has prolonged the programme and increased the errors in daughter estimation.

The main procedure used for measuring radon daughter concentrations was the modified Kusnetz method with the time-factor relation extended to permit half-shift tours and with a correction for thorium B (^{212}Pb) applied when required[11]. Pairs of 10 1/min pumps are used underground for 10 min periods and the glass fibre filters are measured in ZnS drawer counters above ground. Pairs of scintillating flasks, obtained from the Commissariat a l'Energie Atomique, are used for measuring the activity concentration of radon gas.

From approximately 100 comparisons at a variety of locations, a frequency distribution for the equilibrium factor, $\frac{100(WL)}{(pCi/1)Rn}$, was obtained[12]. High values predominated, as one would expect from ventilation conditions, and the distribution was not very wide[11]. These facts enabled the Board to introduce a rule-of-thumb for translating flask results into daughter concentrations at known volumetric flow rates: when the flow rate is less than 500 cfm (0.25 $m^3 s^{-1}$), an equilibrium factor of 1 is assumed; when it is greater, an equilibrium factor of 0.5 is applied. This is simple but crude, and whenever flasks indicate high daughter values, confirmatory daughter measurements are required. As ventilation is improved and radiological control is developed, there will be a need also to improve the interpretation of flask data. One way of doing that[13] is for the inspector to record the age of ventilation air[14] at each sampling location from which value the equilibrium factor can be estimated as for a tunnel. This is more soundly based than the foregoing procedure, but the complicated ventilation patterns in some mines will in turn complicate the interpretation.

Newer Techniques

An instrument which has proved of considerable value recently is the James-Strong radon daughter monitor (RDM). This has been described in detail[15], and it is sufficient to note here that two gross alpha counts in a 5 min counting regime yield working level, RaA concentration, and the ratio of RaC to RaA concentrations. Excellent agreement has been found in repeated intercomparisons with the modified Kusnetz method and the data on individual daughters have proved invaluable when studying air circuits. A small number of RDMs have been produced in the NRPB laboratory, but the possibility of commercial exploitation is being pursued. Some modifications, mainly mechanical, will be necessary.

The combining of respirable dust and radon daughter measurements is an attractive proposition. The method, devised by Ogden[4], and now being refined by him, involves the use of a Type 113A Casella gravimetric dust sampler[16]. This sampler acts by drawing air through a horizontal elutriator, the penetrating respirable dust being collected on a filter for subsequent weighing or compositional analysis. An alpha counting session is inserted first.

Dust sampling is done on a shift-long basis at a known constant aspiration rate, and the instrument is brought to the surface. After a fixed interval of not less than 20 min, two sequential alpha counts of 85 and 105 min are made on an automatic counter. The radon daughter concentration is then calculated using a tabulated set of constants. Thoron (^{220}Rn) daughter concentration can also be calculated.

The radon daughter concentration thus measured is not a shift mean: rather, it is heavily weighted by the ambient conditions toward the end of the sampling period. Detailed field intercomparisons with the modified Kusnetz method have yet to be made, but preliminary results are encouraging. Clearly, the Ogden method offers the mine operator, if obliged to do routine dust sampling, the opportunity to monitor airborne radioactivity as well, provided the appropriate alpha counting equipment is purchased. It is not flexible enough, however, for intensive short-term surveys.

It would be inappropriate to refer to body monitoring for ^{210}Pb using phoswich detectors[17] in this paper, since the potential of the technique is related to retrospective assessment rather than operational control.

Limits and Results

An annual exposure limit of 4 WLM has been adopted in the UK, this being equivalent to the current ICRP recommendations[18] for exposure to radon and daughters. A radiological protection programme is being designed around this limit. In particular, an environmental surveillance schedule linked to the magnitude of the exposure has been devised[19] and some of the data discussed below arise from that scheme. The schedule is shown in Table 3.

Table 3

Surveillance Scheme in Non-Coal Mines

Annual exposure, WLM	Frequency, monthly
Less than 1	12
1 to 3	6
3 to 4	3
More than 4	1*

*until remedied.

A derived concentration limit in unfiltered air of 3×10^{-7} Ci/m^3 radon is set in the new CEC Basic Safety Standards[20]. This limit is appropriate for highly ventilated mines (with an equilibrium factor of 0.1) but is unacceptable for the non-coal mines discussed here.

Some 80% of the non-coal mines have been surveyed at the time of writing. The exposure patterns of UK non-coal miners that have emerged so far may be summarized as in Table 4.

Table 4

Radon Daughter Exposure of UK Non-Coal Miners

Annual exposure, WLM	Men exposed, end August 1976 No.	%
0 to 1	986	49
1 to 4	454	23
4 and more	564	28

This indicates a reduction in the high-exposure category since the end of the first phase, when over 40% were included, reflecting the impact of monitoring, some contraction in the industry, and the fact that second-phase mines generally had lower daughter concentrations. In short, about 50% of the 2000 miners surveyed to date should be regarded as "radiation workers", and this proportion is not likely to change markedly. The mine types involved are tinstone, haematite, fluorspar, gypsum, honestone, fireclay, and ballclay.

It is planned to include such miners in the National Register of Radiation Workers. The register will contain information, supplied by the mine owners, sufficient to identify individual miners. With the aid of National Health numbers, it will be possible to obtain information about the deaths of individuals. Thus, mortality data for these miners could be compared with those of other workers. An exposure recording system will also be set up so that the information in the register can be fully exploited.

At present, exposure assessment is roughly obtained from the surveillance schedule described above. A weighted average annual exposure for all the miners in a particular mine is calculated from concentration and occupancy values for representative locations. This has been called the time-weighted full-shift exposure[21]. Attention is, however, being paid to the need to obtain representative data in an economical fashion, and it would appear that mine owners will need to be involved in routine monitoring to that end.

One question of representativeness that was explored in a small number of mines was whether environmental daughter concentrations were representative of breathing zone concentrations during various operations, especially where pneumatic equipment was being used[10]. This was done by making, simultaneously, environmental measurements at typical locations and breathing zone measurements throughout a number of shifts. The modified Kusnetz method was used for both. Agreement was found to be good. The temporal variability discovered did, however, highlight the difficulties of haphazard sampling.

Problems and Perspectives

There are extremely difficult ventilation problems in a number of non-coal mines. In one case, a complex of 10 or more smaller mines, there are said to be over 100 miles of interconnected roadways and about 20 shafts. Some successful attempts have been made to reduce average radon concentrations below 0.3 WL by increasing or altering ventilation, but the possibility of having to use personal respiratory protective devices, as a temporary measure in some situations, is being faced.

One such device is the Airstream helmet designed at the Safety in Mines Research Establishment[22]; it functions as a safety hat, eye protector, and respirator. Air is forced at 200 l/min through coarse and fine filters by a fan in the helmet and then flows inside a hinged visor. The appliance complies with BS 2091[23] for a Type A dust respirator showing less than 10% penetration. Face-seal leakage is negligible because of the positive pressure, and a protection factor of 10 at least is afforded against respirable dust. This factor would be adequate against radon daughters for virtually all situations in UK non-coal mines. Production models using rechargeable nickel-cadmium cells to drive the fan motor are due in late 1976, and it should then be possible to carry out efficiency tests using radioactive aerosols. Eventually, fan motors will be run from miners' lamp batteries.

Another potential problem revealed during this work was radon exposure of the public from upcast shafts. Some shafts have fairly low superstructures in the middle of mining villages. The increments in radon concentration in the neighbourhood of a number of these shafts following changes in ventilation patterns have been computed; to date, no unacceptably high values have been discovered.

From the point of view of radiological protection, non-coal miners have the highest occupational exposure of any group in the UK in relation to a recommended limit[24]. From the point of view of general health and safety in non-coal mining, however, the situation does not look so serious. This is illustrated as follows. The US epidemiological study of uranium miners[25] yields, on extrapolation, a risk estimate of some 0.3 deaths annually from lung cancer per 1000 miners exposed to 100 WLM. On the other hand, accident statistics[26] for non-coal mines in the UK yield an estimate of two deaths annually per 1000 miners. Further perspective is given to the problem by the incidence of lung cancer among adult males within the UK, that is, 1.5 cases annually per 1000 persons[27]. Narrow concern for the radiological safety of miners must therefore be tempered with broader concern for the other hazards they face.

REFERENCES

1. Bowie, S H U, Ostle, D and Gallagher, M J (1970) Uranium reconnaissance in northern Scotland. Trans. Inst. Min. and Metal, Section B, 79, B180-B182.

2. Michie, U McL (1972) Further evidence of uranium mineralization in Orkney. Ibid., 81, B53-B54.

3. Duggan, M J, Howell, D M and Soilleux, P J (1968) Concentrations of radon-222 in coal mines in England and Wales. Nature (Lond), 219, 1149.

4. Ogden, T L (1974) A method for measuring the working-level values of mixed radon and thoron daughters in coalmine air. Ann. Occup. Hyg., 17, 23-24.

5. National Radiological Protection Board. Unpublished data.

6. Doll, R (1958) Cancer of the lung and nose in nickel workers. Brit. J. Industr. Med., 15, 217-223.

7. Goldman, K P (1965) Mortality of coal-miners from carcinoma of the lung. Ibid., 22, 72-77.

8. Duggan, M J, Soilleux, P J, Strong, J C and Howell, D M (1970) The exposure of United Kingdom miners to radon. Ibid., 27, 106-109.

9. Boyd, J T, Doll, R, Faulds, J S and Leiper, J (1970) Cancer of the lung in iron ore (haematite) miners. Ibid., 27, 97-105.

10. Strong, J C, Laidlaw, A J and O'Riordan, M C (1975) Radon and its daughters in various British mines. NRPB - R39, Harwell, National Radiological Protection Board.

11. Strong, J C and Duggan, M J (1973) The effect of the presence of thoron daughters on the measurement of radon daughter concentrations. Hlth Phys, 25, 299-300.

12. Billard, F, Miribel, J, Madeleine, G and Pradel, J (1964) Methodes de mesure du radon et de dosage dans les mines d'uranium. Radiological health and safety in mining and milling of nuclear materials, Vol. 1, pp.411-423. Vienna, International Atomic Energy Agency.

13. Strong, J C and James, A C. National Radiological Protection Board, Personal communication.

14. Rolle, R (1972) Radon daughters and the age of ventilation air. Hlth Phys, 23, 118-120.

15. James, A C and Strong, J C (1973) A radon daughter monitor for use in mines. Proceedings of the 3rd IRPA Congress, Washington, DC, pp.932-938.

16. Dunmore, J H, Hamilton, R J and Smith, D S G (1964) An instrument for the sampling of respirable dust for subsequent gravimetric assessment. J. Sci. Instrum., 41, 669-672.

17. O'Riordan, M C (1975) The Board's low background facilities. Radiological Protection Bulletin No. 13, pp. 16-18, Harwell, National Radiological Protection Board.

18. ICRP Publication 2 (1959) Recommendations of the International Commission on Radiological Protection, Report of Committee II on Permissible Dose for Internal Radiation. Oxford, Pergamon Press.

19. Purvis, R T. Health and Safety Executive, Personal communication.

20. Council Directive of 1 June 1976. Revised basic safety standards for health protection of the general public and workers against the dangers of ionizing radiation. Official Journal of the European Communities, 19, No.L187, 12 July 1976.

21. Holaday, D A (1974) Evaluation and control of radon daughter hazards in uranium mines. HEW Publication No. (NIOSH) 75-117, Washington, DC, US Dept. of Health, Education and Welfare.

22. Greenough, G K (1974) The dust helmet: protection for head, eyes and lungs. Underground Services 2, No. 5. Essex, Foundation Publications Ltd.

23. BS 2091 (1969) Specification for respirators for protection against harmful dusts and gases. London, British Standards Institution.

24. Duggan, M J, Greenslade, E and Jones, B E (1969) External radiation doses from occupational exposure. Nature (Lond), 221, 831-833.

25. Archer, V E, Wagoner, J K and Lundin, F E (1973) Lung cancer among uranium miners in the United States. Hlth Phys, 25, 351-371.

26. HM Chief Inspector of Mines and Quarries (1975) Report under the Mines and Quarries Act (1954) for 1974. London, Her Majesty's Stationery Office.

27. Office of Population Censuses and Surveys (1975) The Registrar General's Review of England and Wales for the Year 1973, Part 1A. London, Her Majesty's Stationery Office.

PROTECTION OF WORKERS IN RADON-RICH ATMOSPHERES: THE MANDATE
FOR QUICK DETERMINATION OF RADON-DAUGHTER CONCENTRATIONS AND A SOLUTION.

by
james d. shreve jr.
kerr-mcgee corporation
oklahoma city, oklahoma, usa

Introduction

A personnel monitor for radon daughter concentrations in the air inspired by a miner has been a goal of long standing. Despite earnest effort by good people, a monitor that is dependable calibratable, rugged, light; that is acceptably priced; and one that can function under the inhospitable conditions encountered in mines, has been slow to materialize. So much so that the zeal for such an instrument has led to inflated expectation of the benefits, were a satisfactory design achieved. The real perspective deserves to be re-established.

A personal dosage monitor - in radiation safety, film badges for x-ray, gamma ray and neutron exposures are notable examples - totalizes the exposure of one man over weeks or months to some potentially harmful environmental factor(s); a reasonable period to consider is one month. To this can be added one to three weeks for reading the sensitive element(s) and reporting the result(s). Clearly, each result is basically historical. If it can serve to protect at all, the delay in so doing makes the protection indirect at best. The measurement frankly viewed really indicates the degree to which the worker has gone unprotected. For all its import to the long term study of occupational diseases among miners little worth in terms of protection can be claimed for individual personnel monitors.

Patently, personnel monitoring cannot substitute for the day to day vigilance by which the complex ventilation systems of larger underground mines are kept balanced and effective. The clear requirement for immediacy of information just cannot be met. In fact, the current reliance on the Kusnetz method fails in many respects to meet the immediacy. To wait 45 minutes or more for an answer forces inefficient use of monitoring personnel and can lengthen miners' exposure to high radon levels. Slow recognition of higher than normal radon daughter levels delays correction of the condition. Moreover, the efficiency of corrective measures taken cannot be judged quickly; the 45-minute minimum still rules.

An Alternative

First positive efforts to rectify this situation must be credited to Evans & Schroeder and later to Groer, Rolle and Hill. The measurement concepts they developed for what they termed an Instant Working Level Meter (IWLM) were sound enough. That the instruments designed were unable to exploit the concepts reliably was lamentable. Groer eventually solved the basic instrumentation problems but produced a unit too heavy, too cumbersome and too costly for routine in-mine applications. Rolle and Hill did develop practicable schemes and instruments that take 7 minutes or so to measure working level (WL) or 1.3×10^5 mev of alpha energy from RaA through RaC'.

An effort overlapping these endeavors was stimulated by an observation of Adare Hill (then working for Kerr-McGee Corporation in Grants, New Mexico and the Hill mentioned above). Playing with radon-daughter ingrowth calculations, especially the relative build up of alpha and beta activity, he normalized the activities to a unit working level (WL). It was then that he discovered the near-constancy of the sum of alpha and beta activities per WL. At Hill's request, the author made independent calculations and was able to confirm the discovery. Out of the desire to get quick determinations of WL while still exploiting Hill's original observation, the parameters were set at 2 minutes for air sampling followed by a 30-second period for filter transfer from pump to counting chamber; thereafter, alpha and beta counting were to be done for one minute. This made the total elapsed time from start of air sampling to a WL reading, 3.5 minutes.

The Kerr-McGee IWLM

As in the work of Evans, Schroeder and Groer, the principles and analytical base involved were irrefutable. To get the instrument to accurately perform in accord with analytical provisions was less than straightforward. Coping with the gamma sensitivity of the beta sensitive element (both external gammas and those accompanying the beta decay of RaB and RaC contribute spurious signals), accounting for conversion elections which also accompany the beta decay but are absent in the $^{90}Sr - ^{90}Y$ sources normally employed for calibrations and eliminating beta backscatter during calibration took considerable experimentation and cerebration. Ultimately, adequate compensation for these interferences was made; calibration procedures were developed to preserve the required additivity of alpha and beta counts. These techniques plus the appropriate choice of counting geometries and the inclusion of adjustable pulse rate multipliers in the circuitry combined to make the Kerr-McGee Instant Working Level Meter (IWLM) a dependable instrument.

Field Trials and Instrument Evolution

Several versions of the instrument were subjected to many months of intensive use in Kerr-McGee underground mines in New Mexico. The present configuration embodies all suggestions made by the users. As a result the current instrument (now manufactured and sold on a license from Kerr-McGee by MDA Scientific Inc. in the USA as the Model 811 IWLM) is a fully field-worthy device, convenient to carry and use. It weighs under 5kg, contains rechargeable batteries, whose capacity permits operation for a full 8-hour shift. Alpha and beta counts are given in digital form as separate readouts from two LED panels. The decimal point is fixed and an alpha-beta sum of 1.000 denotes 1.00 WL. Both alpha and beta backgrounds are counted for the first minute of the sampling cycle and are available for display for 1 1/2 minutes thereafter; the beta background is a good indicator of external gamma flux. However, recordation of background is not required for WL determination since both backgrounds are subtracted automatically when the filter is counted.

An external pump is necessary and the air sampling rate must be 2.5 l/min for the simple sum of the two readouts to give WL directly.

It follows that the same instrument can be made to serve - particularly where gamma ray backgrounds are small - for quite low radon-daughter concentrations. The flow rate just has to be increased by the appropriate multiple of 2.5 l/min; of course, the readings must be reduced by the same multiple.

IWLM Recycle Option and Kusnetz Reading

There exists an important opportunity for WL reading verification and for a more error-free WL value. Within 30 seconds after the first IWLM cycle ends, the filter holder is lifted from the counting slot. As this 30 second period ends (4-minutes total elapsed time), the instrument is reset to cycle again. After the automatic recount of background which takes one minute, the filter holder is reinserted to be counted in the usual time window of 2.5 to 3.5 minutes in the cycle. Actual total elapsed time, from the start of cycle #1 to the second WL reading is 7.5 minutes. The resulting alpha-beta sum, multiplied by 1.3 to correct for the longer sample decay time, is a check on the first reading and is less sensitive to 'age of air'.

Finally, the filter may be saved to serve as a Kusnetz sample after a wait of 40 or more minutes from the end of sampling - the usual procedure. However, since the filter contains the activity extracted from only 5 liters of air, the reading must be doubled before the Kusnetz factor is applied. For IWLM - Kusnetz method comparisons, this procedure has the evident advantage of performing all assessments on the very same filter, i.e., on activity extracted from the same air parcel.

Inherent Errors of the Method

If the equilibration factor (age of air) were held constant, the IWLM reading - sum of alpha and beta counts - would change directly as WL within counting statistics. However, the alpha-beta sum varies with age of air as Figure 2 shows. Unlike other methods which rely on a single alpha count and indicate nothing about age of air, the IWLM, by observing alpha and beta activities simultaneously, provides a means to estimate age of air. The simplest approach is that of Figure 1 wherein the alpha/beta ratio is employed as the age index. This judgement will be precise to the degree counting statistics permit if there is no preferential plate-out of some radon daughters on the walls or on dust that falls out or if no active particles become attracted to the wall. In short, these distortions of natural ingrowth ratios of daughters, should they occur, perturb the alpha/beta. That such situations are less likely and less severe in active mines that are humid and/or depend on diesel driven equipment, assures that, more often than not, the IWLM count ratio through Figure 1 will furnish an excellent age of air. This in turn promises a sound base for datum correction should the ultimate in resolution be needed.

Notice that Figures 1 and 2 include parallel guidance for the 'second chance' or recycle option. A determination made on recycle is clearly much less open to influence from age of air. For most applications, data so derived can be accepted without concern for corrections.

In Kerr-McGee experience, the technicians who use IWLMs regularly are quick to learn that when the alpha count is more than double the beta count, the air is very 'young' and the WL value is an over-estimation (Figure 2). Likewise equality of the two counts spells old air and values that are low by almost 10%.

One final point: caution is required when IWLM and Kusnetz values for the same mine area are compared. These two techniques agree very well for 20-minute air but for age extremes in either direction from this calibration reference, the sense of the error in one technique is opposite to the other. That is to say Kusnetz readings are low for 'young' air, high for 'old' air. Therefore disagreements appear to be more pronounced than they really are.

Radon Concentration

Another spinoff from measuring both alpha and beta activity is the chance to infer radon concentration. Figure 3 charts one format of this information. Having read alpha and beta from the IWLM, the sum gave the working level; clearly, the alpha-beta difference is equally ascertainable. If a user enters the ordinate at the alpha-beta difference, moves across to the appropriate WL curve, the corresponding abscissa is an estimate of the mean radon concentration.

The dashed radial lines viewed with respect to the difference - WL intercept, i.e., the point found before the radon concentration was read, tell the time or age of air that was necessary for the measured mix of daughters to have grown from that radon concentration.

Conclusion

The elementary observation of near-constancy of the sum of alpha and beta activity on a air sampling filter has led to a novel instrument. By exploiting this observation, a good working level determination can be made in a total time of 3.5 minutes. Second cycling and a total elapsed time of 7.5 minutes permits two determinations on the same filter, the second of which is more indifferent to daughter equilibration factors than either the 3.5 minute value or that defined by the Kusnetz method.

This instrument and auxiliaries, designed by Kerr-McGee Corporation and licensed to MDA Scientific, Inc. for manufacturing and marketing, also permit estimates of external gamma radiation present at the uranium mine site, the equilibration factor of the radon daughters and the radon concentration itself.

Its capacity to reduce exposure of underground workers via quicker detection and correction of high radiation areas is unmatched in a portable device. It has been found to at least double the efficiency of ventilation technicians.

This sum of attributes is unobtainable in any other known instrument marketed in the world today.

REFERENCES

1. Schroeder, G. L. and Evans, R. D., Some Basic Concepts in Uranium Mine Ventilation, Radium and Mesothorium Poisoning and Dosimetry and Instrumentation Techniques in applied Radioactivity, Annual Progress Report from Massachusetts Institute of Technology, MIT-952-5 (May 1968); see also MIT-952-6(1969).

2. Groer, P. G., Keefe, D. J., McDowell, W. F. and Selman, R. F., An Instant Working Level Meter with Automatic Radon-Daughter Readout. Final Report from Argonne National Laboratory (Argonne, Illinois, USA) to U. S. Bureau of Mines, August 1974.

3. Rolle, R., Determination of Radon Daughters on Filters by a Simple Liquid Scintillation Technique, AIHAJ, November-December, 1970.

4. Hill, A., Rapid Measurement of Radon, Decay Products, Unattached Fractions, and Working Level Values of Mine Atmospheres, Health Physics, Vol 28, No. 4, April 1975.

5. Kusnetz, H. L., Radon Daughters in Mine Atmospheres - A Field Method for Determining Concentrations, Industrial Hygiene Quarterly, March 1956.

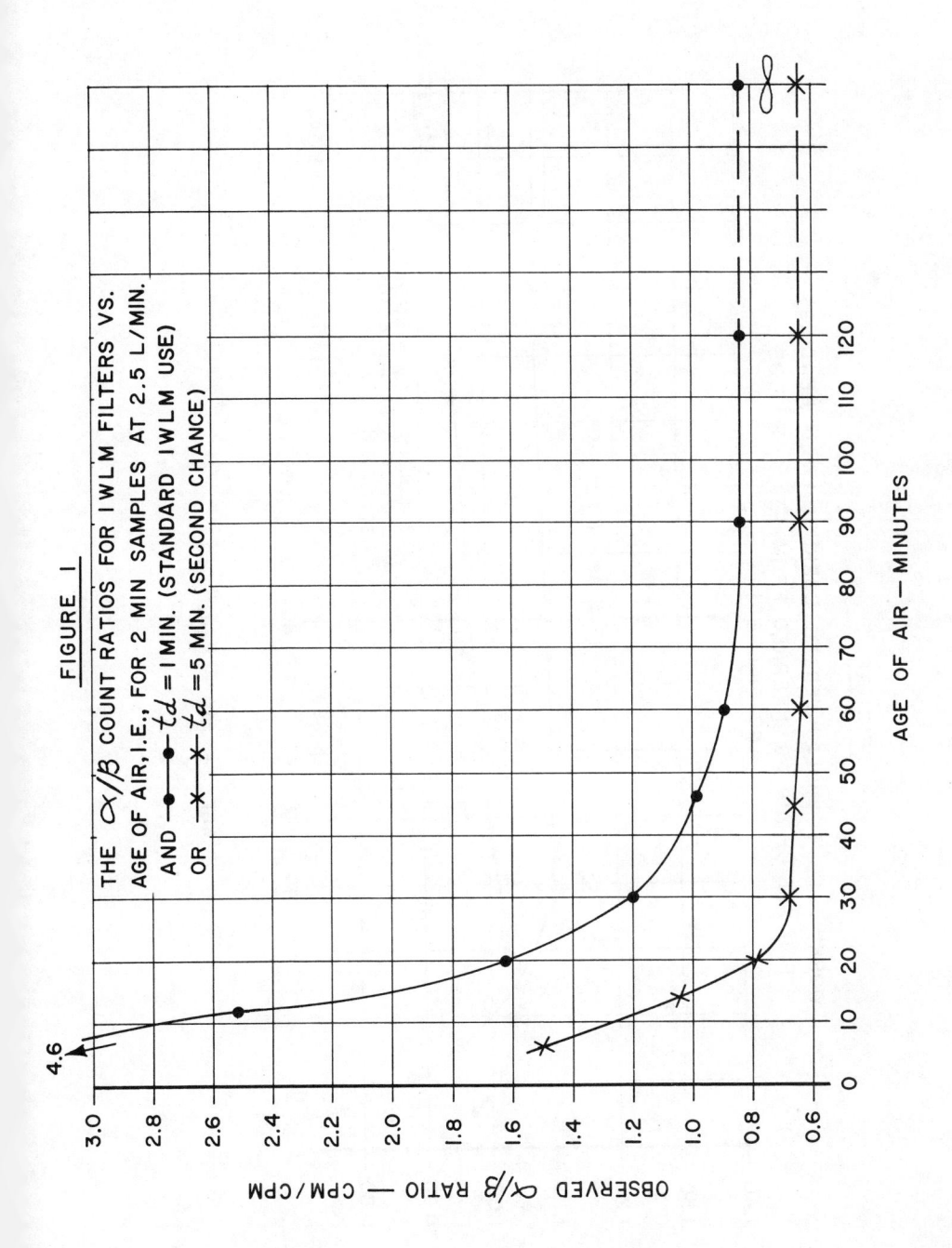

FIGURE 1

THE α/β COUNT RATIOS FOR IWLM FILTERS VS. AGE OF AIR, I.E., FOR 2 MIN SAMPLES AT 2.5 L/MIN.
AND ●——● t_d = 1 MIN. (STANDARD IWLM USE)
OR ✶——✶ t_d = 5 MIN. (SECOND CHANCE)

FIGURE 2

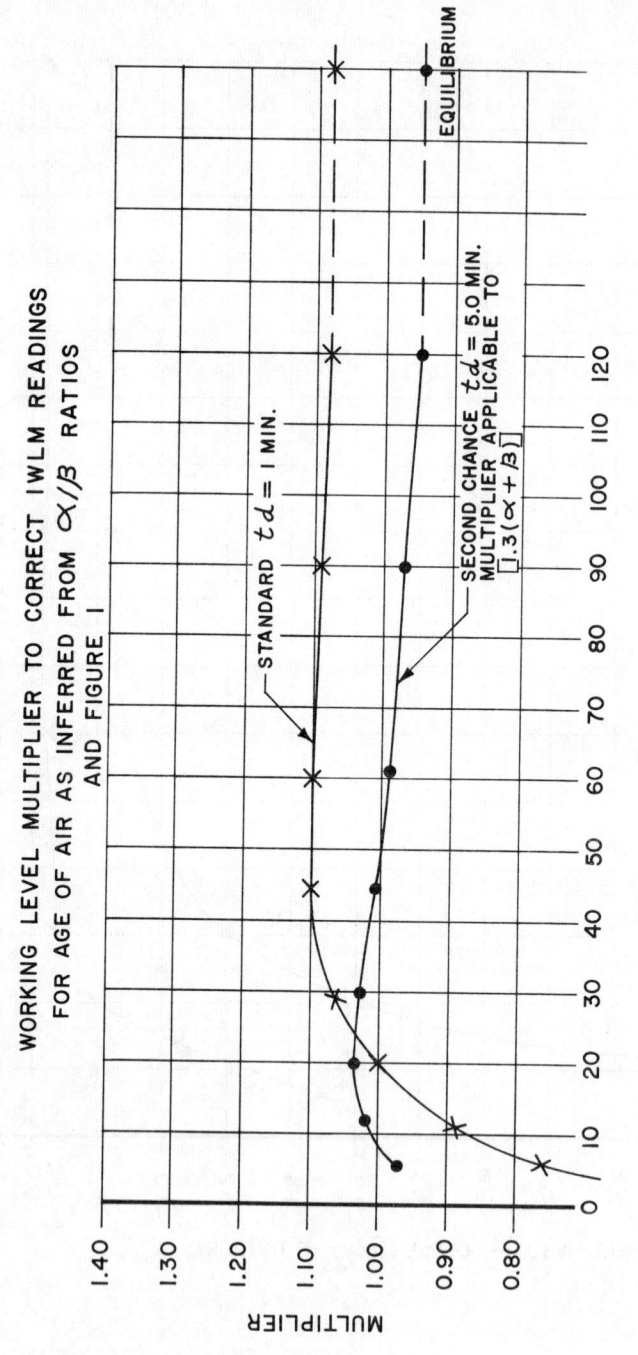

WORKING LEVEL MULTIPLIER TO CORRECT IWLM READINGS FOR AGE OF AIR AS INFERRED FROM α/β RATIOS AND FIGURE I

FIGURE 3

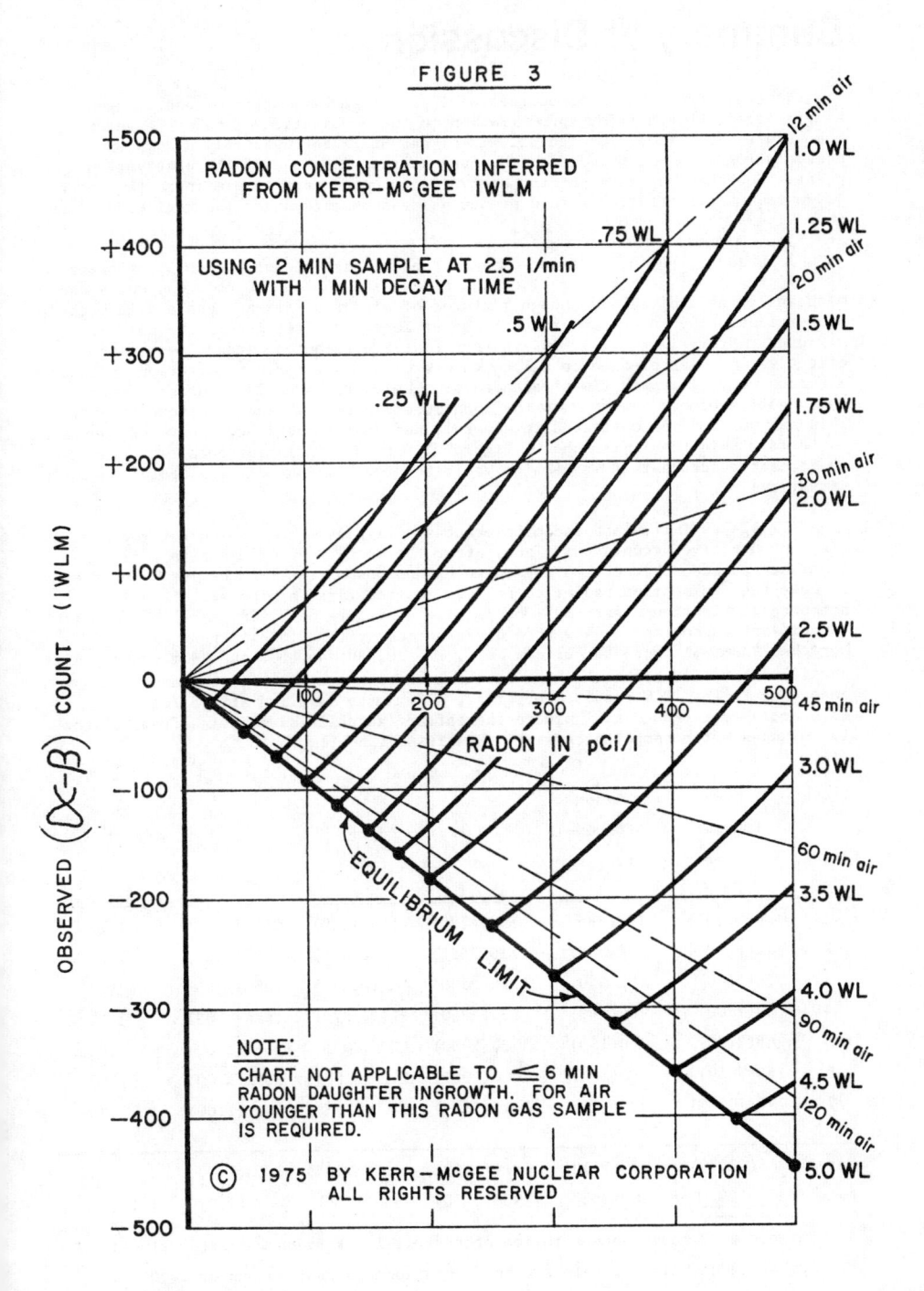

Summary of Discussion

Several persons questioned the seeming oversimplification of assigning a single 'age', WL and radon concentration to the combination of radon daughters encountered in a mine. The actual mix of daughters is invariably a complex accumulation from radon additions by many sources upstream of the observation point. How then can an apparatus analytically based upon simple ingrowth from a given radon concentration afford proper WL measurements under the real conditions in a mine?

In answer, the author assured the attendees that one of the earliest tests of the premise of near-constancy of the alpha-beta sum per WL unit was concerned with analyzing arbitrarily chosen mixtures of air of different ages and in varied proportions. The technique proved to be at least as effective in the case of mixtures. From Figure 2, it is apparent that the technique stands to be more effective, more error-free in all cases where the daughter concentration includes elements whose ingrowth times extended from below to above 20 minutes; positive and negative errors inherent to the constituent elements cancel to some extent. This has not been demonstrated in general, nor does it need be. That the approach is valid within the limits shown (Figure 2) for any single ingrowth time, is sufficient evidence that it remains valid for any combination of singular ingrowth situations.

Air age estimates and radon inferences - Figures 1 and 3 - do not yield to so tidy an intuitive acceptance. Thus, after the meeting, some examples were fashioned to add substance to these two extensions of the IWLM basics. In order to make the explanation easier to follow, imagine first a mine air situation - not unrealistic though extreme - comprising a mixture of 7 air parcels, each with 100pCi/l of radon, and each with a different decay time, a different history. Let the 7 nominal ingrowth periods be 6, 12, 20, 30, 45, 60, and 90 minutes. All partial activities appraised must be those contributed to a filter sample taken according to the IWLM requirement, i.e., a 2-minute sampling at 2.5l/min with each activity reckoned at 1 minute thereafter (actually by counting from 0.5 to 1.5 minutes after sampling ceases). See inset table.

Parcel No	Ingrowth Time (min)	Radon Concentration pCi/l	Activity (dpm) on filter from Alpha	Beta	WL	Alpha to Beta Ratio
1	6(5310)*	100(885)	540(4780)	116(1030)	.113(1.00)	4.66
2	12(6060)	100(505)	710(3585)	277(1400)	.198(1.00)	2.56
3	20(6600)	100(330)	818(2730)	505(1680)	.300(1.00)	1.62
4	30(7200)	100(240)	933(2250)	774(1870)	.414(1.00)	1.21
5	45(8100)	100(180)	1120(2000)	1135(2025)	.560(1.00)	.98
6	60(9000)	100(150)	1280(1905)	1415(2105)	.672(1.00)	.90
7	90(10800)	100(120)	1525(1845)	1795(2175)	.826(1.00)	.85
Total	263(53070)*	700(2410)	6916(19095)	6017(12285)	3.08(7.00)	
Ave	37.6(22.0)**	100(345)	988(2730)	860(1755)	0.44(1.0)	1.15(1.56)

* Parcel age x radon concentration of parcel, i.e., radon-weighted values.
** Sum of radon-weighted values ÷ total of radon concentrations or 2410.
<u>Comparative values from use of Figures 1 and 3</u>. Note that the alpha-beta difference must be multiplied by 0.23 (the IWLM counting efficiency) to transform from

dpm to cpm as required by Figure 3. Thus, 128 dpm indicates a count difference of 29 and 985 a difference of 227.

Ave   41(22)   100(340)

The second example is described by the parenthetical entries in the same table. Again a severe case is elected to test the IWLM concepts. The specifics apply to the same ingrowth times but to 1.00 WL of daughters in each parcel. That is to say the radon concentration is elevated to support 1.00 WL in each of the 7 parcels. To do this the 100pCi/l of radon common to all parcels in the first example is multiplied by the reciprocal of the WL it produced in each case.

The exercise consists of computing, for the mixture in each example, the overall WL (sum of all alpha and all beta activity), the alpha-beta difference, the alpha-beta ratio, the effective age of air and the radon concentration. Then, by use of computed WL and both the alpha-beta difference, the alpha-beta ratio, and Figures 1 and 3, determine comparative values of effective air age and radon concentration. The degree to which computed and graphical values agree will test the dependability of the IWLM for such estimations.

In the first example, WL is commanded by 'old' air, the longer ingrowth times. The second example accents influences from 'younger' air, shorter ingrowth times. Both examples are probably harsher, and broader ranges of, daughter mixes than ordinarily occur in working mines. Yet, they are not impossible. Therefore, it should be reassuring to prospective users of this type of IWLM to see that good estimates of age of air and radon concentration can be realized in routine employment of the instrument.

SUPERVISION OF RADON DAUGHTER EXPOSURE IN MINES IN SWEDEN

J O Snihs and H Ehdwall
National Institute of Radiation Protection (NIRP)
Stockholm, Sweden

Introduction

Since 1969, measurements have been made on radon and radon daughters in Swedish mines. These are ferrous and sulphide ore mines, altogether some 50 mines with about 5000 miners (1976). Sweden has also one uranium mine but up to now this has been an opencast mine and it has only been used on a small scale. However, there are now proposals for extending this mine for the underground mining of about 1300 tons of uranium per year.

Very soon after the discovery of the radon problems in the mines at the end of the 1960s, measurements and surveys leading to research and the introduction of technical improvements were started with the active support of the mining companies and the Swedish Mining Association. The main problems were: first, to develop appropriate equipment for sampling and measurement; second, to find the causes of high radon and radon daughter levels and their variation with locality and time; and third, to develop acceptable methods for the routine measurement of radon daughter exposures and to examine possible health effects caused by earlier exposures.

Special radon regulations were issued in March 1972 (1). In these regulations the maximum permissible annual radon daughter exposure corresponds to a full working-time stay in a radon daughter concentration of 0.3 WL, this can also be expressed as an equilibrium equivalent radon concentration (see below) of 30 pCi/l. A retrospective lung cancer study was started in 1971. The results of this study were published in 1972 (2) and it was concluded that there was a clear excess of lung cancer among the underground miners. The excess was clearly related to the estimated radon daughter exposure (3). Now, in 1976, the radon situation in Swedish mines has been considerably improved although there are still a few mines with excessive radon daughter levels.

Sampling and measuring techniques

We have not developed a personal dosimeter for radon daughters or used such dosimeters in Swedish mines. The estimations of radon daughter exposure are therefore based on representative values for given working areas based on radon daughter measurements or on radon measurements corrected by multiplying by a predetermined equilibrium factor (see below). The latter method is the most common one.

Radon is sampled in evacuated 4.8 or 0.8 litre conventional propane containers which are opened in the mine at the place of interest and then sealed. These containers are then sent by mail to the NIRP in Stockholm and are normally measured within 3 days of sampling. The containers are easy to handle, airtight and sturdy. As the sampling procedure is very simple it can be carried out by untrained persons.

The measurement procedure for radon is to evacuate an 18 litre ionization chamber and connect it to the sampling container. The efficiency of this air transfer is increased by twice adding radon-free air to the sampler and then reconnecting it to the ionization chamber. For a 4.8 litre sampler 91 % and for a 0.8 litre sampler 99 % of the radon is thus transferred to the ionization chamber which is then filled with radon-free air to atmospheric pressure. The amplified ionization chamber current is indicated by a pen recorder, a 1 pCi/l sample in a 4.8 litre sampler giving a net deflection of about 5 mm for the most sensitive range. As a rule two samples can be measured in an hour. The equipment is shown in Fig. 1.

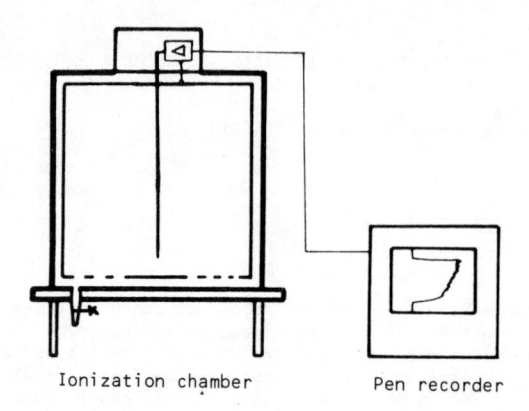

Fig. 1 Laboratory equipment for radon concentration measurements (Constructed by B. Håkansson, NIRP)

Radon daughters are sampled by sucking air through a fiberglass filter using a battery driven pump, the air volume being measured by a volumeter. The filter is then measured with a ZnS detector and the activity calculated using Kusnetz's method (4). By measuring the filter again after about five hours the ThB content of the air can also be estimated.

In order to obtain a rough measure of the activity of the surrounding rock in the mine the gamma dose-rate is also measured at various points. For these measurements we use a 10 litre ionization chamber containing nitrogen at 20 atmospheres. The sensitivity allows the observation of gamma exposure-rates below 5 µR/h.

Continuous measurement of the radon concentration in a mine can be made using the equipment shown in Fig. 2. It consists of a 10 litre ionization chamber, an air pump, a container with a filter and desiccant, and a pen-recorder with a paper speed of 1 inch per hour. Air is continually sucked through the filter and desiccant into the ionization chamber and out via the pump. The air flow rate is 1 litre/min. This equipment runs from the 220 volt line and it is enclosed in a wooden box to protect it from the dust in the mine.

<u>The principles of surveillance</u>

The principles for surveying the radon and radon daughters in the Swedish mines are partly determined by the special circumstances which exist in non-uranium mines. The radon source in the Swedish mines is only seldom a uranium-rich ore. One main source was found to be the large amounts of crushed rock left in the mine which the ventilation air had to pass before entering the working places. By changing the direction of the ventilation air or sealing off the spaces containing crushed rock this source was eliminated. The other main source was found to

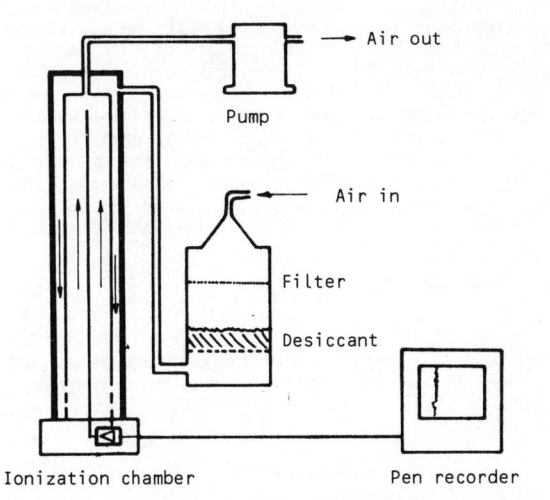

Fig. 2 Equipment for continuous registration of radon concentration in air (Constructed by B. Håkansson, NIRP)

be radon-rich water in the mine. It is often a more troublesome task to eliminate this source and improvement of the ventilation has often been necessary in combination with other measures such as minimizing open water surfaces. However, with knowledge of the main sources and their variation with time and locality the main emphasis in routine measurements should be to check that radon levels are within the expected limits and that there have been no unknown changes in the conditions which affect the radon levels in the mines.

The first measurements, which we have termed "basic" measurements, in a mine are made with the object of finding the main radon sources, of locating the very high or potentially high radon and radon daughter concentrations, for instance spaces with a lot of running water, and of determining the actual radon daughter concentrations at working places. Measurements are also often made of the radon concentration in mine water, of the gamma radiation levels in the mine and of the radium concentrations of the rock. It is then possible to recommend suitable countermeasures if the estimated radon daughter exposures are too high. These measurements are made by the NIRP personnel in close collaboration with protection officers and ventilation experts from mining company. Later "follow-up" and "check-up" measurements are often made by the mining company personnel themselves by taking samples in evacuated containers and sending them to the NIRP for measurement. In some mines these measurements are supplemented or replaced by radon daughter measurements which are performed by specially trained personnel at the mine with their own equipment or by a consultative group organized by the mining companies.

The radon regulations specify that the frequency with which the check-up measurements must be made depends on the radon daughter levels previously found in the mine. If the level is lower than 0.1 WL one check-up every 2 years is prescribed, for 0.1 - 0.3 WL once every year, for 0.3 - 1 WL once every 6 months and for higher than 1 WL once every 3 months is prescribed.

Estimation of radon daughter exposure

In many mines the check-up measurements are only made on radon. The radon daughters are very seldom in equilibrium with radon and we therefore calculate using an equilibrium factor F. This factor is defined as the ratio of the total potential alpha-energy of the radon daughter concentration in air to the total potential alpha energy of the daughters if they were in equilibrium with the given radon concentration. Thus if the radon concentration is C, the product C·F corresponds to a concentration of radon for which the daughters in equilibrium would have the same potential alpha-energy concentration as the actual radon daughter concentration in the air in question. C·F is called the "equilibrium equivalent concentration" of radon and is expressed in pCi/l. An exposure of 1 WLM corresponds to $1.7 \cdot 10^4$ pCi·h/l. The equilibrium factor F is determined by several simultaneous measurements of the radon daughter and radon concentration. Later estimations of radon daughter exposure are based on these values of F and the subsequent measurements on radon. Normally an average value of F is used for all working places in a mine or in parts of a mine. This simplifies the estimation of the radon daughter concentrations from the corresponding radon measurements. Another simplification is the use of so-called "zones". Following the first measurements of radon daughters, the mine is divided into one or several (in practice not more than two) zones as follows:

   Zone I : <10 pCi/l (= < 0.1 WL)
       II : 10 - 30 pCi/l (= 0.1 - 0.3 WL)
      III : 30 - 100 pCi/l (= 0.3 - 1 WL)

By knowing the time spent in the various zones the exposure can be estimated by multiplying the time spent in each zone by the average concentration in the zone, i.e. for zone I: 5 pCi/l (= 0.05 WL), for zone II: 20 pCi/l (= 0.2 WL), for zone III: 65 pCi/l (= 0.65 WL).

It is obvious that these simplified procedures need a good knowledge of the variations and of the factors influencing these variations in the radon concentration and in the equilibrium factor F. These variations are continuously being studied in Swedish mines.

The representativity of a radon measurement with time has been studied thoroughly in several mines by using the continuous measuring device or by taking repeated samples. There may be variations of the radon concentration with season, as shown in Fig. 3, with maximum values in the summer and minimum values in the winter.

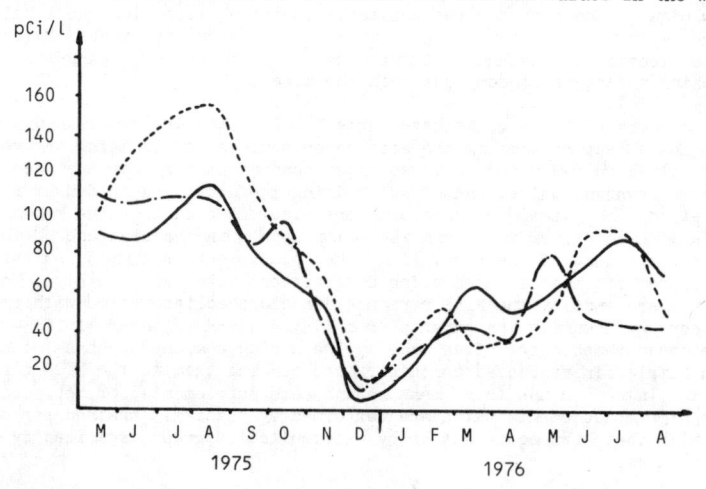

Fig. 3 Variation of radon concentration in three different places in Persberg mine

The reason for these variations is the influence of the outdoor temperature on the ventilation efficiency in the mine. Great variations may be caused by temporary changes such as repair of the ventilation fans, as illustrated in Fig. 4.

The variation of the equilibrium factor F is more complex. The distribution of the average values of F in 37 mines is shown in Fig. 5. The overall average of F in these mines is 0.7 with a standard deviation (S) of 0.14 (20 %). The variation of F in each individual mine is greater. On average S is 37 % (range 25 - 50 %). This variation of F in a mine reflects the different values of F for different parts of a mine. In some mines the value of F decreases as the radon concentration

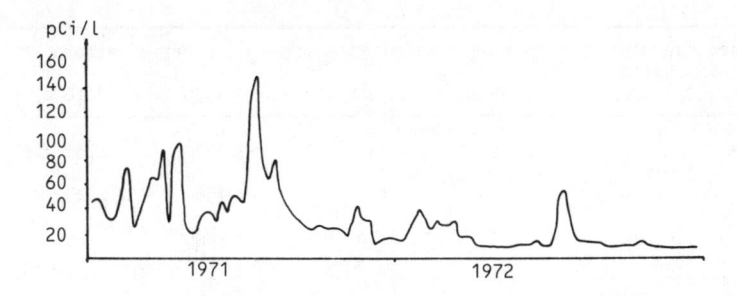

Fig. 4 Variation of radon concentration in Exportfältet mine during 1971-1972

decreases. This is an expected correlation if better ventilation is the only reason for the decreasing radon levels. In other mines there is an opposite correlation; however, this is also explicable in some cases (5). The value of F also varies along the route of the ventilation air. From the entrance to working places to the exit from the mine the value of F may increase from 0.15 to 0.65 (5).

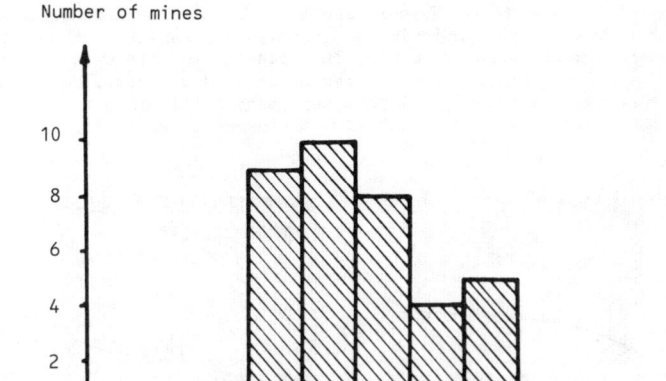

Fig. 5 Distribution of equilibrium factor F in Swedish mines

In conclusion, the variations of radon concentration in a mine with locality and time must be known. These variations can be correlated to known causes. The variation of the equilibrium factor F is more complex and the standard deviation of about 40 % hitherto found for individual mines is to be accepted as an uncertainty of the method. The variations which may occur within a chosen zone lead to a maximum error of 50 %. The overall uncertainty in the estimation of the radon daughter exposure is thus around 60 %. This error is probably small in comparison with all the uncertainties which are associated with the correlation between the radon daughter concentration and the dose to the respiratory region.

The improvement in the radon situation in Swedish mines

Since the first measurements in Swedish mines at the end of the 1960s the radon situation in Swedish mines has been considerably improved, see Table 1 and Fig. 6.

Radon daughter concentration pCi/l	Number of mines 1970	1976	Number of miners 1970	1976
< 10	25	29	1110	2730
10-30	8	12	1560	2345
30-100	18	5	2000	225
100-300	4	0	130	0
Total	55	46	4800	5300

Table 1. Number of mines with the radon daughter levels shown and the resulting number of miners exposed.

In 1970 the overall average of the radon daughter exposure for a miner was 80,000 pCi·h/l (= 4.7 WLM) and in 1976 28,000 pCi·h/l (=1.7 WLM) reflecting the fact that many mining companies have spent millions of dollars on improving ventilation. This is especially the case for the largest mining companies with about 55 % of the employees. Here it should be noted that another reason, perhaps the main reason, for improving the ventilation was the introduction of diesel powered machines in the mines. The reduction of the exposures in the mines of smaller companies has not always been so successful. However, this is compensated to some extent by the fact that many small mines have closed down and that these were often the mines with the highest exposures.

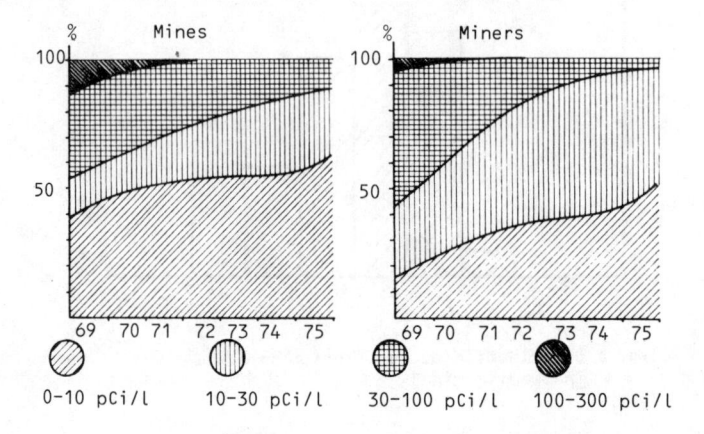

Fig. 6 The percentage of mine and miners in the given radon daughter concentrations during 1969-1975

Anticipated future problems

Even if the acute problems of radon in mines have mainly been overcome, the radon problems remain as a potential risk in the mines and continued checking is therefore needed. New mines and deeper mines will also motivate continued careful consideration of the radon problems.

Paradoxically, the radon problems in the one uranium mine in Sweden, at Ranstad, have hitherto been minor. That is because of the very low emanating power of the uranium ore. Even in unventilated exploratory drifts the radon daughter concentration is less than 1 WL. Prospecting has been carried out for other mines with higher grade uranium ore in the north of Sweden. The water in the areas concerned contains much radon, of the order of 1 µCi/l, and it is therefore to be expected that uranium mines in these areas will have greater radon problems than the mine at Ranstad.

Radon and radon daughter measurements have also been made lately in underground working spaces other than mines, for instance underground military installations, tunnels for telephone cables, electricity and water, and underground hydroelectric power stations. Radon daughter concentrations of 1 - 3 WL may occur in these installations mainly because of the presence of water with high radon concentrations.

References

1. Instructions issued by the Swedish National Board of Industrial Safety, 1972, No 82, Bl 4426. Svenska Reproduktions Aktiebolag, Vällingby, Sweden (in Swedish).

2. K. G. S:t Clair Renard et al., 1972, Lungcancer among miners in Sweden 1961-1968. Gruvforskningen serie B, No 167. Svenska Gruvföreningen, Stockholm, Sweden, (in Swedish).

3. J. O. Snihs, The approach to radon problems in non-uranium mines in Sweden. Proceedings of the third international congress of the International Radiation Protection Association, Sept. 9-14, 1973, Washington, D. C. pp. 900-912, U. S. Atomic Energy Commission, Office of Information Services, Technical Information Center Oak Ridge, Tennessee, 1974.

4. Hearings on Radiation Exposure of Uranium Miners before the Subcommittee on Research, Development, and Radiation of the Joint Committee on Atomic Energy, Part 2, Washington, D. C.: Government Printing Office, 1967.

5. J. O. Snihs, The significance of radon and its progeny as natural radiation sources in Sweden. Proceedings of Noble Gases symposium in Las Vegas, U.S.A. Sept. 24-28, 1973, pp. 115-130. U.S. Environmental Protection Agency.

SOME DATA OF PERSONAL AND AREA MONITORING IN THE SWEDISH URANIUM MINE, RANSTAD

P.O. Agnedal
AB Atomenergi
S-611 01 NYKÖPING   SWEDEN

INTRODUCTION

The uranium ore at Ranstad is found in the upper layer, 3 - 3,5 m of an alumn shale. The content of uranium is about 300 ppm and the total amount that can be mined is about 300 000 tons. The mine is an open pit mine and the mill is constructed for a production of 120 t/y, but the highest production has been about 50 t/y during the years 1966--68. The production was thereafter further decreased and the work at the Ranstad establishment is now directed to find ways how to develop the technicalprocedures in order to produce about 1300 t/y in an open pit mine and an underground mine.

AREA MONITORING

During the years mineing operations have been going on at Ranstad both personal and area monitoring have been performed and the results of these measurements will be briefly reported here. The activity level has been controlled by measurements of radon and daughter products in air external gamma radiation and determination of uranium in urine samples. To a smaller extent personal dosimetry has been used during one year in order to determine external radiation.

The samples for radon and daughter measurements have been taken in areas where high concentrations could be expected. The results are presented in Table 1 and all measurements show very low concentrations. In order to have an idea of the background gamma radiation and also of the gamma radiation in underground mine some measurements have been made and the results are presented in Table 2.

As said above uranium in urine has also been measured in samples from workers handling the dry uranium concentrate and from laboratory personnel. The detection limit for these determinations is 20 µg U/l. Very few samples have had concentration higher than that value and repeated sampling and determinations have shown a rapid decrease of the concentrations. The external dose measured by film dosimeters on workers did not exceed the detection limit which is 25 mrem/month.

As can be seen from these figures given in table 1 and 2 the exposure of workers in the Ranstad uranium mine will be lower than in many other Swedish mines as reported by Snihs at this meeting.

TABLE 1

Measurements of Radon Daughters 1965--74

		pCi/l
Mining Area	Controll room	0.2 - 0.7
	Course crusher area	0.2 - 0.6
	Transport tunnel	0.1 - 0.6
Dressing plant		
	Sink- and floatplant	0.1 - 0.5
Leaching plant		0.1 - 3.6
Tailings area		< 0.3

Radon Measurements 1974

Air in mine and milling area and above tailings < 1 -3 pCi/l

Emanation of radon from the walls in the mine   0.1 - 0.2 pCi/m^2/s

TABLE 2

Gamma measurements External Radiation μR/h	
Above uranium plateau on the alum shale	220
Above restored area on tailings	23
Above area not disturbed in the open pit environment	25
Above tailings	245
At the walls in the tunnel area depending on the distance from the wall	60 - 500

AREA MONITORING IN THE NINGYO-TOGE MINE

Yoshiyasu Kurokawa*
Kenji Nakashima*
Yoshihisa Kitahara*
Ryuhei Kurosawa**
* Power Reactor and Nuclear Fuel Development
  Corporation (Japan)
** Waseda University (Japan)

1. General

In 1954, the Geological Survey Institute of Japan initiated prospecting of uranium resources.

In 1956, the Atomic Fuel Corporation (AFC) was established as a governmental agency to take charge of nuclear fuel development.

In 1967, the AFC's work was taken over by the Power Reactor and Nuclear Fuel Development Corporation (PNC) for the purpose of further continuing the development of technology relating to the new project of ATR and FBR.

Earlier in the summer of 1955, a uranium ore deposit having a grade of over 0.1% U_3O_8 was discovered in the Ogamo area, where located was an abandoned gold mine, in Kurayoshi City, Tottori Prefecture. This was a typical vein-type uranium deposit in Japan. The mine was named "the Kurayoshi Mine," and a drift prospecting was started by AFC in 1956.

At the end of 1955, a new uranium deposit was discovered in the sedimentary stratum running in the border line between Tottori and Okayama Prefectures. This was the first step to the opening of the Ningyo-Toge Mine. The Ningyo-Toge Mine is the largest uranium mine in Japan, and mining, milling and refining are carried out there. This mine is located in a high land 700 to 730 meter above the sea level in the Chugoku mountain range (Fig. 1).

The deposits are sedimentary uranium-deposits which occur in the basal part of the strata - mainly conglomerate and sandstone and partly mudstone - composed of the inland sedimentary rocks and pyroclastics formed in Tertiary to Pliocene with the granites which had been widely distributed in Chugoku mountains in the end of Cretaceous to the beginning of Tertiary as the batholiths.

The main deposits are about 20 km from east to west and about 15 km from south to north, and the estimated amount of uranium

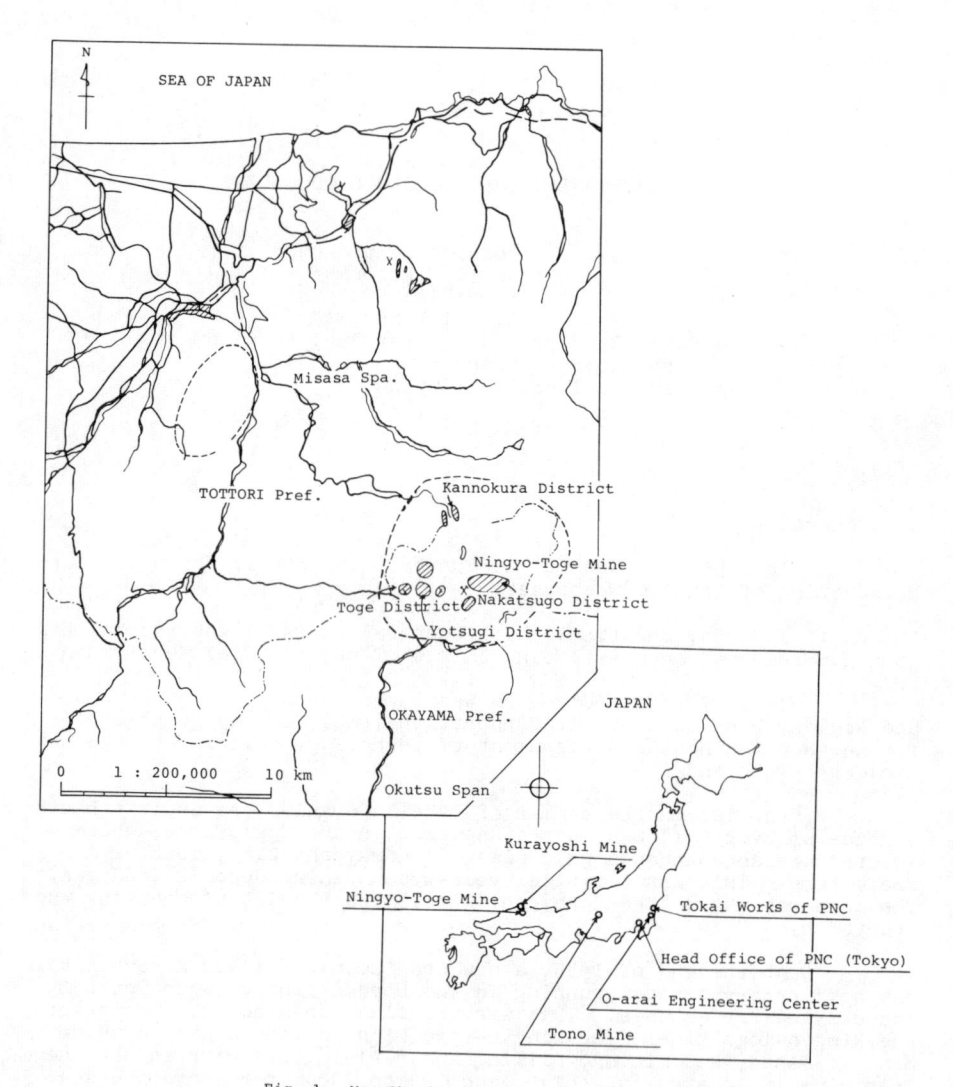

Fig. 1. Map Showing Location of Ningyo-Toge Mine

deposit is about 5 million tons calculated in terms of the mean grade of U_3O_8 0.51%.

The principal ore deposits are those of Toge, Nakatsugo, Yotsugi, and Kannokura, all of district are under prospecting with a considerable length of prospecting tunnels excavated (about 30,000 meters). While actual mining operations of major or minor scales have been carried out in the above mentioned three districts except the Kannokura District.

The ore in the Toge and Nakatsugo District constitutes mainly "Ningyoite" which is the primary mineral, and their grades are 0.09% and 0.58% respectively in terms of U_3O_8. On the other part, the ores in the Yotsugi District consist of the secondary minerals such as autunite, of which grade is 0.15% in terms of U_3O_8.

The ore deposit of the Nakatsugo District consists of comparatively compact rock and because of its small coefficient of diffusion, its radon escape rate is small against its uranium grade, while in contrast to this, the ore deposit in Yotsugi District, which consists of the secondary minerals, its radon escape rate is rather high comparing with the former.

The ore taken out from these mines is carried to the nearby milling plant which has an ore processing capacity of 50 tons/day. Here, the ore is all crushed, and after going through an wet process of sulfuric acid leaching, they are refined into yellow cakes or uranium tetra-fluoride via uranyl chloride solution. (Refer to Fig. 2)

The prospecting tunnels were undertaken by a some companies in the basis of contract, and in the peak time there were nearly 100 mining workers. But after entering into actual normal mining, the mining and milling operations are performed by about 20 mining workers and about the same number of milling workers. Among the mining workers, there are some who have 20 ∿ 30 years of mining experience. Limiting only to the experience in uranium mining, their experiential years may be around 15 years. Since the milling operation was started only in the latter half of 1970, and the milling experiment had been in practice just a few years before the actual milling operation, the real experience of the milling workers is about 10 years.

2. Environmental Assessment

   (1) Ventilation Plan

      i) Underground

      The uranium mines in Japan are subject to regulations prescribing the total doses of both the external and the internal exposures to radon and uranium in the working environment. But for the actual control objectives, the ventilation plans for uranium mines are formulated on the basis of the radon concentration 2.5×10^{-7} µCi/cm³. For the planning of ventilation, some experimental equations of radon emanation rate are employed. For instance, the radon emanation rate from the new excavated face of the mine gallery of the Nakatsugo District adopts 2.0×10^{-1}Cu µCi/m²/min (Cu:U_3O_8 grade per cent), and for the averaged emanation rate for the actual mining area, the expression of 8.6×10^{-2}Cu µCi/m²/min is adopted. In the case of estimation from radioactive survey of boring holes, the expression of $1.6 \times 10^{-2} R_E$ µCi/m²/min is employed (R_E: mR/hr).

      No special consideration is given to internal exposure of mining workers by uranium dust. Generally, a relative humidity of the mine atmosphere is as high as nearly 100% and thus dust generation

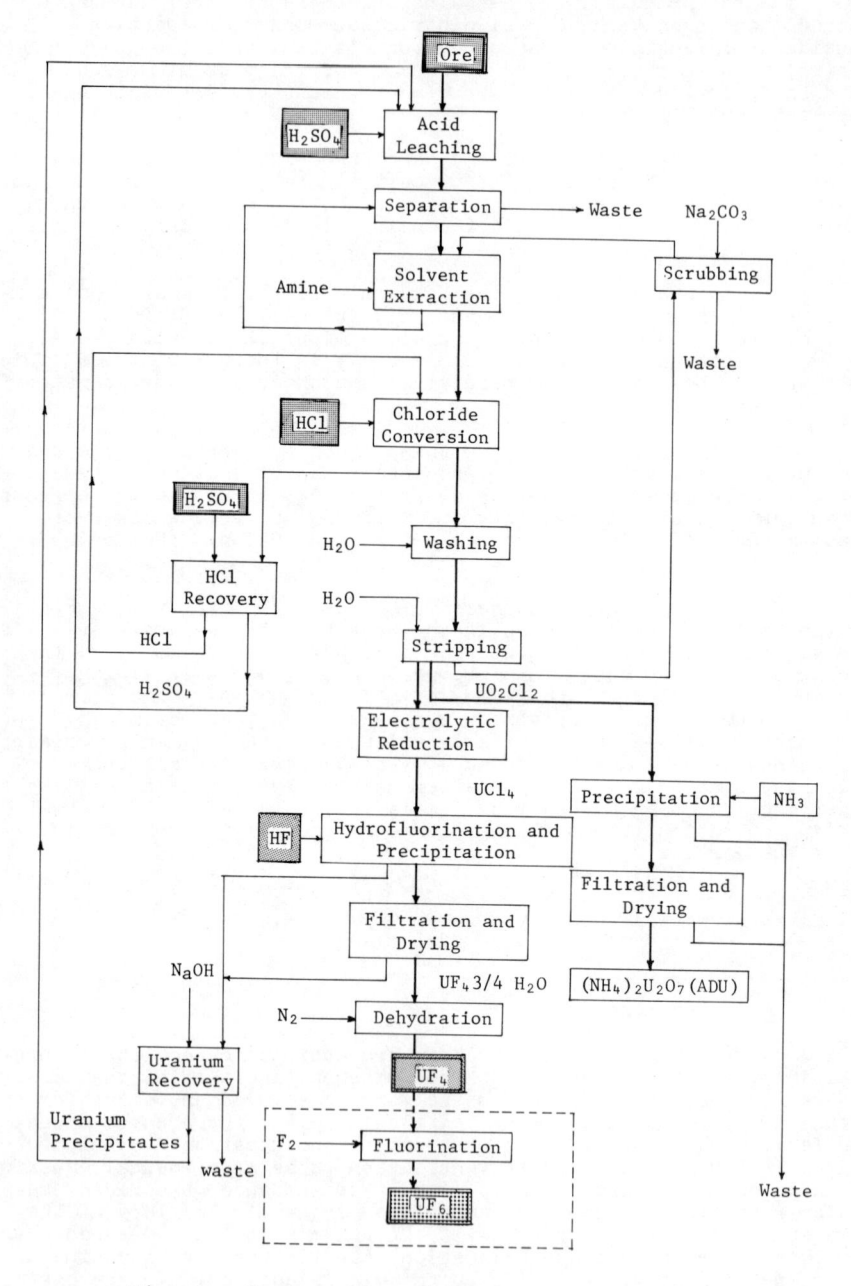

Fig. 2. Process Flow Diagram of Milling Plant

is comparatively small presenting no particular problem.

The principal ventilation method is by a driftway ventilation. The Yotsugi District employs a forced draft system while other districts have adopted an induced draft system. Generally, the latter has more radon emanation rate by about 5 ∿ 10% than the former. The adit of level is used in principle for the air inlet of the ventilation system.

For the ventilation in the case of the drift heading (or in the case of a blind drift), an air is sent by a locally installed air blower via the wind-duct.

ii) Milling Plant

It was difficult to accurately assess the amount of radioactive materials released from the process in the milling facilities making it thus difficult to formulate detailed ventilation plans. But by covering the vessels and containers with sheets as much as possible, and inserting air suction pipes into them to make them negative pressure, the contaminated air was exhausted outside the plant facilities.

(2) Items to be Measured

The assessment of the working area is performed to determine the concentration of the airborne particles which are inhaled by the workers, and at the same time to set up guidelines for planning measures to control the amount of airborne particles as low as possible. The items to be measured for this purpose include, besides the dust, the distribution of concentration and particle size of such substances as uranium, radon, radon-daughters, etc.

The dust is, of course, a cause of silicosis, and the dust in the mine atmosphere constitutes the same composition as that of the ore deposit which forms the mine's ore deposit. The concentration of dust is generally in the range of 0.2 ∿ 0.4 mg/m^3 in most cases, and in many cases, which contained some 0.2 ∿ 0.4% of uranium.

The particle size of dust makes logarithmic normal distribution, and its mass median aerodynamic diameter (MMAD) was 4μm and $\log_{10} \sigma g$ was 0.81.

On the other part, the concentration of the dust in the plant was slightly lower in various processing sections except for the ore crushing section where it was about the same contents as in the mine atmosphere. But, as a matter of fact, in the ADU drying process, the concentration of uranium dust in the air showed a high figure of several 10 μg/m^3. The description of the measuring methods of uranium dust, radon, radon-daughters, etc. is given as following section.

i) Measurement of Uranium Dust

The airborne dust in the mine atmosphere is collected by a high-volume air sampler (300 ∿ 400 ℓ/min) in a short time (about 20 min), while in the milling plant, dust is collected all day long by a low-volume air sampler (about 20 ℓ/min). In the plant, measurement of uranium dust both the inside of the facility and in the exhaust air is conducted periodically once every week. But the measurement of uranium dust in the mine is done only a few times per year. Furthermore, the dust in the ADU drying process is taken by personal dust sampler (2 ℓ/min) in all day.

For the filters, Wattmann No. 41 and Toyo Filter No. 5A are employed, and in-dust uranium analyses are performed by means of radioactivity measurement or chemical analysis (by the fluorimetric method or neutron activation analysis).

ii) Measurement of ^{222}Rn

The sampled air which is removed of the radon-daughters is enclosed into an ionization chamber. Then after having been left standing for 2 ∿ 4 hours, an ionization current of the chamber is measured by a vibrating reed electrometer, and its radon concentration is determined by a calibration table. Both the radon concentration of underground and the plant are regularly checked once every week.

iii) Measurement of Radon-Daughters

In the early period, the concentration of radon-daughters in the mine atmosphere is determined by a method of similar to Tsivoglou Technique. That is, by sampling daughters for five minutes at the flow rate of about 10 ℓ/min with a glass paper filter, and after the sampling, the collected daughters on the filter were counted for 30 seconds every minute by a ZnS scintillation α-counter, and then the mean count rates at 3, 10, and 20 minutes were estimated. Then, three simultaneous equations were formulated and solved to obtain the concentration of each individual radon-daughters. The accuracy of this Tsivoglou method, however, is not so high that the below described weighted least square method is employed at present. This method uses membrane filters (millpore DA:Pore size 0.65 μm). Having taken a sample of airborne radon-daughters for five minutes (or 3 or 7 minutes) at the flow rate of about 13 ℓ/min, after sampling, the collected daughters on the filter is counted six times by the above mentioned α-counter at intervals from 2 to 6 minutes, 7 ∿ 11 min., 12 ∿ 16 min., 17 ∿ 21 min., 22 ∿ 26 min., and 27 ∿ 31 min. respectively. Assuming the respective net counts as $I_1, I_2, I_3, I_4, I_5,$ and I_6, the following observation equations (1) are obtained:

$$\left.\begin{array}{l} 5.19959 \ \eta X + 3.69729 \ \eta Y + 15.9450 \ \eta Z = I_1 \\ 2.04897 \ \eta X + 5.65708 \ \eta Y + 13.3727 \ \eta Z = I_2 \\ 1.15941 \ \eta X + 6.99063 \ \eta Y + 11.2154 \ \eta Z = I_3 \\ 0.953794 \ \eta X + 7.83658 \ \eta Y + 9.40617 \ \eta Z = I_4 \\ 0.935318 \ \eta X + 8.30665 \ \eta Y + 7.88877 \ \eta Z = I_5 \\ 0.952099 \ \eta X + 8.49050 \ \eta Y + 6.61616 \ \eta Z = I_6 \end{array}\right\} \quad (1)$$

where, η = count efficiency

X,Y,Z = collecting rates (dpm/min) of ^{218}Po, ^{214}Pb, and ^{214}Bi respectively.

The concentrations of ^{218}Po, ^{214}Pb, and ^{214}Bi are respectively C_X, C_Y, C_Z μCi/cm³.

Assuming the sampling rate of νℓ/min, it is:

$X = C_X \cdot \upsilon \times 2.22 \times 10^9$. But as each expression of the equations (1) has different grade accuracy, it must be multiplied by the weight of $\sqrt{P}$. $\sqrt{P}$ proportionates to the reciprocal number of the relative error of each count. Therefore, it is expressed as $\sqrt{P} \div 1/(\frac{\sqrt{I}}{I}) = \sqrt{I}$.

$$X_{1.1} \eta X + X_{1.2} \eta Y + X_{1.3} \eta Z = I_1$$
$$X_{2.1} \eta X + X_{2.2} \eta Y + X_{2.3} \eta Z = I_2$$
$$\vdots \qquad (1')$$
$$X_{6.1} \eta X + \eta_{6.2} \eta Y + X_{6.3} XZ = I_6$$

Then, the following weighted observation equations are induced.

$$\sqrt{P_1}\, X_{1.1}\, \eta X + \sqrt{P_1}\, X_{1.2}\, \eta Y + \sqrt{P_1}\, X_{1.3}\, \eta Z = \sqrt{P_1}\, I_1$$
$$\sqrt{P_2}\, X_{2.1}\, \eta X + \sqrt{P_2}\, X_{2.2}\, \eta Y + \sqrt{P_2}\, X_{2.3}\, \eta Z = \sqrt{P_2}\, I_2$$
$$\vdots \qquad (1'')$$
$$\sqrt{P_6}\, X_{6.1}\, \eta X + \sqrt{P_6}\, X_{6.2}\, \eta Y + \sqrt{P_3}\, X_{6.3}\, \eta Z = \sqrt{P_6}\, I_6$$

where, $\sqrt{P_1} = \sqrt{I_1}$, $\sqrt{P_2} = \sqrt{I_2}$ ....

Consequently, the weighted normal equations are as follows:

$$\Sigma(X^2 i \cdot_1 \quad \eta^2 \cdot Pi)X + \Sigma(Xi\cdot_1 \cdot Xi\cdot_2 \cdot \eta^2 \cdot Pi)Y + \Sigma(Xi\cdot_1 \cdot Xi\cdot_3 \cdot \eta^2 \cdot Pi)Z = \Sigma(Xi\cdot_1 \cdot Ii \cdot Pi)$$
$$\Sigma(Xi\cdot_1 \cdot Xi\cdot_2 \cdot \eta^2 \cdot Pi)X + \Sigma(X^2 i\cdot_2 \quad \eta^2 \cdot Pi)Y + \Sigma(Xi\cdot_2 \cdot Xi\cdot_3 \cdot \eta^2 \cdot Pi)Z = \Sigma(Xi\cdot_2 \cdot Ii \cdot Pi) \quad (2)$$
$$\Sigma(Xi\cdot_1 \cdot Xi\cdot_3 \cdot \eta^2 \cdot Pi)X + \Sigma(Xi\cdot_2 \cdot Xi\cdot_3 \cdot \eta^2 \cdot Pi)Y + \Sigma(X^2 i\cdot_3 \quad \eta^2 \cdot Pi)Z = \Sigma(Xi\cdot_3 \cdot Ii \cdot Pi)$$

As the "weight" was different in each measurement, the weighted normal equation (2) had to be prepared and solved by use of a mini-electronic computer in order to determine the radon-daughters concentration.

Apart from this method, the following method is also employed so that radon-daughters concentration can be obtained by a simple calculation in the actual sampling spot:

Transforming each one of the equations (1), the following equations (3) are obtained.

$$5.19959\, \eta X + 3.69729\, \eta Y + 1.59450\, \eta Z = I_1$$
$$4.16218\, \eta X + 20.4843\, \eta Y + 33.9944\, \eta Z = I_2 + I_3 + I_4 = I_{2-4} \qquad (3)$$
$$1.88742\, \eta X + 16.7972\, \eta Y + 14.5049\, \eta Z = I_5 + I_5 = I_{5-6}$$

and the equations (4) are provided for solving the equations (3).

$$(0.300381\, I_1 - 0.234923\, I_{2-4} + 0.220373\, I_{5-6}) \times \frac{1}{\eta} = X$$
$$(-0.00415603 I_1 - 0.0497096 I_{2-4} + 0.1210701 I_{5-6}) \times \frac{1}{\eta} = Y \qquad (4)$$
$$(-0.0342735\, I_1 + 0.0881340 I_{2-4} - 0.0999362 I_{5-6}) \times \frac{1}{\eta} = Z$$

In this method X,Y,Z can be easily obtained. The errors ΔX, ΔY, and ΔZ of X,Y,Z obtained by the method are estimated as follows:

Table 1. Radon-Daughters and Primary Particle Concentration, and WL (Nakatsugo District) (December 12, 1973)

Working		Time	Rn-drs Concentration in $\mu Ci/cm^3$			Primary particle of ^{218}Po	Working Level	Equilibrium Ratio %
			^{218}Po	^{214}Pb	^{214}Bi			
Working area	Scraper operation	09:57	1.24×10^{-7}	2.0×10^{-8}	1.5×10^{-9}	3.0×10^{-8}	0.23	10
	Picking & scraping work	10:39	8.2×10^{-8}	6.4×10^{-9}	5.3×10^{-9}	6.3×10^{-9}	0.14	8
	Same as above	10:59	9.4×10^{-8}	7.9×10^{-9}	3.5×10^{-9}	5.4×10^{-9}	0.15	5.7
	Work stopped	13:06	9.6×10^{-8}	1.1×10^{-9}	2.2×10^{-10}	9.2×10^{-9}	0.11	7.5
	Same as above	13:40	8.2×10^{-8}	6.2×10^{-9}	5.7×10^{-9}	—	0.14	3.2
	Scraper operation	14:15	6.0×10^{-8}	4.3×10^{-9}	4.0×10^{-9}	3.8×10^{-9}	0.10	5.9
	Face arrangement	15:15	8.4×10^{-8}	6.6×10^{-9}	~ 0	2.4×10^{-9}	0.12	6.4
	Rock drilling	15:51	7.4×10^{-8}	6.4×10^{-9}	1.8×10^{-10}	6.9×10^{-9}	0.11	6.7
Ventilation shaft		17:27	1.57×10^{-7}	4.9×10^{-8}	1.9×10^{-8}	1.9×10^{-9}	0.48	12

Equilibrium Ratio = $\dfrac{\text{Measured WL}}{\text{WL of }^{222}Rn:^{218}Po:^{214}Pb:^{214}Bi=1:1:1:1}$

Mean value of $^{218}Po:^{214}Pb:^{214}Bi$ (Working area) : 1.0:0.093:0.039
(Total exhaust) : 1.0:0.31:0.18

Volume of working area: 111 m³

Ventilation rate: 240 m³/min

$$\{(0.0902291_1+0.0551891_{2-4}+0.0485641_{5-6})^{\frac{1}{2}} \times \frac{1}{\eta}\} = \Delta X$$

$$\{(0.00001731_1+0.00247101_{2-4}+0.00146581_{5-6})^{\frac{1}{2}} \times \frac{1}{\eta}\} = \Delta Y \quad \quad (5)$$

$$\{(0.0011751_1+0.00776812_{2-4}+0.00998711_{5-6})^{\frac{1}{2}} \times \frac{1}{\eta}\} = \Delta Z$$

The relative errors of ^{218}Po, ^{214}Pb, and ^{214}Bi concentrations obtained from the equation (4) in the case of 1.0 : 1.0 : 1.0 of ^{218}Po : ^{214}Pb : ^{214}Bi, $\eta = 0.3$, ^{218}Po concentration = 1×10^{-7} µCi/cm³ and sampling rate = 13 ℓ/min are respectively ±9.1%, ±2.7% and ±3.1%, and also they are respectively ±6.0%, ±3.6% and ±6.7% in the case of 1.0 : 0.5 : 0.3.

iv) Determination of Working Level

The working level (WL), when the concentrations of the radon-daughters ^{218}Po, ^{214}Pb, and ^{214}Bi are assumed as Cx, Cy and Cz µCi/cm³ is:

$$WL = (1.34 \times 10^8 \cdot Cx + 6.60 \times 10^8 \cdot Cy + 4.85 \times 10^8 \cdot Cz) \times \frac{1}{1.30 \times 10^2}$$

According to Japanese regulation, ^{222}Rn concentration is used for standard of radiation control. Therefore measuring of WL alone is not used for such purpose. WL is determined by substituting the Cx, Cy and Cz as obtained in the previous paragraph into the above equation.

v) Measurement of Primary Particle Concentration

The primary particle concentration of ^{218}Po is estimated from the relative amount of ^{218}Po deposited to the inner surface of the diffusion battery. The size of the diffusion battery now in use is 3.97mm interval, 28.0mm high and 301mm long, and the number of channels is 7, while the passing velocity is about 1.9 ℓ/min per channel. Not only the primary particles of ^{218}Po but also a small portion of the secondary particles are get deposited to the diffusion battery. But the ratio of the deposited secondary particles never exceed 2% of the total secondary particles.

The relative concentration of primary particles is extremely high so that the uranium mines in Japan use principally electrical machine and no diesel engines which generate a large amount of condensation nuclide. But due to the high ventilation rate, even the existing rate of ^{222}Rn - ^{218}Po is extremely unequilibrium, and the so-called f-value by ICRP is apparently low. The examples of the measured values of radon-daughters concentration, primary particle concentration, WL, etc. are given in Table 1.

vi) Measurement of Particle Diameter of Radon-Daughters

The particle diameters of radon-daughters are measured by use of other type of diffusion batteries. Two types of diffusion batteries are employed. One is of 1.08mm interval, 29.2mm high, and 599.7mm long with 24 channels and its passing flow velocity is 0.58 ℓ/min per channel, while the other one is of an assembly tube type with an inner diameter of 2.0mm, a total length of 1305mm, and 624 channels. Its passing flow velocity is $2 \times 10^{-4} \sim 3 \times 10^{-1}$ ℓ/min per channel. The mean diameter of secondary particles of radon-daughters is measured by use of the former, and the particle diameter distribution is measured by use of the latter.

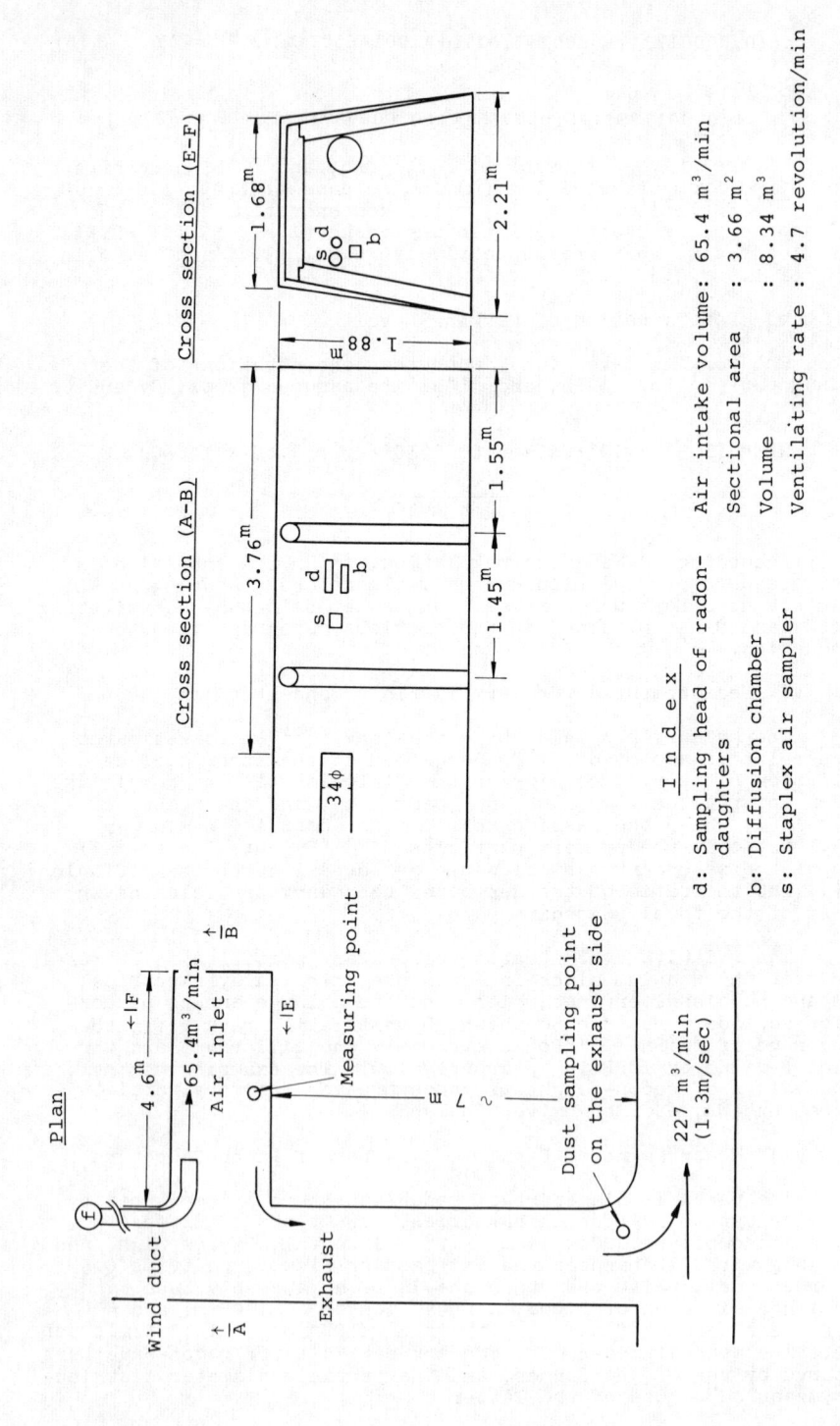

Fig. 3. Measuring Points in the Excavation Headings in Nakatsugo District

From the results of the experiment by use of the former, it was found that the mean diameter obtained from the diffusion coefficient of the secondary particles of ^{218}Po, was 0.03 μm while from the results of the experiment using by the latter, it was determined that the particle diameter is close to the logarithmic normal distribution, and the activity median aerodynamic diameter (AMAD) of ^{218}Po was 0.058 μm (the mean particle diameter obtained from the diffusion coefficient was 0.03 μm the same as above mentioned, and $\log_{10}$ σg was 0.39, and AMAD of ^{214}Pb and ^{214}Bi was 0.13 μm and $\log_{10}$ σg was determined as 0.30.

(3) Results of Measurement

i) Underground

Fig. 3 ∿ 6 represent the examples of a day's time-lapse changes in the progress of excavation and the working faces in the Nakatsugo District. The work cycle involves drilling, dynamite charging, blasting, then scraping and loading ores onto a tub. In the course of the scraping work, localized picking is done using picks to arrange and smooth the cutting face. After scraping, the working faces are reinforced with steel pillars.

The working is by only one-shift, and the mechanical ventilation is stopped during night time. The main fan is set to automatically operate one hour prior to the entry of miners into the adit.

The radon concentration at the working area is $1 \sim 2 \times 10^{-7}$ μCi/cm^3, while uranium dust shows considerably low values.

ii) Milling Plant

The results of measurement in the plant are given in Table 2, Table 3 and Fig. 7. Uranium concentration in the air exists on nearly the same level in the any processing area of the plant except the ADU processing area, in which the concentration of uranium is high. Radon concentration is comparatively high level values in the area of the ore bin. It is, however, almost zero in all the processing area after solid-liquid separation because the radium is removed from the process flow.

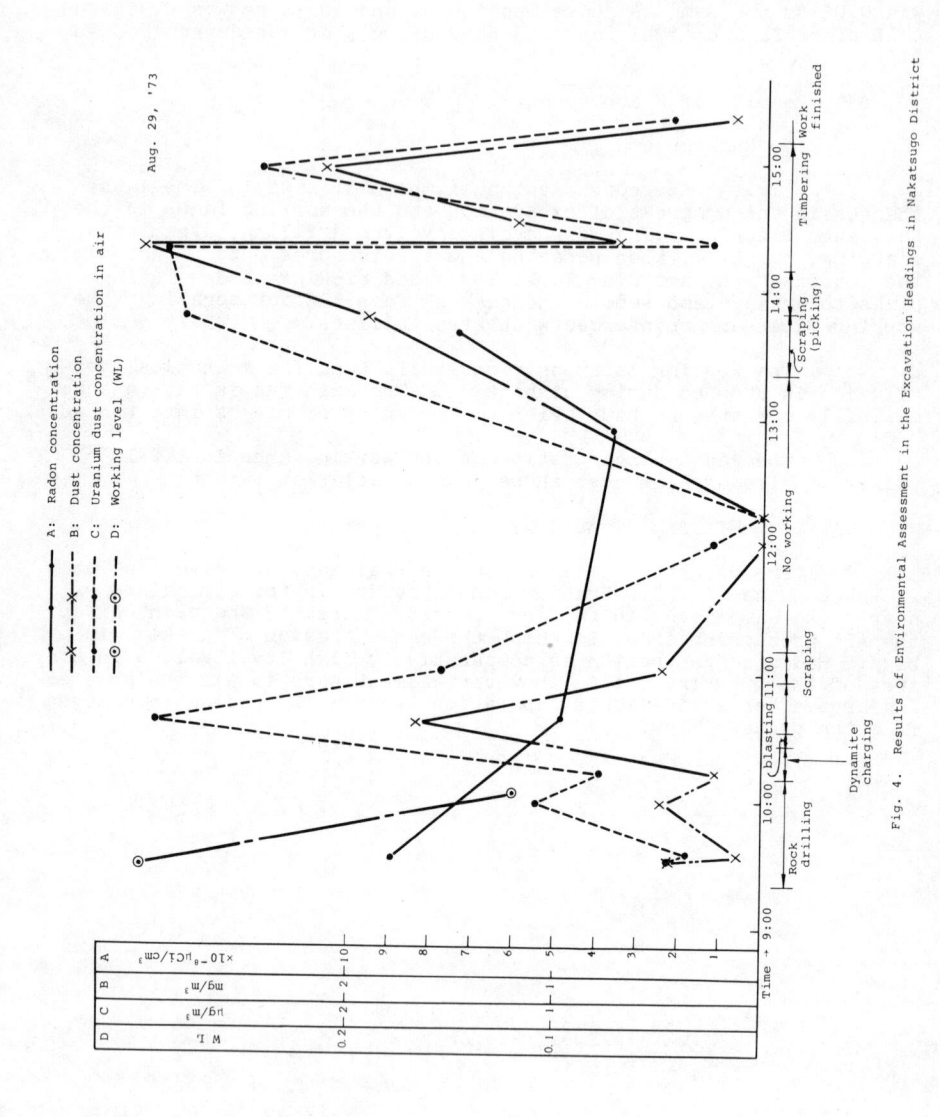

Fig. 4. Results of Environmental Assessment in the Excavation Headings in the Nakatsugo District

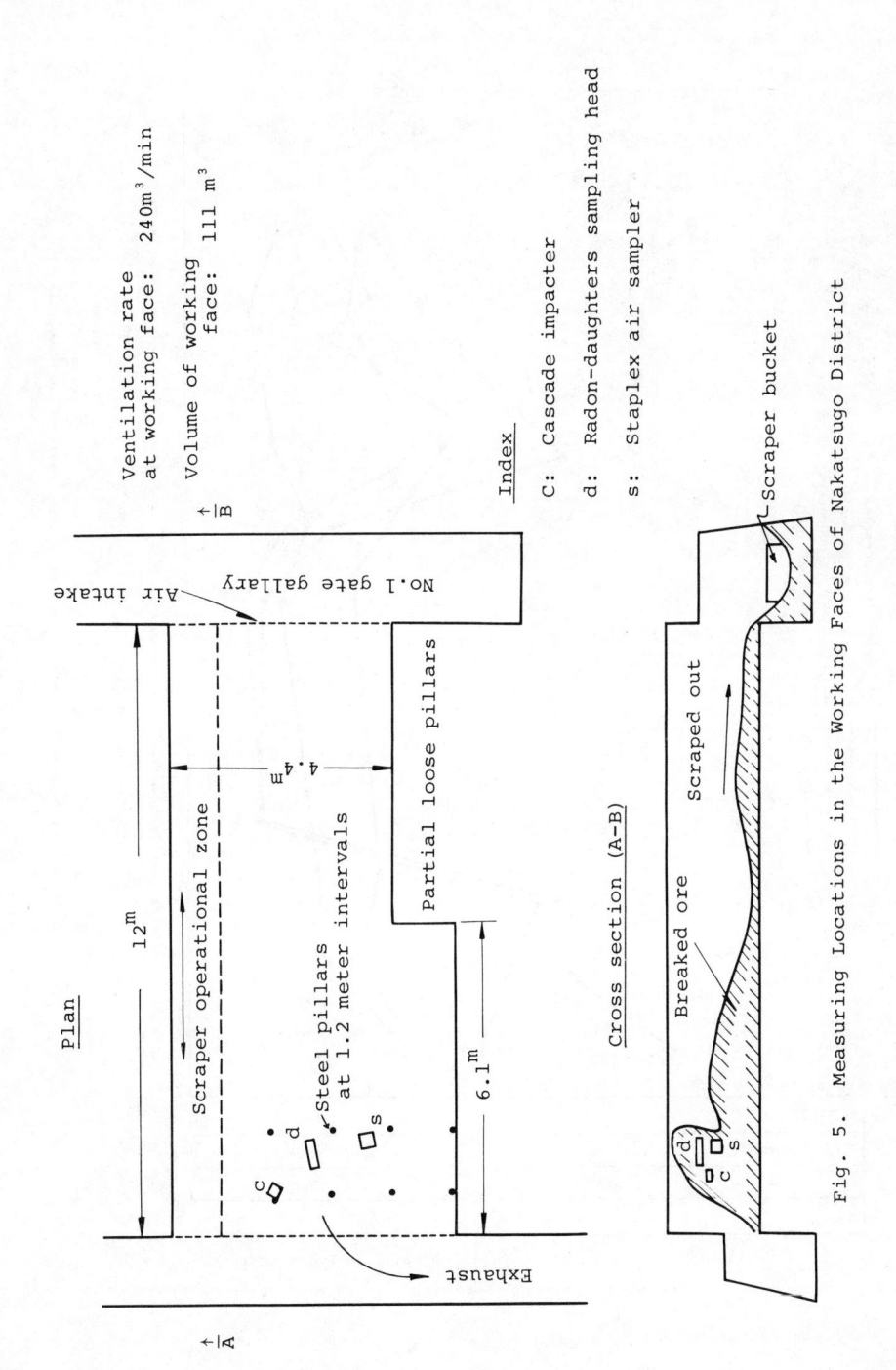

Fig. 5. Measuring Locations in the Working Faces of Nakatsugo District

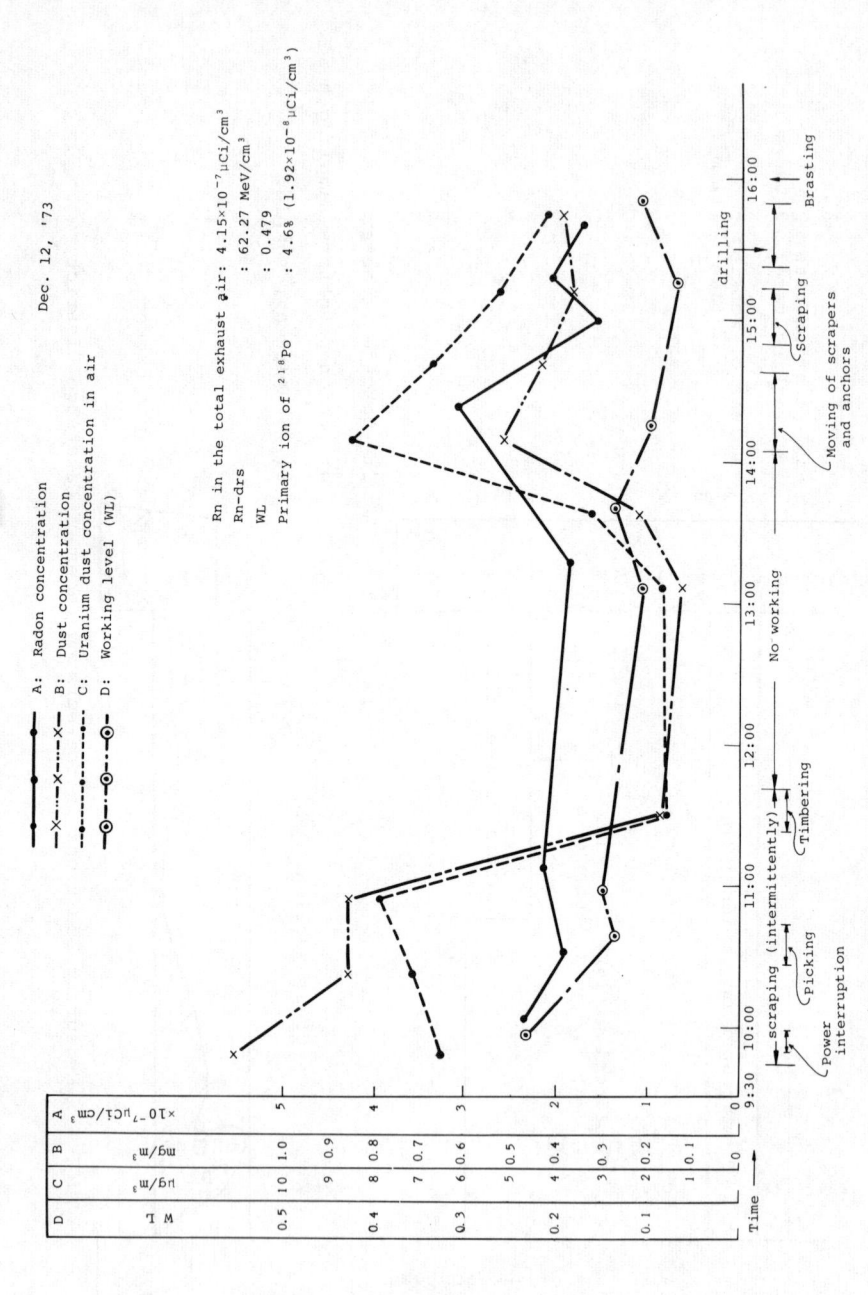

Fig. 6. Results of Environmental Assessment in the Working Faces of Nakatsugo District

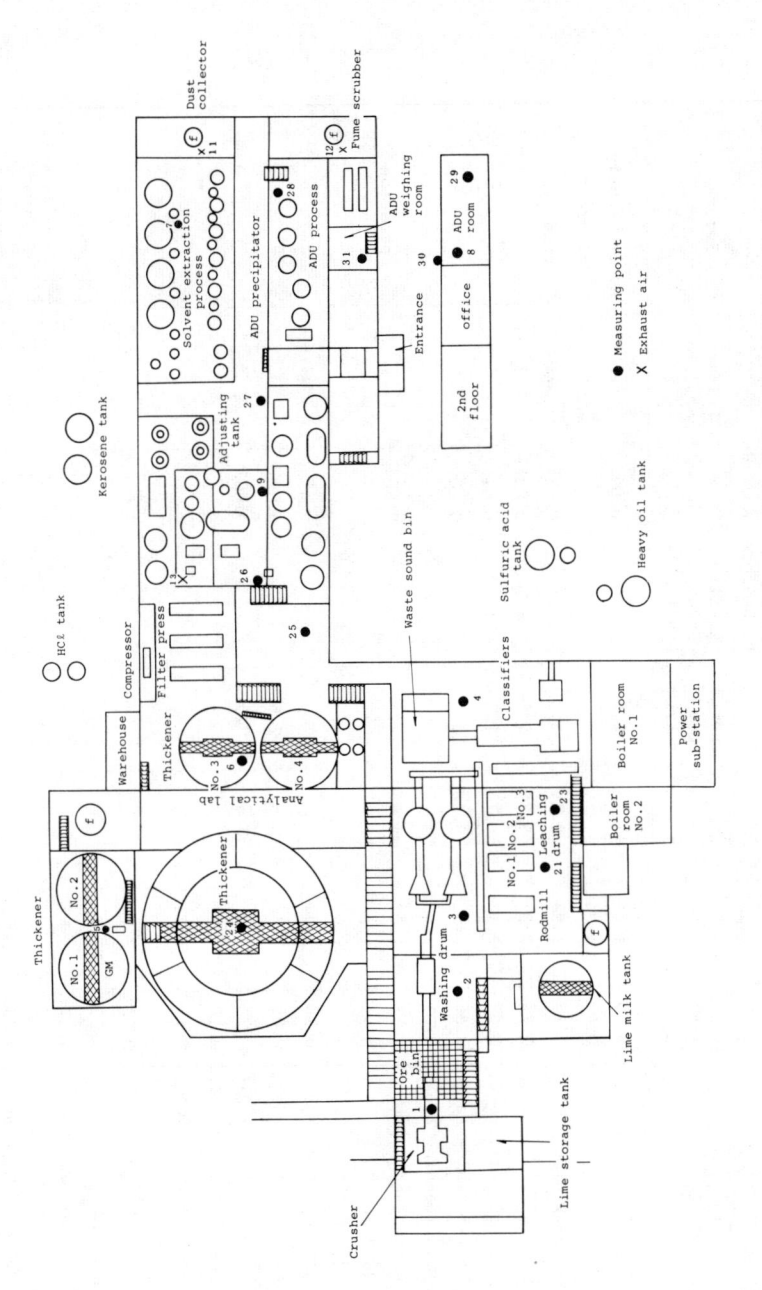

Fig. 7. Measuring Points of Working Area in the Milling Plant

Table 2. Measured Results of Working Area in the Milling Plant
(December 10 - 11, 1973)

No.	Sampling Place	Dust con. mg/m³	U dust con. μg/m³	Rn con. 10^{-7} μCi/cm³	Energy con. MeV/cm³	Rn Daughter Nuclide Equilibrium ratio %	WL	Primary particles %	Remark
1	Ore bin	0.13	1.24	11.1	6.6	4.7	0.051	7.7	No ore supply
		0.43	4.34						Under ore supply
2	Washing drum	0.17	1.98	2.79	3.3	9.3	0.025	3.2	Stopped
		0.75	7.07	5.80	4.8	6.5	0.036	4.3	⎫
3	Drag classifier	0.22	2.13	6.76	1.3	9	0.01	0	⎬
21	No.1 leaching tank	0.25	2.61	9.40	13	11	0.10	7.1	⎭ In operation
23	No.3 leaching tank	0.21	1.45						
4	Waste sand hopper	0.11	1.03	5.46	4.3	6.2	0.003	13.1	
6	No.2 and No.4 thickeners	0.22	1.64						
24	15m chickener	0.10	1.04						
25	Filter press	0.63	3.97	0.6	2.6	8	0.02	9	
26	Pregnant liquid pit			0.3	0.3		0.002		
27	Uranyl chloride pit	0.18	1.55						⎫
28	ADU precipitation tank	0.11	1.33						⎬ ADU work
29	ADU filter press	0.087	2.53						⎭
8	ADU dryer	0.13	72.4						ADU packing
30	ADU room entrance	0.17	66.7						ADU removal from filter press
31	ADU weighing room	0.039	2.62						ADU packing
		1.38	450						

Table 3. Results of Measurement in the Milling Plant (Sept. ~ Dec., 1975)

No.	Sampling Place Name of Place	Dose rate mR/hr Mean	Dose rate mR/hr Max.	^{222}Rn con.×10^{-8}μCi/cm^3 Mean	^{222}Rn con.×10^{-8}μCi/cm^3 Max.	U con.×10^{-12}μCi/cm^3 Mean	U con.×10^{-12}μCi/cm^3 Max.
1	Ore bin	0.32	0.50	4.68	12.50	1.78	5.33
2	Ore washing yard	0.14	0.20	1.93	4.30	1.21	2.47
3	Climber conveyer	0.14	0.18	1.13	1.80	0.69	1.21
4	Waste sand hopper	0.14	0.22	0.32	0.85		
5	No.1 and No.3 thickeners	0.10	0.15	0.25	0.83	1.20	3.08
6	No.2 and No.4 thickeners	0.08	0.08	0.11	0.21	0.48	1.81
7	Solvent extraction room	0.05	0.06	< B.G.	< B.G.	0.35	1.50
8	ADU room	0.08	0.10	< B.G.	< B.G.	1.17	3.12
9	UF$_4$ room	0.06	0.08	< B.G.	< B.G.	0.21	0.32
11	Solvent extraction system's scrubber					0.45	1.43
12	ADU system's dust collector					0.58	0.82
13	UF$_4$ system's scrubber					0.12	0.43

Discussion

Q. The radon concentration listed in the 5th column in Table 2 was high.

A. It is necessary to change from 10^{-7} µCi/cm^3 to 10^{-8} µCi/cm^3 in the 5th column of Table 2.

MEASUREMENTS OF RADON DAUGHTER CONCENTRATIONS IN STRUCTURES BUILT
ON OR NEAR URANIUM MINE TAILINGS*

F. F. Haywood, G. D. Kerr, W. A. Goldsmith,
P. T. Perdue, and J. H. Thorngate
Health Physics Division, Oak Ridge National Laboratory
Oak Ridge, Tennessee 37830, U.S.A.

ABSTRACT

A technique is discussed that has been used to measure air concentrations of short-lived daughters of ^{222}Rn in residential and commercial structures built on or near uranium mill tailings in the western part of the United States. In this technique, the concentrations of RaA, RaB, and RaC are calculated from one integral count of the RaA and two integral counts of the RaC' alpha-particle activity collected on a filter with an air sampling device. A computer program is available to calculate the concentrations of RaA, RaB, and RaC in air and to estimate the accuracy in these calculated concentrations. This program is written in the BASIC language.

Also discussed in this paper are the alpha-particle spectrometer used to count activity on the air filters and the results of our radon daughter measurements in Colorado, Utah, and New Mexico. Results of these and other measurements are now being used in a comprehensive study of potential radiation exposures to the public from uranium mill tailing piles.

INTRODUCTION

Several processes exist for the recovery of naturally occurring uranium. Generally, ore (whose uranium content by weight ranges from a few tenths of one percent to several percent) is transported from mine to mill where it is crushed and leached. The waste material (mill tailings) from this operation is discarded on mill property and is nearly the same volume and weight as the crushed ore entering the mill. The leaching procedure is specific for uranium; therefore, most of the daughter radionuclides including ^{230}Th, ^{226}Ra, and ^{210}Pb remain in the wastes and are left behind once the facility becomes inactive. Uranium mill tailings are sandy in nature, having a particle size which ranges from about -10 mesh (sand) to -200 mesh (slimes). The concentration of ^{226}Ra is highest in the slime fraction, frequently as high as 3000 pCi/g. In the sand fraction, a concentration of 100-300 pCi/g is common. Radon emanates readily from the tailings, and if the tailings are in intimate contact with a structure or stored nearby, radon is likely to accumulate on the inside. The level of accumulation depends in part on the concentration of ^{226}Ra in the tailings, the total amount of tailings involved, and their specific location with respect to the structure.

───────────────

*Research sponsored by the U.S. Energy Research and Development Administration under contract with Union Carbide Corporation.

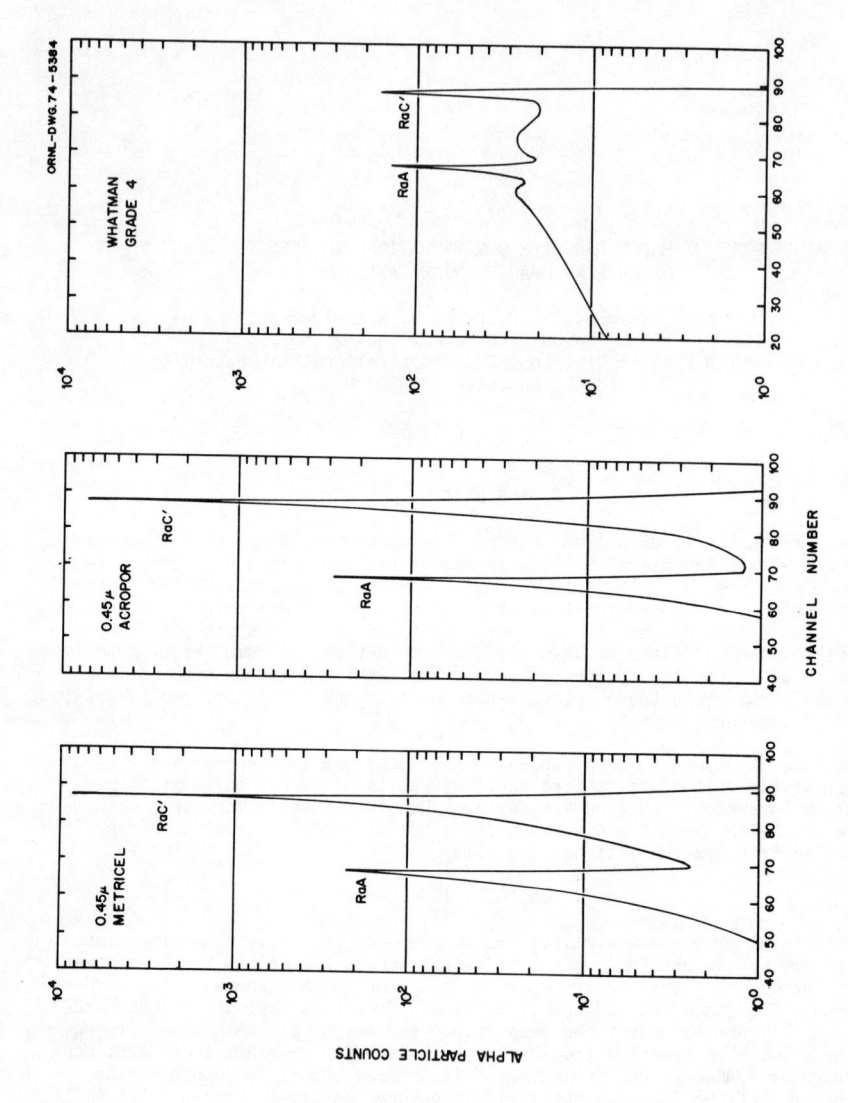

FIGURE 1. Comparison of the Spectra of Radon Daughter Radionuclides Collected on Filters of Varying Composition and Efficiency.

Leaching operations have ceased at a number of mills in the western part of the United States. At several of these sites, tailings have been removed for private as well as public use. Examples of this practice include the construction of commercial buildings on property where tailings were pushed back to make way for the buildings, but were not removed from the ground on which the structures were built, and the use of tailings as a stabilizing material under the floor of new residences or as backfill around basement walls. Because of this practice, people working in the commercial buildings, and those living in residences are subjected to elevated radon daughter concentrations ranging from small fractions of normal background levels to more than one working level* (WL).

The radon daughter measurements discussed in this report were made in Colorado, Utah, and New Mexico. Most of these measurements were made in cases where the proximity of tailings to a structure was well known (in intimate contact with a structure or stored on nearby property). Results of these and other measurements are now being used in a comprehensive study of potential radiation exposures to the public from uranium mill tailings. The evaluation of potential exposures to radon daughters inside structures presents a formidable problem to the dosimetrist.

MEASUREMENT METHODS

Numerous techniques have been presented for the measurement of radon daughters.(1) Our measurement philosophy centers around the sampling of aerosols in air and the use of alpha spectroscopy techniques for determining the concentration of radon daughters.

A modification of the Martz spectroscopy technique (2) has been developed at this laboratory (3) to improve the accuracy and sensitivity of radon daughter measurements. In this modification, radon daughter concentrations are calculated from one integral count of the RaA and two integral counts of the RaC' alpha-particle activity collected on a filter. A computer program, RPCON4, is available which will handle differing air sampling rates, sampling and counting times, and counter efficiencies. This BASIC-language program computes the air concentrations of RaA, RaB, and RaC and estimates the accuracy of these calculated concentrations. (4)

In our alpha-particle spectrometer, helium is flowed between a silicon diode detector and a filter which are separated by a distance of about 0.5 cm.(5) By using helium, the counter can be operated at atmospheric pressure with considerable gain in filter handling simplicity and very little loss of resolution compared to alpha-particle counts made in a vacuum. The resolution of the spectrometer for air samples collected at a flow velocity of about 50 cm/sec with a membrane (Metricel, Gelman GN-6) and glass-fiber (Acropor, Gelman An-450) filter having a 0.45-μ pore size is compared with a more porous filter (Whatman grade 4) in Figure 1. Either the membrane or glass-fiber filter with a medium pore size of 0.45 to 0.80 μ allows easy resolution of the RaA and RaC' alpha-particle activity. At flow velocities of up to 100 cm/sec, these filters are normally more than 99% efficient in collecting aerosols. (6)

For sampling times in the range of 5 to 15 min, one RaA counting interval from 2 to 12 min and two RaC' counting intervals from 2 to 12 min and from 15 to 30 min after the termination of the air sample collection have been found to give a good overall accuracy in the calculated concentrations for a wide range of RaB/RaA and RaC/RaA activity ratios. The starting time of 2 min for the first count is the shortest practicable time for transferring the filter from the air sampling device to the spectrometer, and an ending time of 30 min is standard for techniques of this type. (2,7,8) Air sampling time intervals greater than 15 min were also investigated but were not found to offer any great improvement in the accuracy of the calculated concentrations.

---

*One working level is equivalent to 100 pCi/ℓ radon in full equilibrium with its short-lived daughters, or any concentration of daughters in one liter of air such that the total potential alpha energy is 1.3×10^5 MeV.

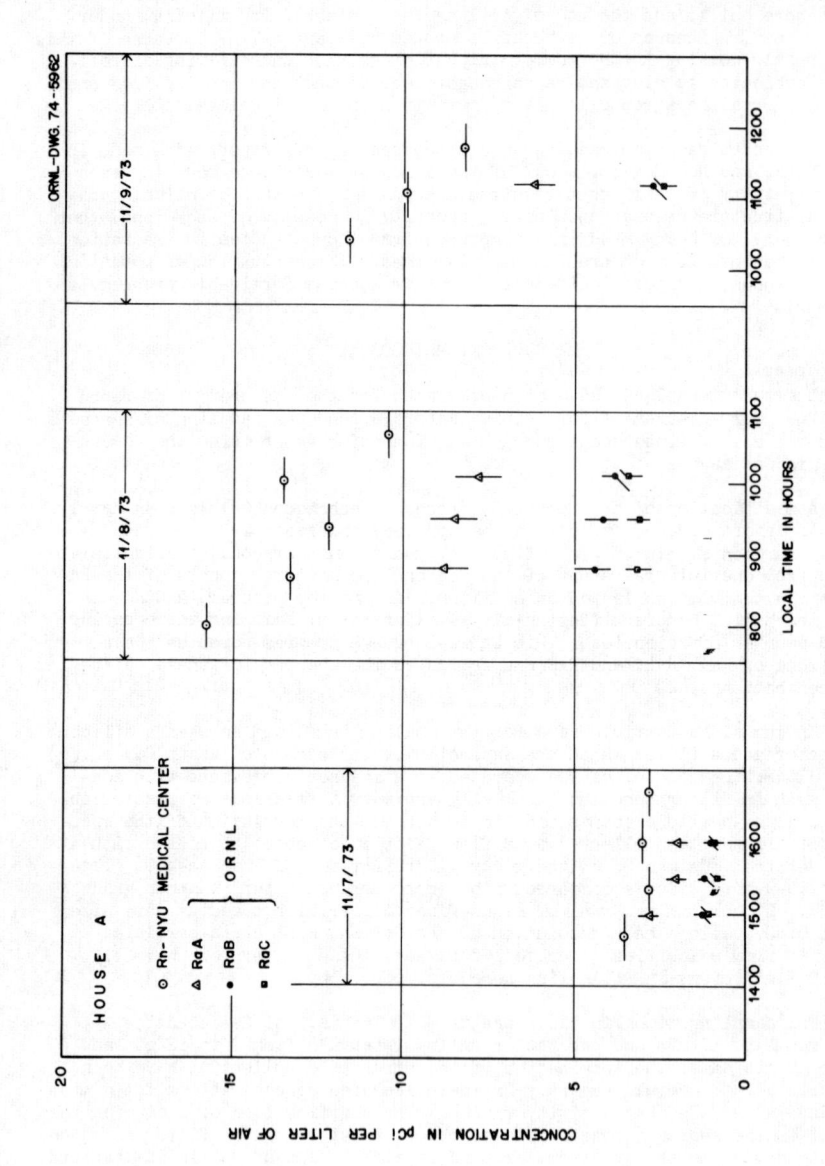

FIGURE 2. Results of Radon and Radon Daughter Measurements in a House in Grand Junction, Colorado, Showing Variation in Concentrations with Time.

APPLICATION AND RESULTS

In late October and November of 1973, the alpha-particle spectrometer and modified spectrometry technique were used to measure radon daughter levels in nine structures in Grand Junction, Colorado, at the request of the USAEC. Tailings from a nearby uranium mill had been used in their construction, but two had undergone remedial action to remove most of the tailings. Because tailings had been used as a fill, the measurements were made in either the basements of the structures or the first floor of slab-on-grade type structures. With one exception, the measurements were made in the living areas of the houses. Examples of the radon daughter measurements in one of the structures are shown in Figure 2. Also shown in the figure are measurements of radon levels made by the Institute of Environmental Medicine of the New York University Medical Center. (9) The radon concentrations from their measurements are plotted at the center of 40-min measuring intervals, indicated by the horizontal bars. Radon daughter concentrations from our measurements are plotted at the midpoint of air sampling times of 5 to 15 min.

The vertical bars on the radon daughter concentrations are an estimate of the probable error in the measurements. These are based on a statistical uncertainty of one standard deviation in each of the three counts of activity on the filter and systematic uncertainties of ± 5% in both the detection efficiency of the alpha-particle spectrometer and the air sampling rate. Air sampling rates were measured with a flowmeter that had been calibrated with a wet test meter, using standard procedures. (10,11) Concentrations of RaA, RaB, and RaC measured in these structures varied from highs of 235 (± 7%), 187 (± 8%), and 156 (± 7%) pCi/ℓ, respectively, in a nonreconstructed house to lows of 0.50 (± 20%), 0.17 (± 35%), and 0.15 (± 30%) pCi/ℓ, respectively, in a school.

Average values for radon daughter concentrations during morning and afternoon periods in some structures in Grand Junction, Colorado, are given in Table 1. The ventilation rates, and therefore the radon and radon daughter levels, of the structures are affected by the types of heating and cooling systems, the opening and closing of doors, etc. Results have been published of some of the New York University Medical Center measurements of the diurnal variations of ^{222}Rn in these and other structures in Grand Junction. (12) Based on their measurements, trends in radon levels spanning the time intervals of our radon daughter measurements in the structures are also indicated in Table 1. Large standard deviations in the average values in Table 1 may be attributed to either rapid changes in radon levels within the structures or large uncertainties in the measurements at the lower radon daughter levels.

With two counting intervals of 2 to 12 and 15 to 30 min used in this work, it is possible to measure equilibrium concentrations of 1 pCi/ℓ each of RaA, RaB, and RaC with relative standard deviations of about 15, 18, and 13 percent, respectively. This assumes an air sampling time of 10 min, an air sampling rate of 10 ℓ/min, and a counter efficiency of 0.20. For these same typical sampling and counting conditions, disequilibrium concentrations of 1, 0.4, and 0.2 pCi/ℓ of RaA, RaB, and RaC can be measured with relative standard deviations of about 15, 26, and 36 percent. If the air sampling time is increased to 15 min and the air sampling rate to 17 ℓ/min, the disequilibrium concentrations of 1, 0.4, and 0.2 pCi/ℓ of RaA, RaB, and RaC can be measured with relative standard deviations of about 11, 18, and 26 percent, and the equilibrium concentrations of 1 pCi/ℓ each of RaA, RaB, and RaC can be measured with relative standard deviations of about 11, 12, and 10 percent, respectively.

In 1975 and 1976, this sampling technique was utilized in radiological surveys in the vicinity of a number of inactive uranium mills in western states. Sampling in these instances was limited to periodic spot or grab samples because of the requirement that the equipment be manned at all times and because only one sampler was available. The objective of this sampling program was directed at the determination of the concentration of radon and its daughters for situations where the proximity of tailings to a structure was well known (in intimate contact with a structure or stored on nearby property).

Table 1. Average and Standard Deviation of Radon Daughter Concentrations Measured in Structures at Grand Junction, Colorado

Structure	Period	No. of Meas.	RaA	RaB	RaC	Average Working Level (WL)	Trend in Radon Concentration
House A	AM	4	7.9 ± 1.2	3.8 ± 0.8	3.0 ± 0.4	0.038	Falling
	PM	3	2.3 ± 0.5	1.1 ± 0.2	1.0 ± 0.2	0.011	Falling slowly
House B	AM	2	40 ± 5	25 ± 6	19 ± 6	0.24	Falling
	PM	2	25 ± 5	19 ± 4	18 ± 3	0.19	Falling slowly
House C	AM	2	1.4 ± 0.4	0.6 ± 0.2	0.5 ± 0.08	0.0062	
	PM	3	0.8 ± 0.3	0.2 ± 0.04	0.4 ± 0.08	0.0035	
House D	AM	2	21 ± 11	9.1 ± 2.6	4.6 ± 1.8	0.085	Falling
	PM	3	28 ± 3	11 ± 0.9	8.4 ± 1.2	0.12	
House E	AM	2	10 ± 4	5.5 ± 2.0	4.6 ± 1.2	0.055	Falling
House F	PM	5	232 ± 22	172 ± 17	145 ±14	1.65	Steady
House G							
Location 1	AM	1	38	15	11	0.16	Rising
	PM	2	28 ± 5	13 ± 0.2	8.2 ± 0.1	0.13	Falling slowly
Location 2	AM	1	42	14	9.2	0.15	Rising
	PM	2	23 ± 7	13 ± 2	6.8 ± 0.4	0.11	Falling slowly
House H	AM	2	6.0 ± 0.7	3.4 ± 0.4	2.0 ± 0.6	0.031	Steady
	PM	4	5.3 ± 0.4	3.1 ± 0.2	2.0 ± 0.3	0.029	Steady
School	PM	3	0.8 ± 0.4	0.2 ± 0.09	0.2 ± 0.1	0.0027	Rising slowly

Table 2. Radon and Radon Daughters Concentrations Measured in Structures at Salt Lake City, Utah

Date 1975	Time	Location	Rn	RaA	RaB	RaC	Working Level (WL)
9-25	11:00	Building (1), Outside		0.43 (19)*	0.23 (36)*	0.17 (34)*	0.0022
9-26	10:40	" , Inside		0.96 (14)	0.17 (75)	0.46 (20)	0.0040
9-26	14:40	Building (2), SW Room		0.78 (15)	0.52 (26)	0.26 (35)	0.0044
9-26	14:40	" "		0.60 (17)	0.30 (34)	0.19 (35)	0.0028
10-2	09:35	" "		3.2 (9)	2.8 (23)	2.1 (20)	0.025
10-2	11:35	" "		2.3 (10)	1.8 (25)	1.7 (19)	0.018
10-2	16:10	" "		1.0 (13)	0.5 (34)	0.6 (21)	0.0061
9-26	11:25	Building (3), Store Room	195.	132. (6)	77. (21)	63. (5)	0.76
10-1	09:43	" "		139. (6)	108. (19)	61. (22)	0.92
10-1	16:05	" "	557.	372. (6)	179. (22)	130. (20)	1.77
9-26	13:00	Building (4), Shop	76.8	49.1 (6)	32.8 (23)	29.1 (18)	0.32
9-26	13:40	" "		16.5 (7)	6.7 (25)	5.9 (19)	0.073a
10-1	10:22	" "		2.2 (10)	0.4 (38)	0.4 (27)	0.0057b
10-1	11:22	" "	6.9	4.9 (8)	0.9 (27)	0.4 (40)	0.011b
10-1	14:50	" "	16.2	9.5 (7)	1.8 (24)	0.8 (34)	0.021b
10-1	15:30	" "		6.4 (8)	1.2 (26)	0.6 (34)	0.015b
10-1	17:10	" "	81.3	43.6 (6)	11.3 (20)	4.2 (32)	0.12c
9-26	16:50	Building (5), Inside		0.28 (24)	0.06 (93)	0.11 (34)	0.001
9-26	17:30	" , Outside		0.34 (23)	0.10 (46)	0.05 (77)	0.001
9-26	18:35	Building (6), Front Room		1.54 (11)	0.49 (31)	0.36 (28)	0.0054
9-27	09:00	Building (7), Store Room		13.2 (7)	9.8 (22)	7.8 (19)	0.092
9-27	10:45	" "		7.7 (7)	5.2 (28)	6.2 (16)	0.057
9-27	11:20	" , Outside		2.1 (10)	1.7 (24)	1.3 (21)	0.016
9-29	12:15	Building (8), Outside		0.68 (16)	0.02(283)	0.19 (26)	0.0015
9-29	14:10	Building (9), S. Basement		0.38 (21)	0.09 (68)	0.14 (33)	0.0014
9-29	14:50	" "		0.61 (17)	0.06(107)	0.17 (29)	0.0015
9-29	15:25	" , Outside Roof		0.4 (21)	0.1 (35)	0.1 (26)	0.0005
9-29	16:45	Building (10), Warehouse	6.6	0.9 (14)	0.7 (28)	0.6 (23)	0.0069d
9-29	17:20	" "		0.8 (16)	0.6 (30)	0.5 (24)	0.0057

* Probable error in percent.
a. Door opened approximately one hour prior to measurement.
b. Doors opened.
c. Sample collected 30 minutes after shop closed.
d. Near open door.

- 225 -

A summary of some radon daughter measurements which were made in Salt Lake City are presented in Table 2. Attention is called to three cases in particular, Buildings 2-4. Building 2 is a sewage treatment plant facility located on property surrounded by the tailings piles from the inactive Vitro Mill. Over the period of these measurements, the working level concentration ranged from 0.003 to 0.02 WL. This room is occupied for short periods each day. Buildings 3 and 4 are commercial structures situated near the boundary of the original tailings pond. Both buildings are constructed over tailings which range to five feet deep and which contain up to 900 pCi/g ^{226}Ra. The resulting working level concentration is highest in the store room (seldom occupied for long periods) of Building 3, reaching a level of 1.77. In the shop of Building 4, the occupancy is higher, and the number of working levels ranges from 0.01 to 0.32. It is noted here that these measurements were made during a period when the temperature was moderate and doors were open frequently. When the building is closed thereby reducing the ventilation rate, the number of working levels will increase. Building 7 is constructed in the same area as 3 and 4, but there are no tailings under the floor. The source of elevated radon levels in this building is the large Vitro tailings pile located at the rear of the building.

During a radiological survey of the inactive uranium mill at Shiprock, New Mexico, hourly measurements of the concentration of radon and progeny were made inside a building now used to house a school for teaching construction equipment skills over a period of 21 hours. The results of this series are presented in Figure 3. Although this building is close to the tailings piles associated with this mill, none are in intimate contact with the structure. Therefore, radon enters the building through ventilation. The maximum working level concentration was 0.021 at 12:15 AM. It is noted that after that hour, the total daughter concentration decreases more rapidly than does radon. A probable explanation for this is the decrease in condensable nuclei as dust settles due to limited movement of air.

CONCLUSIONS

A modified alpha spectroscopy technique has been used for the measurement of radon daughters collected on a high efficiency air filter. Two counting intervals are used providing one measurement of RaA and two measurements, separated by a known interval, of RaC'. It was observed that with this technique, concentrations of 1, 0.4, and 0.2 pCi/ℓ of RaA, RaB, and RaC can be measured with relative standard deviations of about 15, 26, and 36 percent. One limitation of the sampling and counting equipment used in this work is that it is not automatic. It is necessary that the filters and counter be handled manually.

REFERENCES

1. R. J. Budnitz, *Health Phys.* **26**, 145 (1974).

2. D. E. Martz, D. F. Hollman, D. E. McCurdy, and K. J. Schiager, *Health Phys.* **17**, 131 (1969).

3. G. D. Kerr, *Trans. Amer. Nucl. Soc.* **17**, 541 (1973).

4. G. D. Kerr, *Measurement of Radon Progeny Concentrations in Air by Alpha-Particle Spectrometry*, ORNL/TM-4924 (July, 1975).

5. P. T. Perdue, W. H. Shinpaugh, J. H. Thorngate, and J. A. Auxier, *Health Phys.* **26**, 114 (1974).

6. American National Standards Institute, *Guide to Sampling Airborne Radioactive Materials in Nuclear Facilities*, ANSI-N13.1 (1969).

7. E. G. Tsivoglou, H. E. Ayer, and D. A. Holaday, *Nucleonics* **11**, (9), 40 (1953).

8. J. W. Thomas, *Health Phys.* **23**, 783 (1972).

9. H. B. Spitz and M. E. Wrenn, personal communications (1974).

10. American National Standards Institute, *Radiation Protection in Uranium Mines and Mills*, ANSI-N7.1a (1969).

11. R. L. Rock, R. W. Dalzell, and E. J. Harris, *Controlling Employee Exposure to Alpha Radiation in Underground Uranium Mines*, U.S. Department of the Interior, Bureau of Mines (1971).

12. H. B. Spitz and M. E. Wrenn, "The Diurnal Variation of the Radon-222 Concentration in Residential Structures in Grand Junction, Colorado", in the *Second Workshop on the Natural Radiation Environment*, HASL-287, 100-117 (September, 1974).

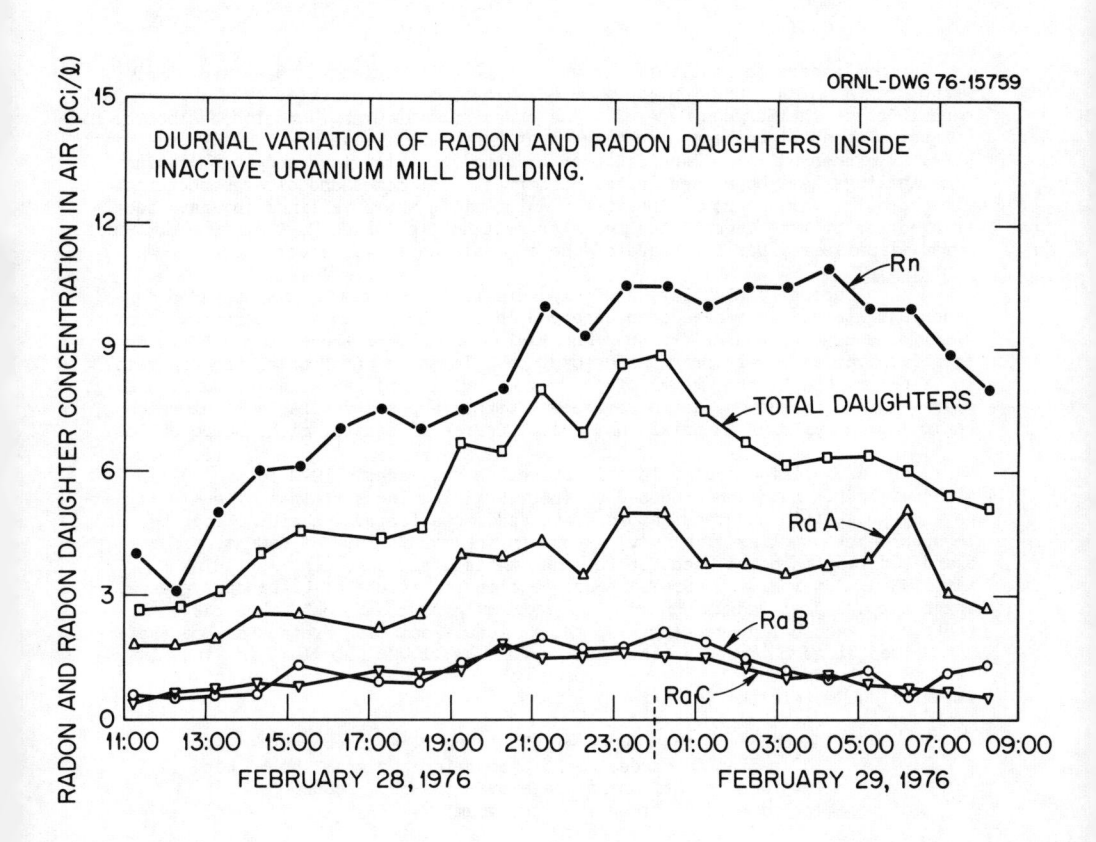

FIGURE 3. Diurnal Variation of Radon and Radon Daughters Inside an Inactive Uranium Mill Building at Shiprock, New Mexico.

Discussion

The discussion following this paper was focused mainly on radon daughter exposures to the general public from building materials. After it was discovered that uranium mill tailings had been used in the construction of homes at Grand Junction, Colorado, the Surgeon General of the United States set up guidelines for remedial action based on levels of exposure to gamma rays and radon daughters in these homes. (a) Those for radon daughters are as follows:

Indoor Radon Daughter Concentration	Recommendation
0.05 or more	Remedial action indicated
0.01 to 0.05 WL	Remedial action may be suggested
0.01 Wl or less	No action indicated

A program of tailings removal has been initiated at homes and other structures at Grand Junction where remedial action was indicated by the recommendations of the Surgeon General. The experience at Grand Junction has not been unique. It has since been discovered that commercial buildings at Salt Lake City, Utah, have been built on land fills of uranium mill tailings and that uranium mill tailings have been used in the construction of homes and other structures elsewhere. In areas around inactive uranium mills where tailings may have been removed for private use, it now seems prudent not to issue a building permit until a radiation survey has been made of the excavated site for a structure.

Abnormally high levels of exposure to either gamma rays or radon and radon daughters in air have been observed in many homes and structures built of granite, concrete, and brick. (b) The radium content of these common building materials varies widely from place to place. In the case of brick and concrete, it is determined by the origin of the raw material components (clay, sand, shales, slags, etc.). Based on past experiences, it now appears desirable to exercise control over building materials where some degree of control might be possible.

A preliminary step in this direction was made in 1972 in the United Kingdom. (c) It had been proposed by their building industry that gypsum obtained as a by-product in the manufacture of phosphoric acid from phosphate rock be used to supplement supplies of natural gypsum in the manufacture of cement, plasterboard ("dry-wall" board used for internal walls), and some other building materials. After giving careful consideration to the use of by-product gypsum, which has a higher radium content than natural gypsum and most other raw components of common building materials, the National Radiation Protection Board of the United Kingdom issued a statement on June 14, 1972. (c) In this statement, it was concluded "that by-product gypsum may safely be used as a building material subject to the following provisos:

"1. Sources of the raw material giving rise to concentrations of radium in excess of 25 picocuries per gram should be avoided so that the average over the whole county shall not exceed 25 picocuries per gram.

---

a. *Natural Background Radiation in the United States*, NCRP Report No. 45, National Council on Radiation Protection and Measurements, Washington, D.C. 20014 (November 15, 1975).

b. United Nations Scientific Committee on the Effects of Atomic Radiation, *Ionizing Radiation: Levels and Effects*, Vols. 1 and 2, United Nations Publications No. E.72.IX.18, United Nations, New York (1972).

c. M. C. O'Riordan, M. J. Duggan, W. B. Rose, and G. F. Bradford, *The Radiological Implications of Using By-Product Gypsum as a Building Material*, NRPB-R7, National Radiological Protection Board, Harwell, Didcot, Berks (December, 1972).

"2. Arrangements should be made for recording the production utilisation of the material and measuring its radioactivity. The information thus obtained should be reported to the Board annually to enable it to carry out periodic assessments of the population exposure."

The above statement provides an example of the desirability of exercising some control over building materials and also the type of controls that might be feasible.

NIVEAUX RENCONTRES DANS LA CONTAMINATION DE L'AIR
EN DESCENDANTS DU RADON DANS LES REGIONS URANIFERES
ET AU VOISINAGE DES INSTALLATIONS D'EXTRACTION
ET DE TRAITEMENT DE MINERAI

J. PRADEL, Ph. DUPORT, P. ZETTWOOG
Commissariat à l'Energie Atomique
Paris, France

RESUME

On présente les résultats obtenus au cours de campagnes de mesures effectuées dans la région de la Crouzille, au nord de la ville de Limoges en France. Cette région est le siège d'une exploitation minière produisant 700 tonnes par an d'uranium. Connaissant la distribution des situations météorologiques dans la région, il a été possible de dresser la carte des valeurs moyennes à long terme de la concentration en descendants du radon sur le site et d'évaluer l'impact des installations d'extraction et de traitement du minerai sur la contamination du site.

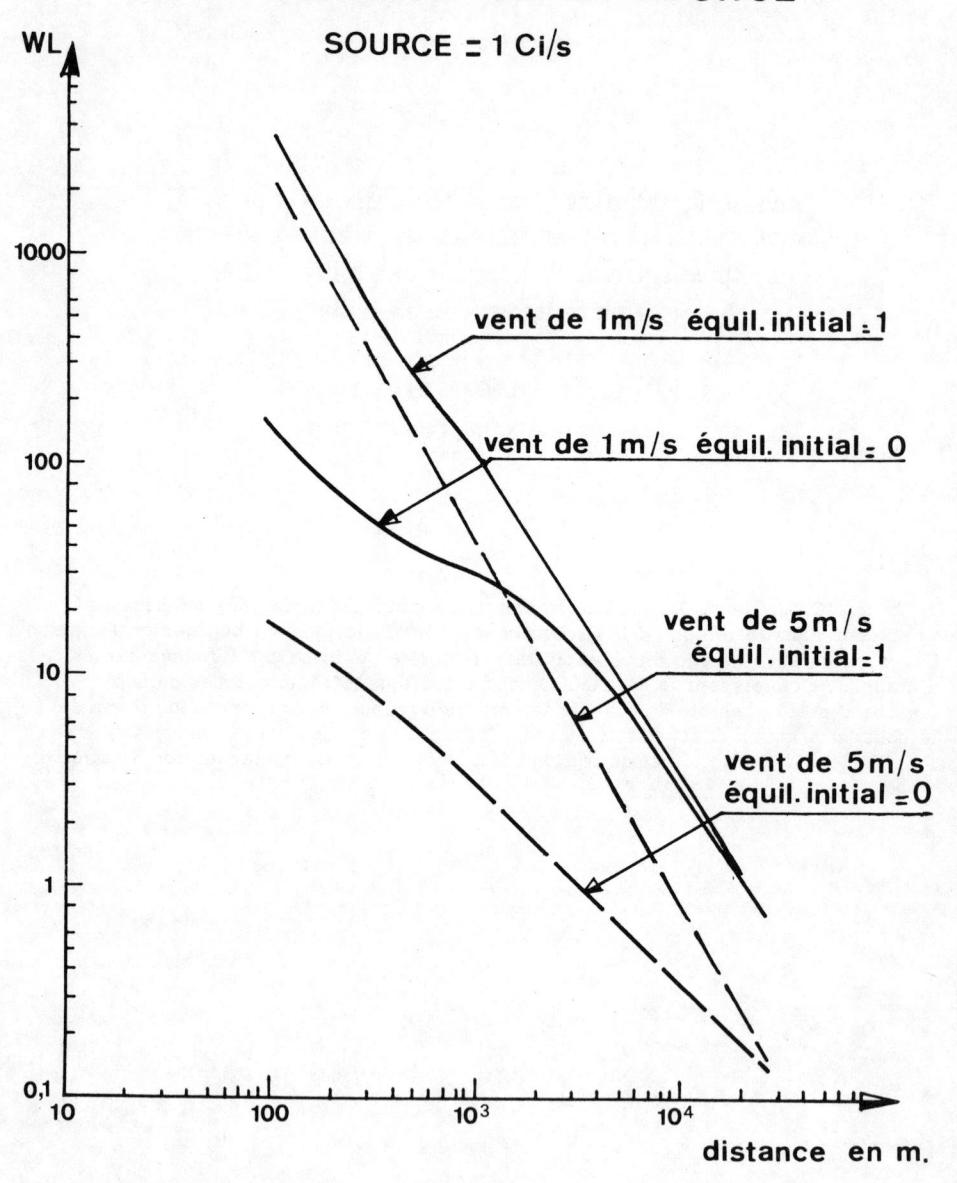

FIG.1

ENERGIE POTENTIELLE ALPHA EN FONCTION DE LA DISTANCE DE LA SOURCE

SOURCE = 1 Ci/s

CADRE DE L'ETUDE

La Division Minière de la Crouzille (près de Limoges, France) est un ensemble producteur de concentrés d'uranium, comprenant des mines souterraines, des carrières, un atelier de stockage et de préparation du minerai et une usine de concentration chimique.

Nous nous sommes attachés à connaître l'impact, sur le niveau de la radioactivité atmosphérique, de la partie de ces installations qui inclut l'aire de stockage du minerai, l'atelier de préparation (broyage, mélange, ...), l'usine de concentration chimique et le bassin d'épandage des effluents liquides et solides dans lequel se sont accumulés les sables radifères depuis 1956.

Méthodologie

Si nous avions une connaissance précise de la source de radon représentée par ces installations, la meilleure manière de déterminer l'impact serait d'utiliser un modèle mathématique de la dispersion atmosphérique sur le site étudié. Dans la mesure où l'on ne s'intéresse qu'à la connaissance des valeurs moyennes à long terme de la concentration des descendants du radon, les méthodes basées sur les modèles mathématiques donneraient en effet des résultats plus précis et à moindre coût que celles basées sur l'expérimentation in situ. Les données de base nécessaires à l'emploi du modèle sont, en dehors des données sur les émissions de radon, les conditions météorologiques et les données sur le relief.

Sur la figure 1, on a représenté les variations de l'énergie potentielle volumique des descendants du radon (exprimée en WL), en fonction de la distance sous le vent dans le cas d'une source ponctuelle de 1 Ci/s de radon et en l'absence de relief particulier. On a considéré deux situations météorologiques, représentatives des conditions de mauvaise diffusion (vent de 1 m/s) et de bonne diffusion (vent de 5 m/s). On a envisagé les deux situations extrêmes de situation d'équilibre au niveau de la source $f = 0$ (radon pur) et $f = 1$ (radon en équilibre avec ses descendants). Dans le cas du radon pur, l'énergie α est nulle à la source, croît et passe par un maximum en s'éloignant sous le vent. A l'aide de cette figure, et connaissant la rose des vents, on peut déterminer les valeurs moyennes à long terme de l'énergie volumique en tout point du territoire. Dans le cas d'une source d'intensité quelconque, les valeurs de WL seraient dans le même rapport que les sources.

L'ennui est que les émissions, ponctuelles ou diffuses, de l'usine, ne sont connues que par valeur supérieure : la source de radon essentielle est constituée par les sables radifères, dont, connaissant la quantité d'uranium extraite, on peut calculer la teneur en radium, donc la production de radon. Il ne s'agit que d'une valeur supérieure, la fraction de radon s'échappant à l'atmosphère, d'ailleurs pratiquement sans descendants, étant faible.

La méthodologie adoptée a donc été la suivante : une campagne de mesures a été effectuée sur le site pour différents régimes de vent à des distances de 2 à 10 km de l'usine. On en a déduit l'accroissement de radioactivité par rapport au bruit de fond naturel. L'utilisation en sens inverse du modèle mathématique permet de déterminer une source équivalente de radon expliquant ce qui a été trouvé sur le site. Connaissant la rose des vents et la source, le modèle mathématique a été réappliqué pour déterminer sur la carte de la région les lignes d'isoénergie potentielle α moyenne.

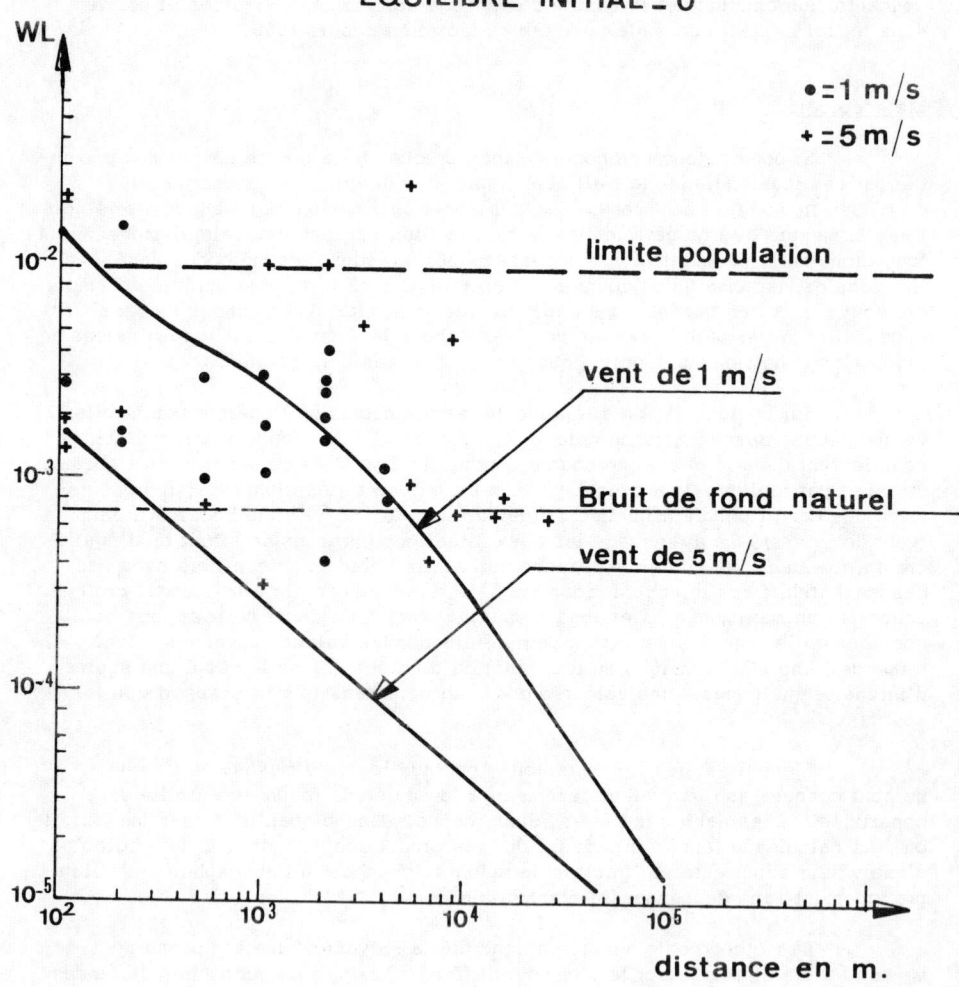

FIG.2

Résultats

Dans le cas de l'Usine de Traitement de la Division de La Crouzille, la quantité totale de minerai traitée a été d'environ 4 700 000 tonnes, d'où l'on tire que la quantité de radium accumulée est d'environ 3 000 grammes, et la source de radon correspondante $6,3.10^{-3}$ Ci/s.

La figure 2 montre la variation de l'énergie potentielle α volumique en fonction de la distance sous le vent (1 et 5 m/s) pour cette source de radon pur. On a fait apparaître la valeur 10^{-2} WL qui est la limite population déduite de la limite travailleur par application du facteur 30 habituel. 10^{-2} WL correspond à 1 pCi/l de radon à l'équilibre. On a tracé également une ligne en tirets horizontale qui correspond au bruit de fond dont la valeur moyenne a été trouvée de 7.10^{-4} WL, des valeurs de 6 à 8.10^{-3} étant possibles.

Enfin, on a porté quelques-uns des 150 résultats de mesures relevés sur le terrain. On a alors recherché graphiquement quelle était la valeur de la source de radon (Ci/s) rendant le mieux compte des mesures de terrain. La valeur de 10^{-4} Ci/s a été adoptée. Cette valeur permet de confirmer que les autres sources (stockage de minerai, usine, atelier) sont négligeables vis-à-vis du bassin d'épandage. On obtient, en ce qui concerne ce dernier, qu'il retient 98,5 % du radon produit. A partir de la rose des vents (figure 3), on a dressé la carte des fréquences de contamination (figure 4), la source étant assimilée à une surface circulaire de 500 mètres de rayon.

Dans la figure 4, la source est représentée par le cercle intérieur hachuré. R_1 et R_2 sont respectivement les rayons correspondant à 10^{-2} et 7.10^{-4} WL dans de mauvaises conditions de diffusivité (fréquence d'occurence : 43 %). Le cercle R_3 correspond à 7.10^{-4} WL, dans de bonnes conditions de diffusivité (fréquence d'occurence 57 %).

Conclusion

Les secteurs où le niveau peut être occasionnellement supérieur à la limite de 10^{-2} WL sont compris dans un rayon de 1 500 mètres autour du centre des installations. Toutefois, ce dépassement n'intervient au pire que 10 % du temps. Il en résulte donc une contamination moyenne de moins de 10^{-3} WL ; c'est-à-dire que, sur l'année, la présence de l'usine a pour effet au plus de doubler la valeur du bruit de fond à 1 500 mètres autour de l'usine.

La contribution de l'usine ne peut plus être distinguée, même sous le vent, des fluctuations naturelles à partir de 5 000 à 10 000 m quelque soit la situation météorologique.

Les mesures d'émission de radon par les sols effectuées sur le site donnent des résultats de l'ordre de 0,1 à 0,5 pCi/s.m^2. Une surface de 1 000 ou 200 km^2 respectivement, est une source équivalente à celle représentée par l'usine.

ROSE DES VENTS SUR LE SITE DE TRAITEMENT

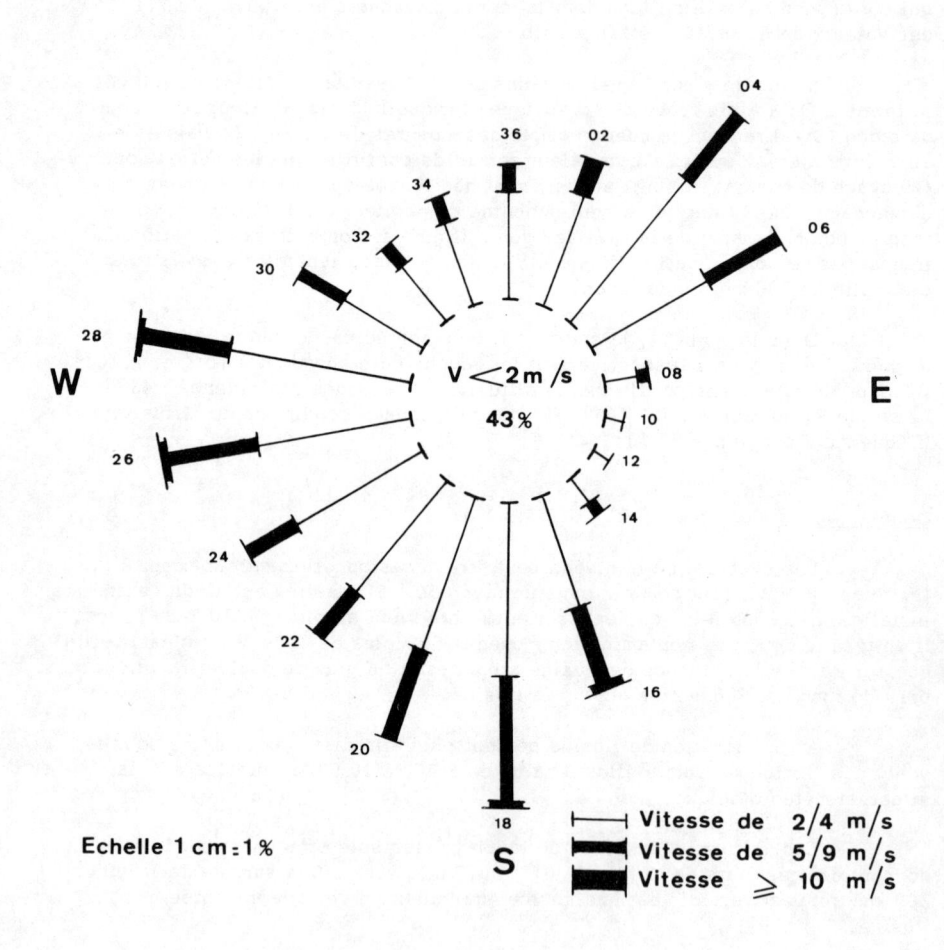

FIG.3

FRÉQUENCES D'OCCURRENCE DE CONTAMINATION PAR SECTEURS DE LA ROSE DES VENTS

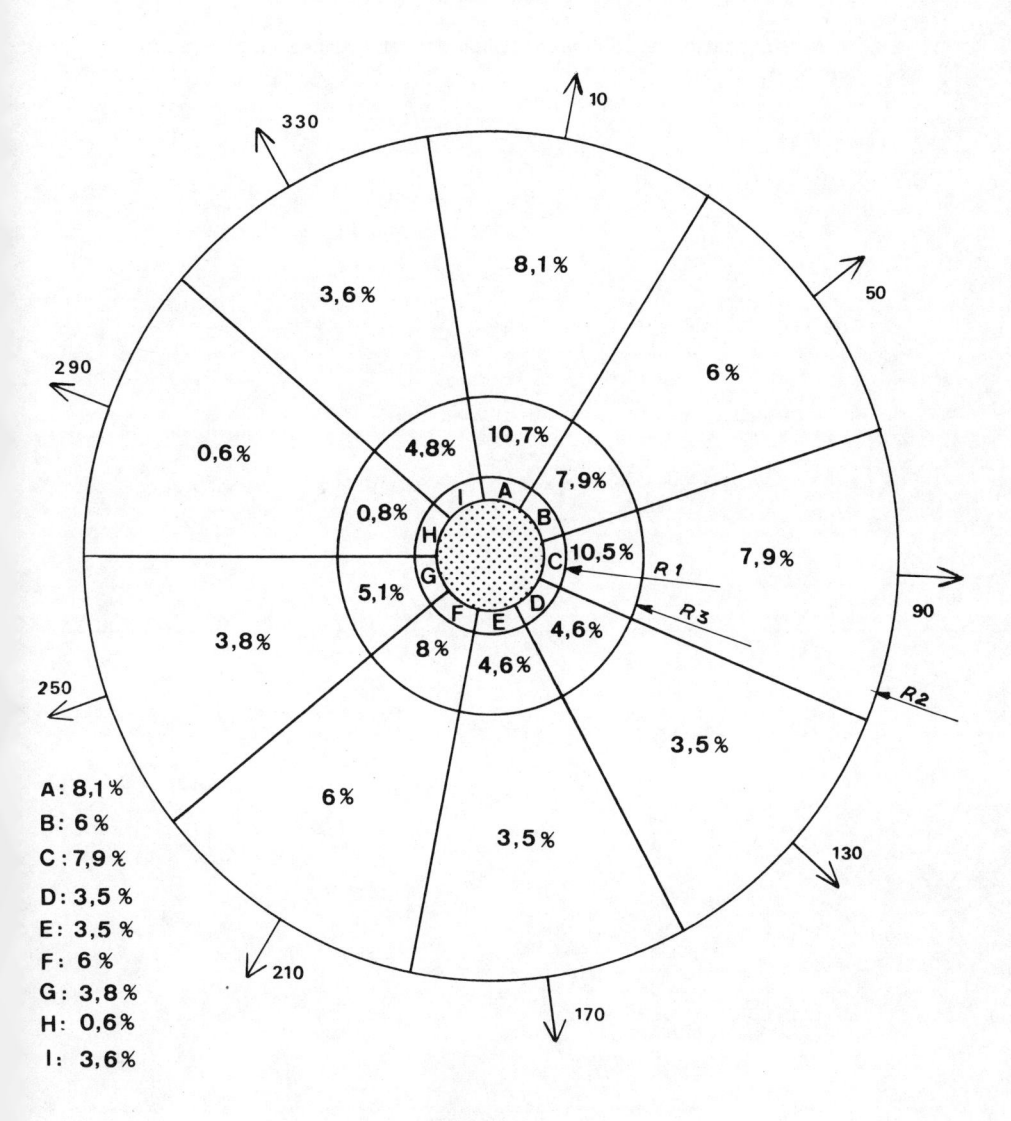

A : 8,1 %
B : 6 %
C : 7,9 %
D : 3,5 %
E : 3,5 %
F : 6 %
G : 3,8 %
H : 0,6 %
I : 3,6 %

FIG. 4

Résumé de la Discussion

- Facteurs pouvant influencer l'importance de la contamination de l'environnement :

    . Etat d'équilibre du radon émis
    . Superficie de la source
    . Influence de l'immersion des déchets radifères
    . Importance de l'émission naturelle des sols.

- Evocation de la contamination de l'environnement sur un site géothermique.

PROBLEMS IN AREA MONITORING
FOR RADON DAUGHTERS

R.A. Washington[1] and J.L. Horwood[2]
[1]Mining Research Laboratories
Elliot Lake, Ontario
[2]Mineral Sciences Laboratories
Ottawa, Ontario

ABSTRACT

Several problems that arise in the measurement of radon daughter concentrations in underground uranium mines are discussed. Preliminary field tests to evaluate sources of error have indicated that the relative standard deviation in the Kusnetz technique varies from 3% to 15%, and that one measurement in twenty may deviate from the true mean by 20% or more. An instant working level meter gives comparable precision, but the result is obtained much more quickly than is possible for the Kusnetz method. A proposed design of an apparatus to provide a stable, reproducible, accurately-known concentration of radon daughters in air for use in calibration is discussed.

I. INTRODUCTION

The problems of area monitoring of radon daughter concentrations in underground mine ventilation systems resemble in many respects those of any industrial on-line quality control system, with two notable differences. The first of these results from the fact that (in distinction from the usual situation in industrial sampling) samples cannot be collected from outside the system. There is, therefore, a significant possibility that the system to be sampled may be disturbed by the presence within the system of the sampler and his instruments. The second difference lies in the fact that there is no reliable method of preparing absolute standards that simulate the real system, for use in instrument calibration.

Effects caused by the presence of the analyst and his instrumentation within the system that is to be assayed are difficult to identify and evaluate. However, the use of correct techniques*, based on good practice, experience, and intelligent identification of potential errors, can minimize or eliminate such effects.

*For example, the sampler should always stand down-wind from the sample pump during sample collection.

Difficulties involved in preparation of standard samples for instrument calibration will be discussed later in this report.

The problems that are common to radon daughter monitoring and most industrial analyses fall into two general classes: sampling problems and measurement problems. Loysen (1) has discussed many of the possible sources of error both in sampling and in measuring, in determining radon daughters by the Kusnetz method (2), which is commonly used for routine measurements in underground uranium mines. Breslin (3) has estimated the overall variance found in routine working level data obtained in uranium mines. Other reports (4, 5) have evaluated the errors in the Kusnetz method and in a little-used alternative technique, the modified Tsivoglou procedure. However, all these reports have dealt with data that were obtained in U. S. mines. Little or no information has been available on the sources of error and the precision and accuracy of data obtained under typical Canadian mining conditions.

This report presents some preliminary results which indicate the overall variance in working level that can be found in Canadian mines using the routine Kusnetz method, and a comparison with data obtained using an "Instant" Working Level Meter. The data are also used to make a preliminary estimate of the validity of working levels determined at the sampling frequency used in the routine mine ventilation control program. Finally, a discussion is presented of the problem of instrument calibration using standard samples, as mentioned earlier.

II. VARIANCE IN KUSNETZ DATA

It has been stated that two types of problems occur in radon daughter monitoring, sampling problems and measurement problems. Because the system to be assayed is gaseous, and the material to be determined is an aerosol, the sampling problems are rather more severe than is usually the case with liquid systems. Thus the difficulty of accurately measuring the volume of air sampled is a major consideration; for a liquid system, this would be a minor (probably, insignificant) problem. A major source of error in the measurement lies in the accuracy of calibration of the detector. These, and other sources of error, are discussed by Loysen (op. cit.).

The total variance in the data includes all of these instrumental variances, plus the systematic variance (i.e. the variance in the radon daughter concentration) which is of concern. In order to establish the latter, both the former (total variance and instrumental variance) must be known. The factors contributing to the instrumental variance can be measured in some cases and estimated with reasonable accuracy in others*. The total variance can be calculated for the system simply by taking a large enough number of measurements. The data given in Table 1 represent a

---

*It should be possible to estimate the instrumental variance with reasonable accuracy by repeated measurements at an underground location where there is a minimal likelihood of changes in the working level (e.g. in a worked-out, unventilated stope), provided a "dust" source (e.g. cigar or cigarette smoke) is provided as a substrate for radon daughter deposition.

TABLE 1

SITE NO. 1
CONDITIONS:
AIR FLOW 70-100 fpm; ~64K cfm; TEMP 11-12° C; R.H. 65-82%; LIGHT DIESEL SMOKE; JUMBO DRILLING; SCOOP TRAMS; LIGHT TO HEAVY TRUCK TRAFFIC.

DATE	$\overline{WL}$	S(WL)	S($\overline{WL}$)	RANGE	n	OBS S(WL)/$\overline{WL}$	OBS S(WL)/$\overline{WL}$	OBS RANGE/$2\overline{WL}$
9/12	0.451	0.052	0.014	0.18	12	0.12	0.031	0.20
10/12	0.537	0.052	0.013	0.19	16	0.10	0.024	0.18
11/12	0.585	0.081	0.022	0.30	13	0.14	0.038	0.26
7/1	0.581	0.056	0.015	0.21	14	0.10	0.026	0.18
8/1	0.598	0.050	0.016	0.16	10	0.08	0.027	0.13
12/1	0.608	0.056	0.017	0.21	11	0.09	0.028	0.17
13/1	0.620	0.047	0.014	0.14	11	0.08	0.023	0.11
14/1	0.595	0.044	0.013	0.16	12	0.07	0.022	0.13
15/1	0.605	0.060	0.017	0.22	12	0.10	0.028	0.18

SITE NO. 2
CONDITIONS:
AIR FLOW 1200-1400 fpm; 85K-100K cfm; TEMP 10-14° C; R.H. 70-80%; HEAVY DIESEL SMOKE; JUMBO DRILLING; LIGHT TRUCK TRAFFIC.

22/1	0.369	0.020	0.006	0.05	10	0.054	0.016	0.068
23/1	0.353	0.012	0.004	0.04	12	0.034	0.011	0.057
27/1	0.370	0.020	0.006	0.07	13	0.054	0.016	0.095
28/1*	0.379	0.032	0.009	0.14	13	0.084	0.024	0.185
29/1*	0.384	0.029	0.009	0.09	11	0.076	0.023	0.117
30/1*	0.334	0.018	0.006	0.06	10	0.054	0.018	0.090

* NIGHT SHIFT

TABLE 2

Date: 29/1   Site no. 2   Detector Efficiency   E = 0.35 ± 0.03

SPLE	TIME	FLOW RATE	COUNT RATE	KUSNETZ FACTOR	WL
1	2000	10.0 ± 1.25 ℓ/m	480 ± 10 m^{-1}	106 ± 5	0.43
2	2030	10.5 ± 1.25	502 ± 10	"	0.43
3	2100	"	466 ± 10	"	0.40
4	2130	"	462 ± 10	"	0.40
5	2200	"	432 ± 9	104 ± 5	0.38
6	2230	"	422 ± 9	106 ± 5	0.36
7	2300	10.0 ± 1.25	420 ± 9	"	0.38
8	2330	"	410 ± 9	"	0.37
9	2400	"	379 ± 9	"	0.34
10	0030	"	368 ± 9	92 ± 5	0.38
11	0100	"	384 ± 9	106 ± 5	0.35

$\overline{WL}$ = 0.384   S(WL) = 0.029   S($\overline{WL}$) = 0.009

preliminary effort to determine the total variance in the WL at two locations in an underground ventilation system. Measurements were made by the routine Kusnetz method (2), as described in Appendix 1. Samples were taken in the return air at intervals of 15 to 30 minutes over 3 to 6 hours during working shifts. Both day and night shifts were tested.

Data for one shift (selected at random) are given in Table 2. These data indicate relatively little variation in WL during the shift. Data obtained on other shifts, however, indicate a much greater variation (e.g. Site no. 1, 11/12, Table no. 1).

These results indicate that the range of values of WL observed at a fixed location varies widely, from 10% to 50% of the mean, and that the relative standard deviation $(S(WL)/\overline{WL})$ varies correspondingly from about 3% to 15%. There is no experimental evidence to suggest that the variance due to the sampling and measuring instruments was not constant throughout the tests. It therefore appears that the observed variation in the relative standard deviation reflects an instability in the actual working level. However, a preliminary application of Svedecor's F-test (6) suggests that only the extreme values of $S(WL)$ differ significantly, in both sets of data. It will be necessary to accumulate much more data for a thorough statistical analysis before any absolute conclusions can be reached.

A chi-squared test (6) applied to the data for site no. 1, excluding those obtained on 9/12 (which appear to differ significantly from the rest on the basis of a t-test*) indicates that the data fit a normal distribution. However, this test is not conclusive, because the possibility of a log-normal distribution is not excluded. Again in this case, only tentative conclusions can be drawn until more data are collected.

Excluding the data for 9/12 (for the reason given above) the grand mean, WL, for site no. 1 was estimated to be 0.588. The standard deviation, $S(WL)$, was found to be 0.060 and the standard deviation of the mean, $S(\overline{WL})$ to be 0.006. It can therefore be estimated that a single observation has a probability of about 0.32 (one in three) of deviating from the true mean by more than about 10%, and a probability of about 0.05 (one in twenty) of deviating by more than about 20%. In other words, under the sampling scheme presently used in the Elliot Lake mines (a single sample in every work-place every month, on the average), about 5% of the measured values will be in error (high or low) by 20% or more.

There is a significant risk in this situation that an erroneous decision could be made. On the one hand, men may be exposed at levels in excess of the limit when the measurement gives a value below the limit. On the other hand, costly improvements to the ventilation system may be instituted because the measurement gives a result above the limit when the true value in fact falls below the limit.

Moroney (6, p 210 ff) has given a simple method of sequential sampling to provide a better estimate of the true value. Applying his procedures, it can be shown that the number of samples

---

*Comparing the data for 9/12 with those for 10/12, the estimated t was 4.30, indicating a statistically significant difference at a confidence level of > 99.9%.

taken in each work place at each visit should be 2. Using proper
control charts, it will then be possible to decide with confidence
whether the exposure limit has or has not been exceeded, or whether
further testing is required. This should not result in a great
increase in cost or labour of sampling, as a very large part of the
total time required in sampling underground is taken up merely in
travelling to and from the sampling sites.

III. THE "INSTANT" WORKING LEVEL METER

It has long been the hope of all those responsible for
industrial analysis that a miraculous instrument would be developed
which would give an instantaneous, accurate, and precise (i.e.
reproducible) estimate of the value to be measured. Uranium mine
ventilation engineers are no exception, and in response to their
desires several "instant" working level meters (IWLM) have been
developed over the last 8 or 10 years. One of the most recent of
these instruments was designed and built at Argonne National
Laboratory (ANL) by Groer et al (7) under contract from the U.S.
Bureau of Mines (USBM). Through the courtesy of Dr. Groer of ANL
and R. Droullard of USBM, an instrument was made available for
testing in the uranium mines at Elliot Lake.

The operation of this instrument has been fully described
elsewhere (7) and it is not necessary to repeat it here. It will
suffice to say that an air sample is drawn through an Acropor
filter at a flowrate of 11 lpm for 2.00 minutes. The radon
daughters on the filter are then measured by counting the alpha
particles emitted during 2.00 minutes by RaA and RaC' using a
solid state (Si) alpha detector and 2-channel pulse height
analyzer, while the beta particles emitted by RaB and RaC are
counted using a plastic scintillator. The beta/gamma background
(recorded in a preliminary 1 minute measurement) is subtracted,
and the data are fed to hard-wired logic circuits which calculate
the total radon daughter concentration in WL, and the concentrations
of RaA, RaB, RaC, and RaC' ($\equiv$RaC) in pCi/ℓ.

The instrument is powered either from a 110 v, 60 Hz, 1 ϕ
supply or from built-in batteries that also serve as a background
shield for the beta/gamma detector. The total time required for
a measurement is 5 minutes, and the only manipulations required of
the operator are (1) loading of the filter, (2) pressing a button
to start a measurement, and (3) recording the results which are
displayed by means of LED's.

The major limitations on this IWLM in its present form
are its weight (approx. 40 lbs or 18 kg) which precludes its use as a
portable instrument for routine measurement, and a severe restriction
on the number of samples that can be taken when using battery power.

Nevertheless, underground tests in a variety of locations,
over a wide range of working levels have indicated that the results
using the IWLM agree fairly well with those obtained for duplicate
samples measured by Kusnetz' technique (Figure 1). A regression
analysis of the data on the IBM 360 computer gave the results in
Table 3.

Although the tests are not yet complete, it is possible
to draw some preliminary, tentative conclusions. First, it appears
that the precision of the IWLM may be somewhat better than that of
the Kusnetz in the range above 0.1 WL. However, below 0.1 WL it
appears that the IWLM becomes rather more erratic, probably because
of the small number of counts recorded by the alpha and beta detectors
and the correspondingly large relative standard deviations.

FIGURE 1: Comparison of Working Levels Measured by ANL Instant Working Level Meter and by Kusnetz' Method.

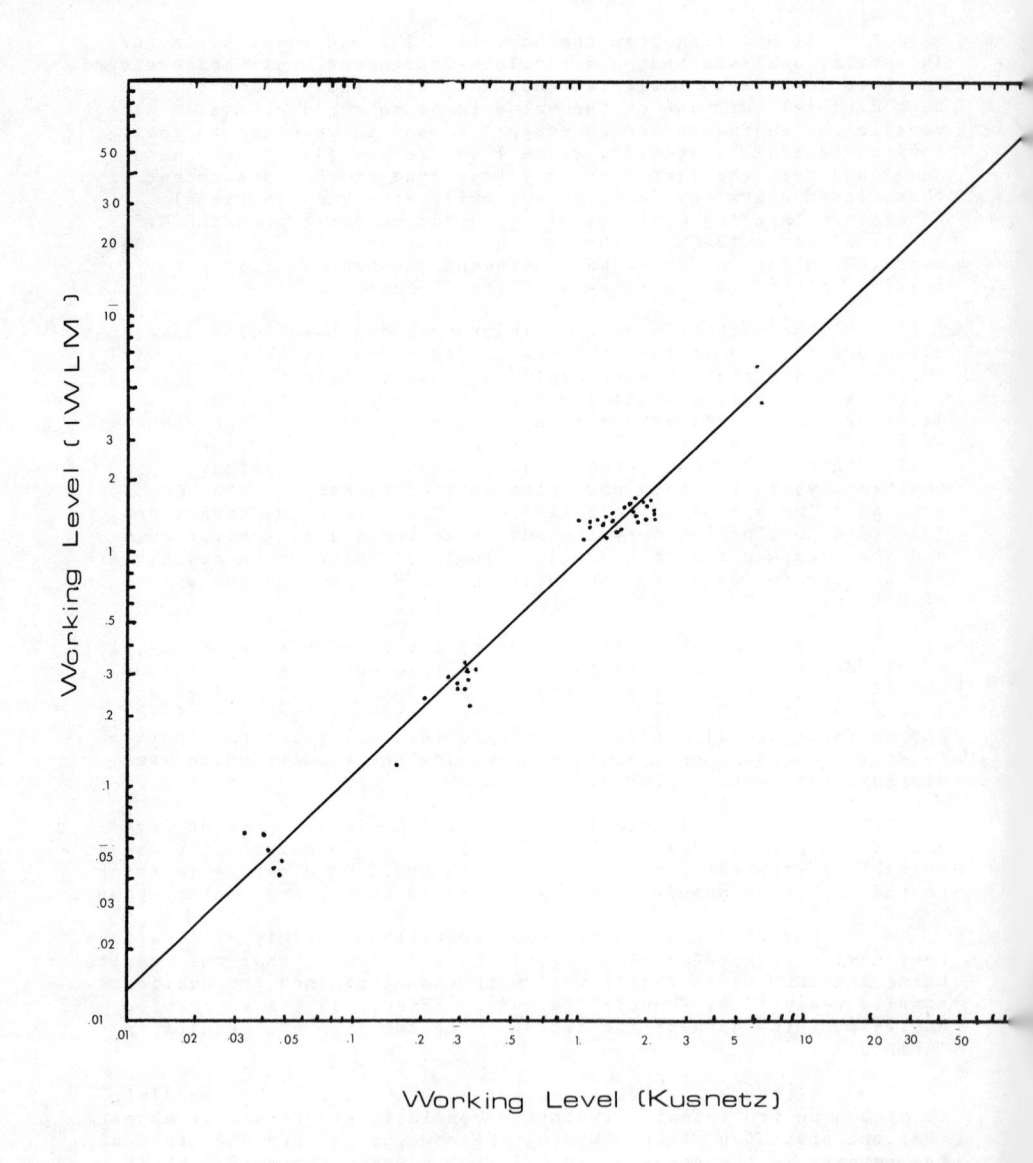

TABLE 3

Regression analysis of comparative working level data by Kusnetz' method and by IWLM:

Intercept:	0.08446
Regression Coefficient:	0.80655
Std. Error of Regression Coefficient:	0.03431
Computed T-value:	23.90594
Correlation Coefficient:	0.96323
Standard Error of Estimate:	0.30162
Standard Error of Intercept:	0.06171

Because no absolute method for measuring radon daughters is available, no absolute conclusions can be drawn about the accuracy of either method. Nevertheless, the fact that the methods appear to give results that are in agreement tends to indicate that neither method is subject to any severe bias.

Field tests are continuing, and the results will be reported, together with complete statistical analyses, as soon as the tests are completed.

1V PROBLEMS IN DETECTOR CALIBRATION

The usual method for calibration of a measurement system is based on preparation of a standard sample of known concentration using measured amounts of pure materials. Unfortunately, it is extremely difficult, for a number of reasons, to prepare reproducibly standard samples of airborne radon daughters whose concentration can be accurately known. It has, therefore, been necessary to employ other radiosotopes (usually Am^{241}) to prepare standards for the calibration of alpha detectors for use in measuring radon daughters. However, it is known (8) that the counting efficiency of Zn S (which is the most widely-used scintillator for alpha detection) for Am^{241} is probably not identical with that for RaA, and that it is even less likely that the efficiency for RaC' is the same as that for Am^{241}. Thus it would be preferable to have a standard prepared using RaA and RaC'.

Some of the difficulties involved in preparing a reliable standard of RaA and RaC' are:

(1) Ensuring that the radon emission from the source (sample of uranium ore of known weight and uranium content, or known weight of radium) is complete;

(2) Ensuring that the radon daughters are in radioactive equilibrium with the radon gas;

(3) Ensuring that the radon daughters are not removed from the system by any means other than radioactive decay;

(4) Provision of a stable, reproducible aerosol as a substrate for radon daughter deposition to facilitate sample collection and reduce the possibility of undetected losses (item 3);

(5) Ensuring that the volume of the aliquot collected from the system is accurately known;

(6) Ensuring that the collection of the airborne daughters from the volume sampled is 100% efficient.

The proposed facility, which is described below, represents an effort to reduce to a minimum the errors resulting from the first four items above. It will consist of a steel drum containing an accurately weighed quantity (250 to 300 kg) of uranium ore which has been carefully assayed. A stream of air (about 1 ℓ/min) will pass through the ore and into a second large steel or aluminum drum (vol. ca. 200 ℓ) containing a source of condensation nuclei (e.g. a hot nichrome heating element). This will provide time for the growth of radon daughters into radioactive equilibrium with radon, and an aerosol on which they can deposit. The air from this growth drum will be drawn through a filter and recirculated to the ore barrel by means of an oil-less rotary vane pump. If it is necessary or desirable, a third, larger, walk-in chamber could be installed in the line between the second drum and the filter and circulating pump. Auxiliary equipment (probes for sample collection, flowmeters, manometers, circulating fans in the growth chamber, check valves for admission of make-up air during sampling, etc.) will be provided at suitable points in the circuit.

It is hoped that this system can be developed into a reproducilbe source of accurately known concentrations of radon daughters. It may even be possible to vary the concentration of daughters by adding fresh air to the system at known rates and bleeding off corresponding amounts of mixture.

V SUMMARY

The sources of error in measuring airborne radon daughters in uranium mines have been examined qualitatively and semi-quantitatively by a number of authors, but there is relatively little quantitative data available, particularly for Canadian mines. A preliminary effort has been undertaken to collect information on the total variance in measured values in typical Canadian mines, and the initial results indicate that the relative standard deviation can vary widely from one location to another. The maximum observed was about 15%. Using this information, it is shown that it would be preferable to collect duplicate samples in the routine ventilation surveys (rather than single samples as at present), and to use properly designed control charts to determine compliance or lack of compliance with the exposure limits.

Comparative tests of an IWLM showed good agreement with the routine method, and indicated slightly better precision. However, more data are required before any firm conclusions can be reached.

Field tests are continuing in an effort to determine the causes of the observed variations, and to establish more clearly whether there is an advantage to be gained from use of an IWLM. At the same time it is planned to try to develop an accurate and reproducile method of generating radon daughters for instrument calibration.

APPENDIX 1

Kusnetz Method for Measuring Working Level

1. A sample of airborne dust was collected by drawing air (using a battery-powered rotary vane pump) through a 25 mm diameter Millipore AA membrane filter (pore size 0.8 μm). The flow rate (11 to 15 ℓ/m) was measured using a Dwyer model VFA Visi-Float flowmeter. Sampling time was 3.00 minutes. The flowmeter accuracy was estimated by the manufacturer to be 5% FS.

2. The sample was allowed to decay for approximately 60 minutes; the decay time, t_1, was measured (from the end of sampling) to the nearest minute. At the end of the decay period the alpha-emission rate of the sample was measured using a calibrated (see para. 4) alpha detector and sealer.

3. The working level was calculated from the formula

$$WL = \frac{C}{V \times E \times K \times t_2}$$

where C = no. of counts observed by the alpha detector in t_2 minutes,

V = volume of air sampled, litres (product of observed flow rate, ℓ/min. and sampling time, min.)

E = calibration factor for the alpha detector (ratio of the observed counts/min. for the Am^{241} standard to the known disintegrations/min.), and

K = Kusnetz' correction factor (2); this factor was estimated to be

$$K = 230 - 2 (t_1 + 0.5\ t_2) \quad (40 \leq t_1 + 0.5\ t_2 \leq 70)$$

$$K = 195 - 1.5 (t_1 + 0.5\ t_2) \quad (70 \leq t_1 + 0.5\ t_2 \leq 90)$$

4. Calibration of the alpha detector was done using a standard Am^{241} source prepared by evaporating an aliquot of standard Am^{241} solution (Amersham-Searle no. S7/34/56/γ, 0.8748 ± 0.0044 μCi in 0.5 N, HNO_3, diluted to 10.00 mℓ with 0.5N, HNO_3) on a stainless steel planchet. The diameter of the deposited source was limited to the diameter of the dust sample on the membrane filter, in order to keep the same source/detector geometry for samples and standard.

REFERENCES

1. Loysen, P. Errors in measurement of working level; Health Physics 16, 629, 1969.

2. Kusnetz, H.L. Radon daughters in mine atmospheres: a field method for determining concentration; Am. Ind. Hyg. Assoc. Quart. 17, 85, 1956.

3. Breslin, A.J. Proposed simplification in the method of monitoring uranium miners' exposures to radon daughters. Technical Memorandum 68-7, Health and Safety Laboratory, U.S. Atomic Energy Commission, New York Operations Office; April 1968.

4. Breslin, A.J. and Wainstein, M.S. Characterization of uranium mine atmospheres; progress report 2. Technical Memorandum 68-4, Health and Safety Laboratory, U.S. Atomic Energy Commission, New York Operations Office, March 1968.

5. Thomas, J.W. Determination of radon progeny in air from alpha activity of air samples; U.S. Atomic Energy Commission report HASL-202; Health and Safety Laboratory, New York, Nov. 1968; Determination of the working level of radon daughters by the modified Tsivoglou method. Technical Memorandum 71-15, Health and Safety Laboratory, U.S. Atomic Energy Commission, New York Operations Office, July 1971; Measurement of radon daughters in air by alpha counting of air filter, U.S. Atomic Energy Commission report HASL-256, Health and Safety Laboratory, New York, April 1972.

6. Moroney, M.J. Facts from Figures, 3rd ed., Penguin Books Inc., Baltimore, 1956.

7. Groer, P.J., Keefe, D.J., McDowell, W.P. and Selman, R.F Rapid determination of radon daughter concentrations and working level with the instant working level meter, International Symposium on Radiation Protection in Mining and Milling of Uranium and Thorium, Bordeaux, France, Sept. 1974.

8. Handbook on calibration of radiation protection monitoring instruments, Technical report series no. 133, International Atomic Energy Agency, Vienna, 1971.

Discussion

A Continuous Monitor for Underground Ventilation Systems

R.A. Washington, Research Scientist

In response to the interest generated at the recent NEA Symposium on personal dosimeters and area monitors for radon and radon daughters by a brief mention of the continuous ventilation monitor system which is being developed at Elliot Lake jointly by Rio Algom Mines Ltd., Mining Research Laboratories and Conspec Controls Ltd. the following brief description is provided.

The system consists of a network of sensors located strategically in the mine ventilation system, various signal transformation and data transmission devices underground and on surface, and a Nova computer on surface to monitor the sensors, record and process the data, and provide an output to a teleype printer/tape punch. The sensors in the system at present include anemometers (various types), fire detectors (ionization type), thermocouples, and detectors for radon, carbon monoxide and methane. It is planned to install barometric pressure sensors, detectors for sulphur dioxide, hydrogen sulphide, oxides of nitrogen, oxygen, and relative humidity in the near future. Continuous sensors for dust and radon daughters are under development for installation later this year.

The system operation is as follows. At intervals of approximately 2 minutes, the output of each sensor is monitored by the computer and the data is processed to provide a running average, which is stored in memory. If the response of a sensor is observed to be outside the preset limits for more than a predetermined number of accesses, an alarm message is immediately printed to identify the sensor, whether it is high or low, the date and the time. The location and status of each sensor (green lamp if within limits, red lamp if without limits) is also displayed on a mimic board to enable ready identification of the location of any problem.

At the end of each shift, a shift-end report is printed automatically giving the shift average value for each sensor, its identification, and the upper and lower limits. Each day, the cumulative averages are also punched on tape, and this tape can be fed back to the computer each month to generate a month-end report of average underground conditions.

Other functions are also available to the ventilation engineer (eg. the printing of the current value for any selected sensor), and additional functions could be added as necessary by modifications to the program. It must be emphasized that the system is, as yet, in a very early stage of development, and may undergo considerable modification before its final form is reached. Additional information may be obtained from this laboratory, or from the research contractor (Conspec Controls Ltd., 44 Martin Ross Ave., Downsview, Ontario M3J 2K8). A full report on the development to date is in preparation, and should be published shortly.

PRESENT PRACTICES OF THE DEPARTMENT OF NATIONAL HEALTH AND WELFARE
FOR THE AREA MONITORING OF RADON AND DAUGHTER PRODUCTS

H. Taniguchi, Ph.D.,
Nuclear Safety Division,
Radiation Protection Bureau,
Health Protection Branch,
Department of National Health and Welfare,
Ottawa, Canada.

Introduction

This communication outlines the present practices of the Department of National Health and Welfare for the area monitoring of radon and daughter products. The measurements have been largely directed towards specific studies pertaining to occupational and environmental health and not to any extent for the routine monitoring of uranium mines.

This Department first became involved with the measurement of radon and daughter products in mine atmospheres in 1958. At that time, A.J. de Villiers and J.P. Windish[1] expanded an on-going study of lung dust diseases among miners in Newfoundland to include radiation exposure in the fluorospar mines of St. Lawrence. An average radon daughters concentration in the range 2.5 to 10 Working Levels was found, even though the uranium content of the ore was less than 0.005% U_3O_8. It was concluded that radon was not being emanated directly from the exposed rock walls, but, that it was being released from mine waters into which radon had dissolved at more distant locations.

In subsequent years, measurements of radon and daughter products have been carried out in other non-uranium mines, including a columbium ore mine and a tin ore mine. In these mines, also, elevated radon daughter levels were found to be associated with high levels of radon dissolved in mine waters. Moreover, the ventilation systems were not specifically designed to cope with radon daughters because the presence of radon was not anticipated.

At the two active uranium-producing mines at Elliot Lake, Ontario, intercomparison measurements have been carried out with the operators directly in the mines and mills[2]. These field studies were undertaken to evaluate the quality of the monitoring data being generated by the mines for the calculation of cumulative occupational exposures to radon daughters.

Recently, the same techniques have been applied to measure radon and daughter products in the homes and buildings of a community located near a major uranium ore-body. In this study, it was found that below ore-grade levels of uranium in crushed development rock used as fill and in driveways generated sufficient radon to accumulate in the homes. In addition, small, natural local occurrences of radioactive material appeared to contribute in some locations.

The procedures used in the foregoing examples are briefly described in the following sections.

Present Practices

A. Radon Gas

Radon gas concentrations are determined by the Lucas chamber method[3]. In the original design, a 5 cm diameter by 6 cm long metal cylinder with a hemi-spherical top and a quartz window on the bottom served as a chamber for radon. The active volume was about 100 ml. The inside of the metal portion was coated with zinc sulphide (silver-activated) scintillator. A stopcock capable of maintaining a vacuum was sealed to the top for emptying and filling the cell.

At the desired sampling location, the stopcock of a pre-evacuated cell is opened to admit air through a filter until atmospheric pressure is reached. After the radon daughters have fully grown in, the scintillation rate is measured with a multiplier phototube and scaler-timer.

During field studies, a 2.5 cm diameter phototube in a portable, self-contained battery-operated detector is used. In the laboratory, a 5 cm phototube is used in a system having a printing scaler and timer. Any desirable counting period up to 999.9 minutes may be selected. At the end of the pre-selected time, the total counts accumulated are printed out, the scaler is reset automatically and the counting cycle is repeated until stopped by the operator. The total elapsed time is also registered to measure the in-growth period for the radon daughters.

The radon cells are calibrated by de-emanating known amounts of radon from standard solutions of radium-226 obtained from the U.S. National Bureau of Standards. The average efficiency observed was 5 cpm per pCi of radon-222 after equilibrium with daughters has been reached. When new, the background of the cells averaged less than 0.1 cpm.

B. Short-lived Radon Daughters

Short-lived radon daughters concentrations are determined by the Kusnetz field method[4], modified by using a scaler-timer in place of a ratemeter to measure the alpha count rate. In this well-known procedure, the air to be sampled is drawn through a filter at a known flow rate for a known time. After allowing the collected radon daughters to decay for a period of 40 to 90 minutes, the alpha count rate is determined with a zinc sulphide alpha scintillation detector. The pulsed output from a multiplier phototube viewing the scintillations is amplified and recorded on a scaler-timer. The detector is calibrated using a source prepared from a standardized solution of americium-241.

The collection system presently in use consists of a rotary-vane vacuum pump driven by a 6 V D.C. motor. Two rechargeable lead batteries in parallel allow at least twenty 5-minute duration samples to be collected. A rotameter-type flowmeter, calibrated against a wet-test meter, is used to measure the flow rate. A flow rate of 10 liters per minute is usually chosen to give a large volume for better counting statistics, especially for samples from homes and buildings. For these studies, the need for minimum size and weight is not as pressing as it is for use in mines.

The filter used is a 25 mm diameter membrane type having a pore size of 0.8 μm. The filter holder itself is readily detachable, permitting a number of them to be pre-loaded for rapid use in the field.

The possible sources of errors in this method have been reviewed by Loysen[5]. In his review of available instrumentation, Budnitz[6] quotes studies which confirm that with appropriate attention to the factors, the method is intrinsically accurate and sensitive.

C. Other Natural Radionuclides

The low-level environmental radioactivity laboratory of the Radiation Protection Bureau in Ottawa is well equipped to carry out radioanalytical procedures for the long-lived precursors of radon and for the long-lived daughter products from the decay of radium-C. Specifically, these include uranium, thorium-230, radium-226, lead-210, bismuth-210 and polonium-210. The instruments on hand include low-background alpha and beta counters, NaI(Tl) scintillation gamma spectrometers, high resolution Ge(Li) gamma spectrometers, an alpha spectrometer, a fluorometer and a liquid scintillation counter. A vacuum system for concentrating radon from air, water and exhaled breath for subsequent measurement in Lucas cells is available. These laboratory procedures have provided useful analytical data to support studies of environmental levels of radon and daughter products which are of interest to public health.

Recent Studies

Recent studies of radon and daughter products in non-uranium mines have shown that surprisingly high levels can accumulate. In a columbium mine, the concentration of daughter products ranged from 3.4 WL at an ore ramp to 19 WL at a new face and 29 WL at the end of a cross-cut. These levels were greatly improved by the judicious use of ventilation.

In a tin mine, which is presently in a stand-by state with only a sump pump operating pending further development, Working Levels started at 1.1 WL half-way into the entrance incline and progressively increased to 5 WL, 11 WL, 47 WL and 63 WL at the farthest point from the entrance.

Radon gas samples were also taken at the same time and locations. This allowed the calculation of the ratio of (100 x WL) to radon concentration in pCi/l. This fraction of theoretical equilibrium ranged from 0.3 in the entrance incline to 0.98 in the interior. This confirmed the lack of any air movement and clearly pointed to the need for substantial ventilation.

As a contrast to the situation in these non-uranium mines, some findings from a systematic survey of a uranium mining community are summarized. Approximately 20% of the 500 homes and buildings were found to have radon daughter Working Levels over an arbitrary value of 0.03 WL. At least 11 were over 1 WL, with the highest over 2 WL. The source of the radon was attributed to the use of below-grade mine development rock which was crushed and used as fill for driveways, sidewalks and other construction purposes. Contribution from small natural occurrences of uranium-containing outcroppings were also not precluded.

It was of interest to measure the fraction of equilibrium values of Working Levels which were reached in these homes. This fraction ranged from 0.3 to 0.5 in the basement of the majority of those homes which contained measurable amounts of radon. These ratios imply a reasonable circulation of air, partly due to the air intake for oil burners of furnaces which were in operation during the measurement period.

Summary

The present practices of the Radiation Protection Bureau for the measurement of radon and daughter products have been briefly described. For radon gas, the Lucas chamber method is in use. Short-lived radon daughter products are determined by the modified Kusnetz method. These field methods are supported by radioanalytical procedures carried out at the environmental radioactivity laboratory.

Some recent studies using these methods have been briefly summarized. Concentrations of daughter products up to 29 WL were found in a columbium mine and 63 WL in a tin mine under development. The level of radon daughters in some homes in a uranium mining community ranged up to 2 WL.

References

1. de Villiers, A.J. and Windish, J.P., British J. Indust. Med. 21, 94 (1964).

2. Windish, J.P. and Taniguchi, H., Report to the Division of Mines, Ontario Ministry of Natural Resources (1974).

3. Lucas, H.F., Rev. Sci. Inst. 28, 680 (1957).

4. Kusnetz, H.L., Am. Ind. Hyg. Quarterly 17, 85 (1956).

5. Loysen, P., Health Phys. 16, 629 (1969).

6. Budnitz, R.J., Health Phys. 26, 145 (1974).

Discussion

A question was raised on whether the mines mentioned in this presentation were normally wet or dry. Both the columbium mine and the tin mine were very wet, with mine water coursing through most drifts. Water from a drift face in the columbium mine had 220 nanocuries of radon-222 per litre. In the tin mine, 60 nanocuries per litre was measured. Concentrations of 4.2 and 12.8 nanocuries per litre have been reported for the fluorospar mine in St. Lawrence, Newfoundland.

It was also noted that dry mines can also accumulate high levels of radon and daughter products, as found in the United Kingdom.

LIMITATION OF EXPOSURE TO THE COMBINED RISKS FROM RADON DAUGHTERS, GAMMA RADIATION AND OTHER RADIOLOGICAL HAZARDS

W. R. Bush
Atomic Energy Control Board, Canada

Although gamma radiation and other radiological hazards in uranium mines are generally considered to be negligible compared to the hazard from radon daughters, gamma radiation may be the predominant hazard when mining high grade ore. We can expect, during the next few years of Canadian mining, to encounter a wide variety of conditions that will fall between the two extremes, from gamma negligible to gamma predominant. It is therefore necessary to develop a formula for dealing with the combined hazard from radon daughters and gamma radiation, and perhaps from other radiological hazards as well.

Confining ourselves for the moment to the combined hazard of radon daughters plus gamma radiation, the simplest formula would combine the risk to the lungs from the radon daughters (for which the maximum permissible annual exposure is 4 WLM) and the gamma radiation (for which the maximum permissible dose to the lungs is 15 rem/yr):

$$\frac{WLM}{4} + \frac{\gamma}{15} \not> 1 \quad (1)$$

The maximum permissible annual exposure to radon daughters would then be $4 - \frac{4}{15}\gamma$, where "γ" is the gamma exposure in roentgens or rems. The use of this formula would not have a significant effect on the current exposure limit for radon daughters until the annual gamma exposure approached 2 rem or so, which might be a desirable feature. However, the formula would not be appropriate for application to miners with gamma exposures approaching the 5 rem/yr limit because it would permit an annual exposure of 2.7 WLM in addition to the maximum permissible whole-body exposure of 5 rem. This would probably not be acceptable under present Canadian regulations.

In order to more fully account for high gamma exposures, one could use a formula such as:

$$\frac{WLM}{4} + \frac{\gamma}{5} \not> 1 \quad (2)$$

The maximum permissible WLM exposure would then be $4 - \frac{4}{5}\gamma$.

A major disadvantage of this formula is that it would virtually never permit annual exposures as high as 4 WLM, whereas it is implicit in the 4 WLM limit that some gamma exposure is acceptable in addition to an annual exposure of 4 WLM.

It may be necessary to select a middle course between the two approaches outlined above. Let us assume that the miner exposures on which the epidemiological studies were based included 1 rem of gamma radiation for every 4 WLM. This leads to a very simple formula:

$$\text{WLM} + \gamma \not> 5 \text{ (but if } \gamma < 1, \text{ WLM} \not> 4) \quad (3)$$

By this formula, if γ = 1 rem, max WLM = 4
2　"　　"　"　= 3
3　"　　"　"　= 2
4　"　　"　"　= 1
5　"　　"　"　= 0

I suggest that the possible error in assuming 1 rem of gamma dose for every 4 WLM is more than compensated for by the mathematical simplicity of this formula. To support this rather simplistic approach, I remind you that mathematical simplicity was the primary reason for the international acceptance of the linear, non-thresold, non-repair theory instead of more biologically logical but mathematically difficult theories.

The basic formula, $\text{WLM} + \gamma \not> 5$ may have to be modified in practice to account for other radiological hazards, and this could perhaps be done as follows:

$$\text{WLM} + \gamma \not> 5 - x \quad (4)$$

where $x = \dfrac{U - U_o}{MPC_u} + \dfrac{Th - Th_o}{MPC_{Th}} + \dfrac{Rn - Rn_o}{MPC_{Rn}}$

where U, Th and Rn are measured concentrations of uranium, thorium and radon in air, and U_o, Th_o and Rn_o are representative of the concentrations that existed at the time of the exposures on which the derivation of the 4 WLM limit was based.

This sort of compensation for the presence of high concentrations of U, Th or Rn would obviously be inadequate at concentrations approaching the respective MPC values, but perhaps it will prove to be suitable at the concentrations to be encountered in well-ventilated mines. This remains to be determined as measurements in high-grade mines become available. I would caution against a more complicated formula unless field measurements prove this to be necessary, because the epidemiological and dosimetric bases of the 4 WLM limit are simply too weak to justify a very complicated correction for radiological hazards other than gamma radiation. This is not to say that all the radiological hazards should not be measured; they should be measured, at least annually, in order to regularly confirm that they are in fact negligible.

The above proposal is not presently part of the Canadian exposure guideline system, but is offered here as an indication of the direction in which the Canadian regulatory position might evolve.

Discussion

The discussion is summarized below in point form.

i. Equation (1) conforms to the critical organ concept presently recommended by the ICRP.

ii. Equation (2) is more-or-less in accord with the additivity of risks concept that is expected to be recommended by the ICRP within the next year or so.

iii. Gamma radiation should be ignored if the annual exposure is only a small fraction, say 1/3 or 1/10, of the maximum permissible dose.

iv. If the WLM limit is reduced to account for gamma exposure, why not reduce it further to account for the annual X-ray examinations required of miners, and why not reduce it even further to account for the risks of exposure to diesel fumes and silica?

v. It is very difficult to meet the 4WLM annual limit in some mines, and it might not be possible to comply if the limit is reduced to account for gamma radiation. Moreover, there is no practical way to protect against gamma radiation in mines.

Mr. Bush replied that the traditional means of protecting against gamma radiation (time, distance and shielding) are expected to be economically viable in the high-grade mines where gamma radiation is the dominant hazard. Vehicle cabs can be shielded, ore in open pits can be shielded with sand, and techniques can be developed to increase worker-to-ore distances and to reduce exposure times.

ASSESSMENT OF AIRBORNE RADIOACTIVITY IN ITALIAN MINES

G. Sciocchetti

Dipartimento Radiazioni, Centro Studi Nucleari Casaccia, CNEN
Rome (Italy)

Abstract

The present paper reports on the activity carried out at the Comitato Nazionale per l'Energia Nucleare in the field of radiological exposure existing in some Italian mines. The activity included:

- a national survey of airborne radioactivity in mines selected on a geographical and geological basis;
- an investigation on the mine environments to identify dosimetric parameters and to plan monitoring programs;
- the development of measuring methods and techniques and of a particular instrumentation.

The data on the national survey, the measuring methods and techniques are briefly illustraded, while the preliminary results, for which further study is in progress, are discussed.

I - Introductory remarks.

The Comitato Nazionale per l'Energia Nucleare (CNEN) started since 1970 a program aiming at the investigation of the possible radiological exposure existing in some Italian mines. The purpose of the study was that of obtaining a general overview of the radiological situation of large groups of persons chronically exposed to low levels of radiation. As it is well known, the radiation exposure is caused by the inhalation of radon and Rn daughters always present in variable concentrations in the mine environment. The radon gas is the decay product of Ra-226 of the Uranium series, always present in variable amounts in rocks. Its emanation is depending on the geolithological characteristics and on the mechanical conditions of the rocks, in addition to the Uranium content of the ore. For instance, the radon emanation from granitic rocks is remarkably higher than that from limestones.

The tectonic faults and the ground waters containing the dissolved gas are specific vectors for radon. These factors can assume a particular importance in the presence of low grade ore and mainly in non-uranium mines. The radon concentration in the air depends also on the atmospheric conditions, particularly on the ventilation rate and on the type of mining operations.

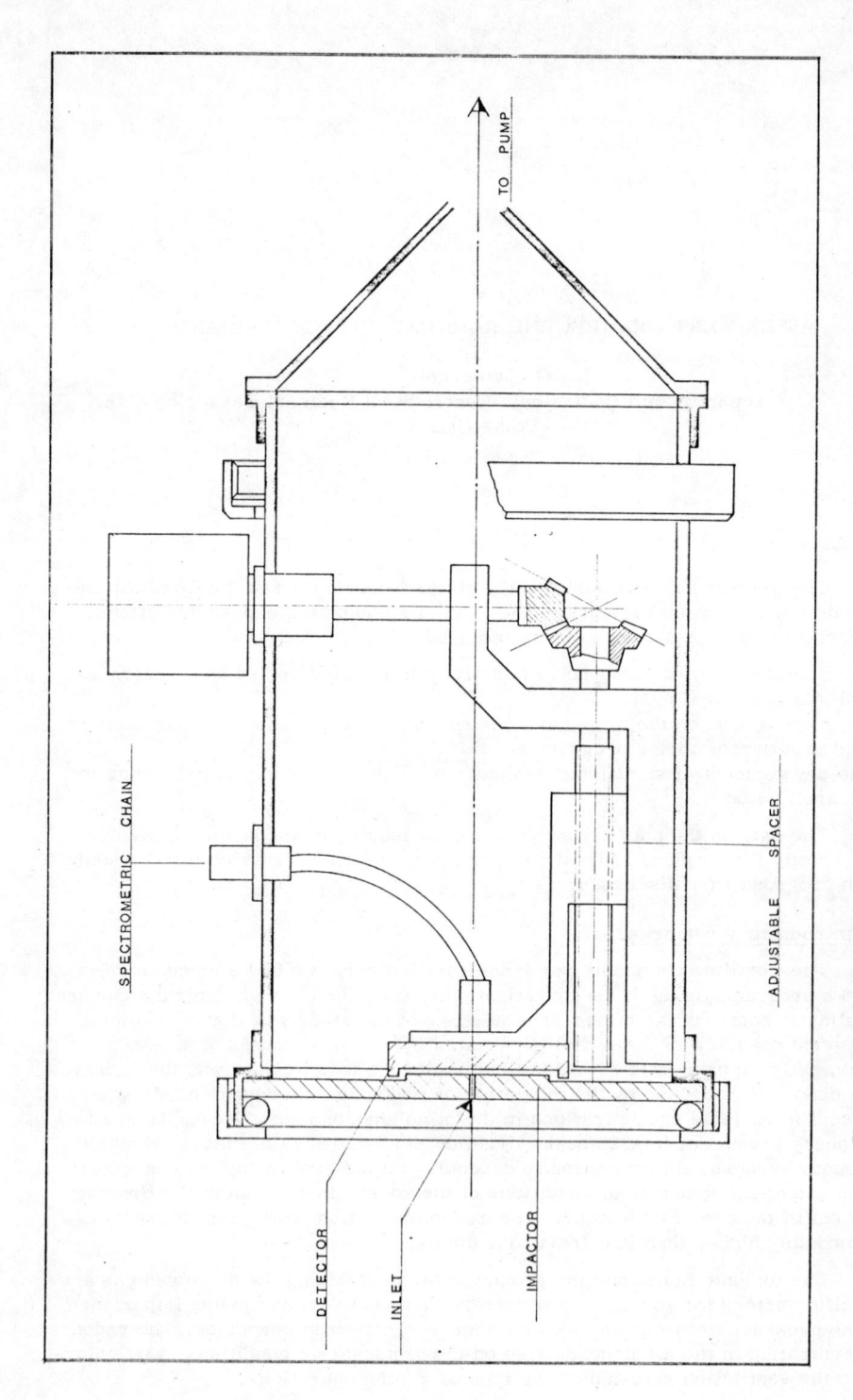

Fig. 1 - RaA ultrafine particles measuring device.

The ventilation rate is also important in determining the radiological exposure, because it affects the equilibrium degree between radon and its daughters.

The program included a national survey on airborne radioactivity in underground mines. The survey was planned according to geographical and geological criteria, so that the correlation between geological characteristics and radon concentration could be investigated.

Further investigations are in progress on environmental parameters, among which, for instance, the atmospheric conditions of the mines able to affect the dosimetric factors (equilibrium factor, uncombined fraction of radon daughters, etc.).

The aim of the whole study consists in a basic approach for assessing the individual dose of miners and for planning operational procedures for the radiological surveillance in non-uranium mines.

In this paper preliminary data and results are briefly summarized.

2 - Measuring methods and techniques

All the measurements were carried out with the supervision and the technical assistance of the Radioactive Gases Unit of the CSN-Casaccia (CNEN), where standardized dosimetric techniques and specific instrumentation were on purpose developed.

The preliminary assessment of the air quality in the mine environments was accomplished through a radon concentration survey. The measurements were carried out with scintillation chambers, coated with a ZnS (Ag) layer.

The radiation exposure of miners was evaluated from the environmental data, determining the equilibrium degree of radon daughters. The measurements were routinely carried out with the "total alpha energy method" of Markov (1). The method does not require single determinations of the concentration of each Rn-daughter and is very rapid; in fact, it implies only one counting of the filter activity in the interval ranging from the 7th to the 10th minute after the end of a five minute collection of the air sample. This method, furthermore, is able to determine the RaA concentration in three minutes.

The accuracy and precision of the Markov method, used for determining the "equilibrium equivalent Radon" EER, were tested against the Tsivoglou method and the spectrometric method. The tests confirmed the reliability of the method selected for the routine monitoring.

Some physical characteristics of the mine atmospheres are also being investigated, with some emphasis on the diffusion coefficient of ultrafine particles and on the so-called "unattached" fraction of radon daughters (2).

The "unattached" fraction of Po-218 (RaA) was determined with a special spectrometric technique. The instrument (Fig. 1) is essentially a modified version of the round-jet impaction stage described by Mercer (3), in which the Po-218 ions plate the active surface of a Silicon barrier detector, instead of a metal disc or a filter surface. The direct measurement of the sample collected on the detector surface results in an increased sensitivity, owing to the short life of RaA, and in a higher resolution, due to the lacking filter absorption.

3 - Survey and Sampling Standardization

The mines selected for the survey were representative of the geographical and geological features of the Italian territory. The different mineralizations and the type of material mined were also taken into account.

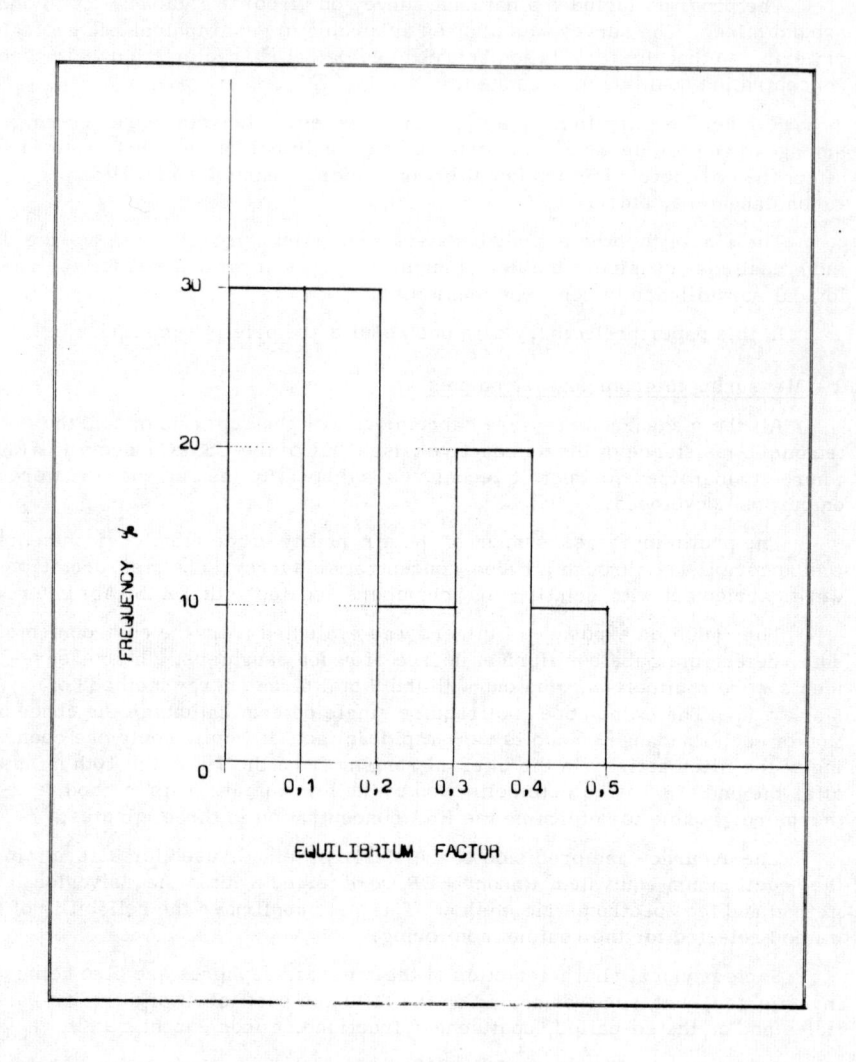

Fig. 2 - Distribution of equilibrium factors.

Eighteen mines with different mineralizations were examined. In these mines various materials, such as fluorspar, iron, coal, sulfur, are extracted (4). The spatial and temporal variations in the concentration of radon and Rn-daughters were observed in representative places. The sampling pattern, carried out according to the representativeness of the inhaled air, was standardized regardless of the type and the size of the mine under examination.

Sampling spots were chosen to evaluate the influence of the location, as main galleries, drifts, etc. and the influence of the type of mining operations, as drilling, blasting and loading, etc.

The temporal variations of the airborne radioactivity were studied in fixed locations in the main gallery of mine A, where crosscuts connecting various active areas are generally converging.

4 - Results

In Table I the results of the measurements carried out in eighteen underground mines throughout Italy are reported. The values represent the maximum and mean concentrations of radon found out in the atmosphere of the mines.

The mines were divided into three groups, according to the geological characteristics of the Italian territory. The first two groups include the areas of the Alps, of Sardinia and of Apennines.

The Uranium prospecting areas are grouped separately: the high values of radioactivity, in fact, are due to the presence of a relatively high degree ore in well defined deposits.

It is possible to observe a clear correlation between the radon concentration levels and the orogenetic belt, in addition to the influence of the type of the host rocks or gangues peculiar for every ore body. In fact, the mineralizations in the Alpine range and in Sardinia, related to the pre-Alpine orogeneses (mines A, B, C, D, E, G and H of Group I of the non-Uranium mines) show remarkably higher Rn concentrations. The values of mine F, in particular, may be explained by the substantially higher content in U_3O_8 of the lignites of the Sulcis area (Sardinia) (5).

The relatively lower values of mines G and H of the first group could be explained on the basis of the host rock. Similarly, the remarkably lower values of mine Q could be explained with the presence of a gangue of prevailing quartz type.

The mineralizations of the Apennines, Group II, related to the Alpine orogenesis show remarkably lower Rn concentrations (mines I, L, M, N, O, P, and R).

The preliminary data showed that some high values of Radon concentration required further investigations to assess the potential radiation exposure.

The values of radon and Rn-daughters concentrations expressed in Working Levels are reported in Table II. Fig. 2 shows the values of the equilibrium factors obtained in various conditions and for different working areas. The diagram demonstrates that radon has never been found in equilibrium conditions with its daughters and that these non-equilibrium conditions cover a wide range; however, they are all lower than 0.5. The data reported in this paper are non-homogeneous, even if the partial accumulation of values near the origin seems to confirm the practical possibility of measuring radon for an operational monitoring.

A mean equilibrium factor F could be established for each mine with better accuracy and on sounder bases.

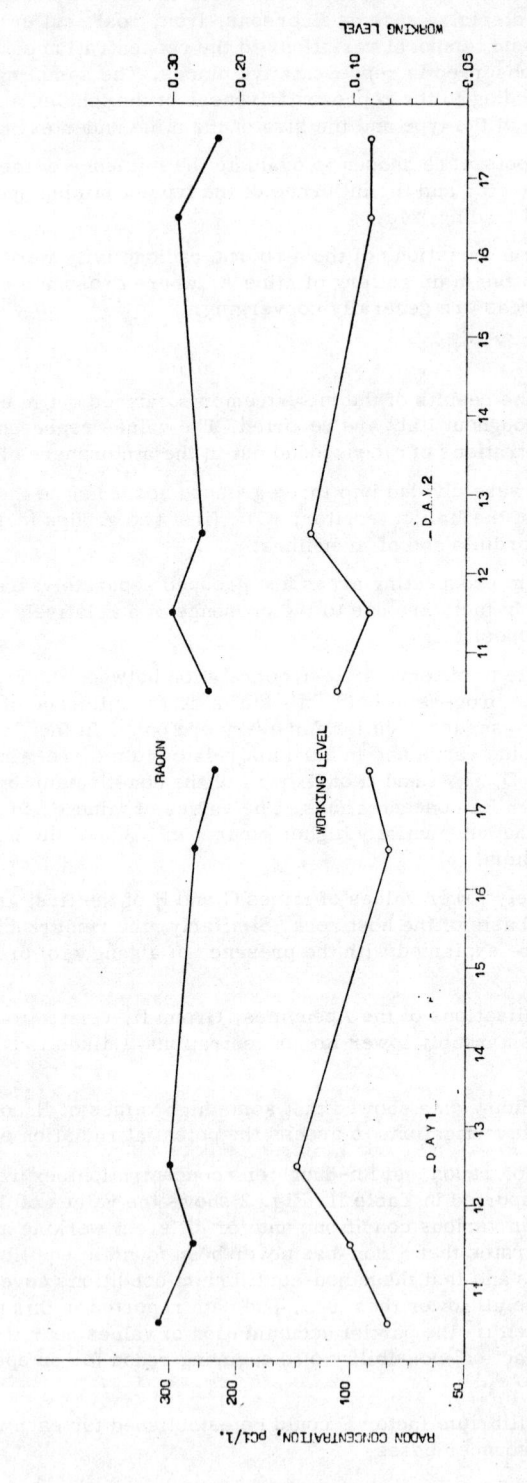

Fig. 3 – Variation of Radon concentration and working level with time at fixed sampling locations.

A systematic study on the radiological characteristics of the mine environment is underway. The parameters considered would be, among others, the atmospheric stability, and the various dosimetric factors, such as the equilibrium factor F, the ratios of the equilibrium among Rn-daughters, the free fraction f, etc. (6). The study aims to obtain criteria to assess the individual radiological exposure, necessary for planning a monitoring programme and for epidemiological studies.

An underground monitoring station was set up in mine A, where the highest radioactivity levels were found.

The sampling location was in the main gallery, in order to get mean values of the radon and Rn-daughters concentrations, representative of the mine environment. This selected area was characterized by a high ventilation rate. The data presented were collected in different working hours for two days. In Fig. 3 the temporal variations of Rn-concentrations and of Working Levels and in Fig. 4 the ratios between RaA, RaC and Rn are plotted. In Table III the variations of the equilibrium ratios of the Rn-daughters are reported, together with the values of the equilibrium factors.

The preliminary data show interesting features, even if further investigations are needed to confirm them. For instance, the atmosphere in selected places shows some stability degree, characterized by the reproducibility of diurnal cycles of Rn concentrations, of Working Levels and of the ratios of Rn-daughters concentrations. Moreover, the data show that radon concentrations are somewhate more stable than Working Levels. The variations in the ratio between RaC and Rn are remarkably lower than those of RaA to Rn ratios Rn-daughters equilibrium ratios present only slight variations. The collected values of equilibrium factors shows a narrow distribution, which may assume some significance for the methodology of radon monitoring to assess the radiological exposure.

The determinations of the unattached fraction of Po-218 (RaA) presents some difficulties, because the low levels of airborne radioactivity existing in nonuranium mines. Consequently, a particular technique was developed in our laboratory providing the direct measurement of unattached Po-218 ions during the collection of samples.

The preliminary results seem to confirm the results recently obtained in France (7) (8), in Japan (9), USA (10) and USSR (11), that is to say that the mean value of 0.10 postulated by ICRP Committee II (1959) Report is somewhat high and that it is more likely that the value may range around 0.03. Our preliminary investigations resulted in a value on the unattached fraction of RaA ranging from 0.02 to 0.04.

The external radiation exposure was also measured with a Spiers ionization chamber. Generally it resulted equivalent to the local background, with the exception of two Uranium mines and some Sicilian mines where higher levels have been found.

5 - Conclusions

The preliminary results reported in the present paper confirm that the mine environment, as defined in Appendix B of ICRP Publication 2, is characterized by "a variable, but essentially continuous contamination of the working place as a result of normal operations".

The contamination levels depend on several parameters, among which, the emanation rate of radon from the mine walls and rocks. For low level environments, such as in nonuranium mines, the role of the geological and the

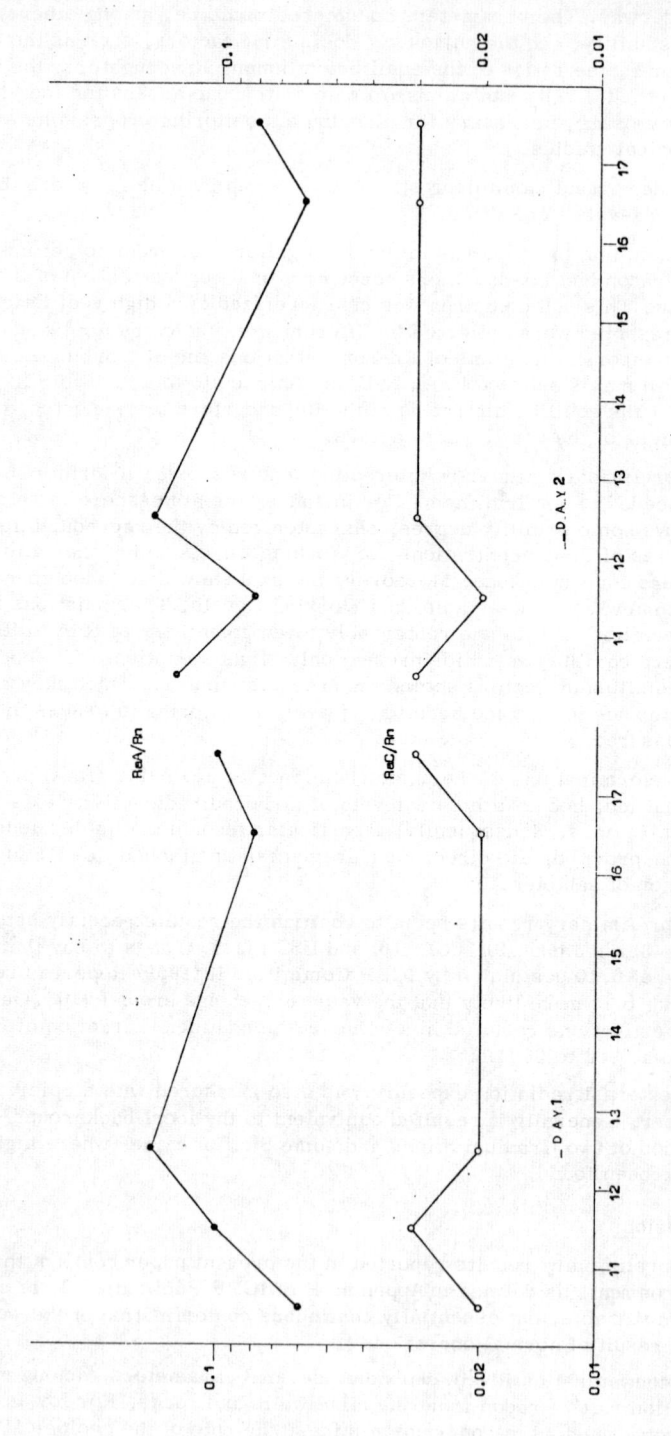

Fig. 4 — Variation of Radium A and Radium C to Radon concentrations ratios with time at fixed sampling location.

physical characteristics of the rocks and the mineralizations should be adequately evaluated mainly in view of the planning of general surveys and monitoring programs.

The assessment of the individual exposure at low dose levels for rather large groups of persons, as it is the case in nonuranium mines, could be carried out through an adequate standardization of the dosimetric factors for each single mine equilibrium factor, unattached fraction, equilibrium ratios and so on, including specific parameters of the ventilation system.

In conclusion, the possibility of radiation exposure in mines has to be taken in account in addition to other hazards which are considered as predominant in mining work. In fact, concentrations of radon and Rn-daughters, at not negligeable levels, have been found also in nonuranium mines.

Consequently, at least from the radiation stand point, the classification of mines into mines with radiological hazards and mines without radiological hazards, would be more significant than the classification into Uranium and nonuranium mines.

Acknowledgements

The Author wants to thank:

Prof. C. Polvani, for suggesting the subject of the present work, Dott. F. Breuer, for his scientific assistance, Mr. F. Scacco (CNEN) for the supervision and technical assistance in the monitoring program, Dott. P. Nascimben (CNEN) for his geological advice, and Ing. A. Galati of Italian Department of Mines for his cooperation and help.

REFERENCES

(1) K.P. Markov, N.V. Ryabov, and K.N. Stas, Atomnaya Energiya 12, 315 (1962).
(2) G. Sciocchetti, and F. Scacco, Unpublished Report, CNEN (Italy).
(3) T.T. Mercer, and William A. Stowe, Health Phys. 17, 259 (1969).
(4) A. Bottino, G. Lembo, F. Scacco, G. Sciocchetti, and S. Tagliati, Proc. Region.Conf. on Radiation Protection, Jerusalem, 5-8 March (1973) Vol.I Israel Atomic Energy Commission YAVNE, 1973.
(5) P. Nascimben, L'Industria Mineraria, anno XXI, I-27 (1970).
(6) G. Sciocchetti, and F. Scacco, Unpublished Report, CNEN (Italy).
(7) J. Pradel, Proc. of Symp. on Radiation Dose Measurements, Stockholm (1967), 527, ENEA, Paris (1967).
(8) A. Chapuis, A. Lopez, J. Fontan, and G.J. Madelaine, Health Phys., 25, 59 (1973).
(9) W. Fusamura, and R. Kurosawa, IAEA, SM 95/26, Vienna (1967).
(10) B.F. Craft, J.L. Oser, and W. Norris, Am. Ind.Hyg.Ass.J., 27, 154 (1966).
(11) A.S. Serdjukowa, and E.I. Savenko, Izvest. Vyss.Uchel.Zav.Razv.Moscou II, 100 (1968).
(12) G. Sciocchetti, and F. Scacco, Unpublished Report, CNEN (Italy).

Table I - Radon concentration in eighteen nonuranium mines.

Mine	Mineralization	Type of material mined	Radon, pCi/l Max.	Radon, pCi/l Mean
Group I - Alps				
A	Impregnation massive ore body	Blende-Galene	800	480
B	Small dikes and massive ore in schists	Siderite	500	280
C	Massive ore	Magnetite	370	190
D	Dikes in lim. stratified	Fluorite	300	270
G	Serpentinites with schistose limestones	Magnetite	120	70
H	Dikes of quartz in granites	Fluorite	100	70
Sardinia				
E	Dikes in granites	Blende-Galene	200	110
F	Layers in limestones	Lignite	200	80
Q	Dikes with gangue of quartz	Blende-Galene	20	10
Group II - Apennines				
I	Limestones stratified interbedded with clay rhaetian limestone	Cinnabar	60	30
L	Cretaceous limestone	Bauxites	35	30
M	Rhaetian limestone	Cinnabar	30	20
N	Gypsum and marls associated with sypsum	Sulphur	30	15
O	Siliceous sedimentary	Manganese	20	15
P	Granited at the bottom limestone at the top	Phyrite	20	15
R	Dikes in clay	Sulphur	10	10
Group III - Uranium prospecting areas				
S	Ore bodies in silicified tuff	Uranium oxide	700	600
T	Ore bodies in tuff and lacustrine sediments	Uranium oxide	240	150

Table II - Radon concentration and working levels in seven mines

Mines	Radon concentration, pCi/l		Working level	
	Max.	Mean	Max.	Mean
C	370	190	0.60	0.40
G	120	70	0.30	0.20
I	60	30	0.01	0.01
L	35	30	0.15	0.12
O	20	15	0.01	0.01
P	20	15	0.07	0.05
S	700	600	0.85	0.75[+]

[+] Closed mine.

Table III - Equilibrium factors and Rn-daughters ratios variation at fixed location during two days

	Radon concentration pCi/l	Equilibrium factor	Equilibrium ratios RaA : RaB : RaC
day 1	316	0.025	1 : 0.38 : 0.34
	260	0.038	1 : 0.29 : 0.26
	300	0.046	1 : 0.30 : 0.16
	255	0.031	1 : 0.34 : 0.30
	232	0.038	1 : 0.34 : 0.31
day 2	237	0.046	1 : 0.29 : 0.26
	300	0.030	1 : 0.34 : 0.26
	252	0.051	1 : 0.30 : 0.19
	290	0.031	1 : 0.42 : 0.48
	226	0.039	1 : 0.44 : 0.40

Discussion

- Chairman

Would you like to comment on the build-up of background on the Silicone detector, page 261 if there was one, and number 2, if there was, what did you do, did you throw it away or did you clean it?

- Dr. Sciocchetti, Italy

The alpha background of the Si-detector was lower than 0.1 cpm for the line of 6.00 MeV of RaA. For the second question: We used an Ortec R-series ruggedized detector. The build-up of the background, which raised essentially with the half-life of Radium F, could be avoided periodically cleaning the detector surface.

- Dr. Snihs, Sweden

On page 265 you say the unattached fraction of Radium A ranged from 2% to 4%; is that in relation to Radon or to Radium A?

- Dr. Sciocchetti, Italy

The values are related to Radon concentration.

- Mr. Kitahara, Japan

Please could you describe the background, which means the last line, the full type background.

- Dr. Sciocchetti, Italy

During the measurements a contribution from alpha-emitters, essentially Radium C, was observed for the Radium A line. It was different from mine to mine, according to the radioactivity level and the equilibrium degree of Radon daughters. The value was always lower than 1% of the integral of Radium A line.

CONTINUOUS AIR MONITORING USING BETA-GAMMA DETECTION FOR AREA MEASUREMENTS OF RADON DAUGHTERS

H.M. Johnson
Whiteshell Nuclear Research Establishment
Pinawa, Manitoba, Canada

The operational health physics monitoring for airborne activity at a nuclear research establishment has been accomplished by a beta-gamma detection system. Of course, in this context, such a system primarily monitors for airborne fission products. However, it has shown significant sensitivity for radon-daughter measurements. The system provides continuous recording, an adjustable alarm and the opportunity of preserving an archival sample. It can operate in the presence of a significant gamma radiation field.

Briefly, the system consists of a filter holder in a lead castle, an end-window geiger tube, commercial electronics and a commercial air-pump. The paper filter is held in a brass-holder inserted within a cylindrical lead castle 9 cm in external diameter and 15 cm in height. The cylindrical castle fits onto a 3 cm thick lead base. The end-window geiger tube is inserted into the castle bore and is connected to the scaler. The scaler incorporates the necessary high voltage supply and has an adjustable alarm level. The output is recorded on a strip chart recorder. The air pump achieves a constant flow rate of 0.6 m^3 per hour. At some locations the system provides airborne monitoring in a gamma radiation field of up to 10 mR/h.

Although this system was not explicitly developed for radon-daughter air monitoring, it operates effectively to give an alarm warning when this activity increases. Of course, unless the equilibrium fraction is known, the monitored activity does not necessarily give an accurate measurement of dose absorbed. However, it has things in its favour: its efficiency is reasonable, it will provide a warning when operationally significant levels are exceeded, the sample volume is large and the filter can be preserved for 210-Pb analysis later.

The system has, for example, successfully monitored the variation in radon-daughter activity in office building air when the ventilation system has been shut down as part of an energy conservation program.

Such a system in its present form would cost approximately 1500 dollars.

Although the design is currently bulky, the point to be made is that beta-gamma detection in this manner appears to be useful for continuous area monitoring.

Discussion

The discussion of this paper commenced with the comment that a short, rapid increase in airborne activity would result in a prolonged, elevated measurement by this type of area monitor.

This criticism was answered by the argument that the instrument was designed as an integrating monitor. The air pump provided proportional sampling and the data of the monitor were proportional to the integrated exposure of a worker to airborne activity for the work period. The concept to which this instrument was applied was that of providing a warning when the integrated exposure to airborne activity had exceeded the limit set for the work period. The alarm would then sound, for example, to declare the area a respirator area. According to law in the Province of Ontario, respirators must be worn in areas where the annual exposure of a miner cannot be limited to 4 W.L.M. This type of instrumentation could provide a warning when the airborne activity exceeded that equivalent to 0.3 W.L. After each shift the filter would be changed either manually or by an automatically advanced filter tape mechanism.

Additional comments referred to other monitoring instrumentation used elsewhere in the nuclear industry. The point was made that instrumentation in the nuclear fuel cycle beyond the mine was well developed. A knowledge of this instrumentation was necessary in the mining context because such instrumentation appears capable of adaptation to meet the conditions and the needs of the mine.

Session V

Panel discussion on area monitoring

Chairman - Président
W. JACOBY
(F.R. of Germany)

Séance V

Table ronde sur la surveillance de l'atmosphère

Summary of Discussion

W. JACOBY (F.R. of Germany)

After an introduction by the Chairman, the Group discussed the general aspects and problems of area monitoring in uranium and other mines. The main topics of the discussion concerned :

- Objectives of area monitoring in mines
- Quantities to be measured and required types of instruments
- Emission monitoring and environmental aspects
- Problems of organisation and training.

In the following the main statements and conclusions of this Panel discussion are summarized.

Objectives of area monitoring in mines

In the general radiation protection philosophy the purpose of area monitoring is the control of radiation sources and radiation fields, to obtain the informations which are required for the evaluation of radiation safety.

Transferring this concept to the special problems in uranium and other mines with a high radon content, the objectives of area monitoring in these mines are the detection of the main radon-sources and the assessment of the activity distribution in the mine's atmosphere. This objective is strongly connected with the necessity to find out efficient and economic radiation protection measures, especially with respect to the layout of the ventilation system. Area monitoring should give us also an early warning, if abnormal changes in the activity level are occuring (alarm-monitoring).

The other main objective of area monitoring is to obtain quantitative informations about the radon-exposure of miners. Until now no radon-monitors were available. Several types of these monitors are now in development or are ready for testing under routine conditions. It was the general opinion of the participants, however, that area monitoring remains an important information source for the estimation of the radon-daughter-exposure of miners. This aspect of area monitoring must be taken into account in the selection of the location and the frequency of measurements and the quantities to be measured.

Finally it was pointed out that area monitoring in mines can provide us with data to estimate the emission rate of radon from the mines shafts into the atmosphere.

Quantities to be measured and required types of instruments

The quantities to be measured in area monitoring are the activity concentration or the potential α-activity concentration (in WL-units) of the short-lived radon-daughters and/or the activity concentration of the radon-gas itself. For routine monitoring the additional measurement of the unattached fraction of radon-daughters does not seem to be necessary, because the available data indicates that this fraction will be rather small in most working areas of mines ; to confirm the general applicability of this statement however additional measurements of this fraction should be done in characteristic areas of some mines on the basis of research projects.

At the present time routine area monitoring is done by spot-check measurements, mainly using the KUSNETZ-method. It was mentioned during this meeting that such spot-check data can lead to relatively large errors in the assessment of the average activity level and integral personal exposure of miners during the whole working period. The higher the activity level is, the more frequent such measurements should be done. Under these circumstances the recently developed instant WL-meters allow a considerable improvement in area monitoring.

However it was pointed out during the discussion that in addition continuous sampling methods should be applied in area monitoring. For alarm-monitoring and ventilation control continuously operating instruments with a direct readout are needed.

In this context suggestions were made for the development of automatic control systems with data transmission to a control center. This system should involve detectors for activity and air flow to control the correlation between the activity level and the ventilation rate. If necessary the control of dust, toxic gases and vapors can be included in such an automatic system.

Summing up, it was stated that the present methods and techniques of area monitoring in mines have to be improved to meet all demands and objectives. In many cases we are still far away from recommending the most suitable method and defining exactly the desired types and properties of monitoring instruments.

Emission monitoring and environmental aspects

With regard to national regulations on the environmental impact from nuclear installations there is some interest to estimate the environmental exposure due to the release of radon from the exhaust air of uranium mines. The emission quantity to be used for the environmental impact assessment is not the Rn-concentration in the air but the average emission rate (for example in Ci per year). This quantity can be determined for example by continuous radon-detectors in the main exhaust shaft.

Environmental impact studies for uranium mines are prepared in several countries. Preliminary rough estimates indicate that the resulting environmental exposure will be small compared with the natural background level of radon and its daughters inside and outside houses. It was also mentioned that the population exposure from tailings will be more important.

Organisation and Training

The evaluation of the radiation hazards in mines and the possible protection procedures depend strongly on the knowledge of the possible radon-sources and of the air distribution in the mine. Due to the complex structure of the mine's atmosphere, experienced persons for radiation monitoring are required. Especially in small mines it is difficult to get qualified persons for this job.

It was strongly pointed out that a good cooperation between the production staff, the radiation protection people and the ventilation engineers is necessary, to optimize the mine's operation from the radiological and economical standpoint. The arrangement of this teamwork in all phases of planning and operation is mostly a matter of the mine's management.

Similar aspects are valid for the supervision and consultation by the official inspectors. In several countries special training courses for radiation inspectors are performed or will be started. It was the general opinion that especially the cooperation between radiation inspectors and the mine's ventilation inspectors should be improved.

Session VI

Summary, conclusions and recommendations

Chairman - Président
P.E. HAMEL
(Canada)

Séance VI

Résumé, conclusions et recommandations

SESSION VI - ROUND TABLE

P.E. Hamel, Chairman
Atomic Energy Control Board
Ottawa, Canada

It is recalled that the purpose of the meeting is to discuss and this we did extensively, existing practices in Member countries. There are 10 Member countries represented, plus three international organizations: IAEA, ICRP and our host organization, NEA. Besides existing practices we also considered recent developments in measuring techniques and interpretation of measurements for dose evaluation. Another important purpose was to contribute solutions to problems in the field of exposure to radons in underground mining operations. To this end, we had four excellent papers that helped to focus our discussion and no doubt these will prove very valuable in the proceedings.

The paper by Dr. Fry on radon and its hazards is, in my mind, an excellent mise-au-point of what is known of radon and its rate of production from natural sources or from manmade activities. It also contains a review of what is currently available to radiation protection specialists in the field of dosimetry and also to others in the field of epidemiology. It also raises the question of protection standards and their bases and suggests that four working level months per year could be one such basis. Secondly, reference is made to total cumulative exposures and based on epidemiological data available from the United States - 800 working level month total might give rise to an acceptable level of 120 working level months below which there is no unacceptable excess lung cancer. Both points could be challanged in our discussion.

In his paper, Dr. Jacobi reviewed techniques for interpretation of measurements in uranium mines and dose evaluation in relation to biological aspects. This is related to the ICRP recommendations which are not necessarily final. The general conclusion indicated amongst other things, that uranium miners might be the most exposed group of workers in the nuclear industry, and this being the case, the investigation programmes should be commensurate to the risk. Estimates are more for long-term cumulative individual exposure. It is still an open question as to whether it should be the total exposure or the rate of exposure. It is also indicated that ICRP is reviewing and may be publishing its recommendations in the not-too-distant future. This may be six months to a year.

Certains résultats biologiques expérimentaux sont maintenant disponibles pour permettre une première tentative d'établir une relation dose/effet. Les études qui ont été faites sur le rat par l'équipe du docteur Chameaud et celle du docteur Lafuma au Commissariat à l'Energie Atomique, semblent indiquer qu'en effet il peut y avoir un grand intérêt à étudier l'animal et établir une analogie entre les effets des produits de filiation du radon chez l'homme. Et encore là, on retrouve des valeurs qui ne sont pas étrangères à celles qui ont été obtenues par les études épidémiologiques, c'est-à-dire dans l'ordre de grandeur de 100 à 200 working level mois. Il semble intéressant de continuer ce genre d'étude et une proposition a été faite à ce sujet. C'est un sujet qui intéresse le Canada et au cours de la semaine, j'ai reçu diverses expressions d'intérêt de la part des représentants.

Finalement, le docteur Ham, mercredi matin, nous a rappelés que les problèmes chez les mineurs dépassaient maintenant le cadre des centres d'études et étaient mis en évidence chez le public et devenaient un problème politique. Même si on ne peut pas mettre ces aspects en équation, les réactions sont tout à fait réelles et on ne peut pas les négliger. J'invite maintenant les commentaires des participants.

M. PRADEL, France

Monsieur le Président, je suis un peu gêné pour faire cette synthèse, parce que j'ai constaté qu'on avait eu souvent des discussions un peu semblables dans les différentes sessions sur des sujets qui ne me paraissaient pas très nettement délimités et il est possible que j'aborde des sujets qui ne soient spécifiquement de ma session, et je voudrais que vous m'en excusiez. Ce que j'ai préparé et que je vous propose c'est un peu ce que je retire moi-même de cette réunion. Je ne vous demande pas d'être tous de mon avis, je crois que ce ne serait pas possible et on a bien vu qu'on n'avait pas un avis unanime sur la question. Mais si cela vous intéresse, je peux vous donner plutôt mes réflexions sur la réunion. Personnellement, j'ai retenu qu'il était souhaitable de développer l'expérimentation sur l'animal pour mieux connaître l'effet des faibles doses cumulées et des co-facteurs. En effet, l'expérimentation sur l'animal me paraît être la seule méthode qui puisse permettre d'obtenir des résultats rapides dans ce domaine.

Ensuite, j'ai retenu d'améliorer la dosimétrie individuelle au moyen de dosimètres individuels déjà mis au point. Ceux-ci permettent d'accroître nos connaissances en matière de dosimétrie, en permettant une mesure précise des working level mois grâce au dosimètre radiothermoluminiscent et en partant de mesures précises des concentrations en radium A, radium C, des working level avec les dosimètres à traces. La dosimétrie individuelle est nécessaire pour effectuer des enquêtes épidémiologiques sérieuses dans le domaine des faibles expositions cumulées. Les résultats à mon avis seront utilisables rapidement pour améliorer les évaluations dosimétriques soit des enquêtes en cours, soit des enquêtes futures. Il doit résulter de l'utilisation de ces matériels une économie en matière de surveillance qui devrait pouvoir compenser, au moins partiellement, l'investissement nécessaire pour leur acquisition.

Troisième point, les normes basées sur les résultats d'enquêtes épidémiologiques doivent être en accord avec les méthodes de dosimétrie utilisées pour ces enquêtes. Ou sinon, elles doivent être adaptées, si les méthodes actuellement utilisées sont différentes.

Quatrièmement, on peut envisager l'utilisation d'un réseau de détection, mais il me semble préférable de limiter son objectif au contrôle du bon fonctionnement de la ventilation. On peut envisager simplement de détecter l'arrêt des ventilateurs, l'ouverture intempestive de portes d'aérage ou toute autre anomalie de ce genre. Un détecteur placé au niveau des rejets dans l'atmosphère et mesurant l'activité en radon ou ses descendants, peut aussi aider à déceler des anomalies dans le fonctionnement de la mine. Mais ces détecteurs ne peuvent se substituer aux techniciens qui doivent fréquemment inspecter les différents lieux de travail pour vérifier les conditions de travail et faire apporter, si besoin est, par les responsables de l'exploitation avec lesquels ils sont en contact fréquent, les modifications nécessaires.

Cinquièmement, les doses d'irradiation externes peuvent être élevées dans certaines mines. Si l'on décide de cumuler ces risques au niveau du poumon, il faut, premièrement, faire des recherches sur la granulométrie, la rétention et le métabolisme des poussières à vie longue afin de préciser le risque et deuxièmement, mieux faire apparaître les effets additifs de l'irradiation externe du corps entier et de la contamination interne pulmonaire. Car la limitation à 5 rem pour les poumons aurait des conséquences très importantes pour l'exploitation des mines à teneur élevée, teneur supérieure à deux ou trois pour mille.

Sixièmement, l'étude de la pollution de l'air dans l'environnement autour des mines et usines doit être effectuée. Mais les contrôles réguliers ne devraient pas se révéler indispensables, car la dilution qui intervient dans l'atmosphère est telle que les rejets effectués à des niveaux faibles ne dépassant pas le working level, les effets ne sont décelables qu'au milieu du bruit de fond naturel, qu'à quelques centaines de mètres de distance. Enfin, il faut effectuer la mesure des concentrations en radon et descendants dans les différents types d'habitation, avant de fixer des limites d'activité pour les matériaux de construction ou des limites pour l'activité de l'air. Car la fixation de limites en dessous des valeurs rencontrées, quelques dixièmes de picocurie par litre, dans les maisons construites en matériaux couramment utilisés, conduira à des dispositions inapplicables risquant de déconsidérer les règlements de radioprotection. Voilà, M. le Président, mes conclusions personnelles.

Mr. ROCK, U.S.A.

Dr. Pradel, you mentioned the limitation to 5 rem for the lungs, shouldn't it have been 15 or did I misunderstand?

Mr. PRADEL, France

J'ai bien dit 5, parce qu'hier dans la discussion, il a été question de ramener à 5 rem au niveau des poumons. C'est M. Bush qui, dans sa formule, a 4 WLM plus gamma = 5. Cela revient à 5 rem au niveau des poumons. Alors c'est contre cela que j'attire l'attention des autorités.

MR. BUSH, Canada

I gave one formula for adding the contribution to the lungs from gamma, where I had $\frac{gamma}{15}$ + working level months, and another one where there was whole body risk $\frac{gamma}{5}$ + lung risk = $\frac{wlm}{4}$.

MR. PRADEL, France

Je suis d'accord avec la première formule, mais je ne suis pas d'accord, personnellement, avec la deuxième, ou en tout cas j'estime que pour adopter la deuxième il vous faut des justifications sérieuses. Il faut être bien conscient des complications que cela entraînera au niveau de l'exploitation.

MR. O'RIORDAN, United Kingdom

Can I just observe that when you do make observations of natural levels of airborne radioactivity in houses you find such levels that you would be no nearer a decision at the end of the day. In fact, levels would be so high that they would tend to lead to confusion.

MR. PRADEL, France

Là aussi, je pense que c'est notre rôle d'informer les gens qui prennent ces décisions des complications que leurs décisions vont apporter. Si vous dites qu'il ne faut pas qu'il y ait plus de 3 ou 1 pCi/ℓ dans les maisons, moi je suis obligé de dire à mes autorités, il y a peut-être 100,000 maisons ou 1 million de maisons à détruire en France. Je crois qu'il est de notre devoir de les avertir de cette situation.

CHAIRMAN

I wish to invite comments on the discrepancies that seem to exist between the results from different individual dosimeters or equipment used for individual dosimetry. It seems difficult to conclude without further critical look at the value of such results.

MR. ROCK, United States

All I'd say is that the dosimeters are available, they are workable, the thing that we've got to overcome is the political aspect to get these things into actual use. By this I mean that we've got to have laws, we got to have compensation for over-exposures and this sort of things.

CHAIRMAN

Your point is well taken, Mr. Rock, but I wasn't thinking of those aspects as much as which dosimeters. Perhaps the answer is simply not one but several, but once a dosimeter is used or a dosimetry system established, it has to be used consistently from then on to make it possible to relate with other results that might be generated.

DR. BRESLIN, United States

I agree with you completely and I wonder if it would be useful to consider some kind of a systematic programme of intercalibration or of intercomparison or both involving various candidate dosimeters, and possibly even area monitoring equipment. It occurs to me that our host agency might have the prestige and the interest in perhaps trying to organize such an undertaking which would involve I imagine, a considerable amount of effort.

CHAIRMAN

Could I obtain other views on the proposal by Dr. Breslin to promote a systematic programme for testing or intercomparison of individual dosimetry systems.

MR. ROCK, United States

This is fine but we are actually comparing the same thing. He's doing the same thing we're doing, we have standard chambers, of course, we're going a little further, in that we are going in mines and making comparisons, but, certainly we are measuring the same thing, and if we're co-relating with known values, an answer is an answer and we can describe it in terms of physical units. I'm not worried about his readings if he's not worried about mine. What I'm worried about is the lack of readings that we have now.

CHAIRMAN

Mr. Rock, may I ask you whom you have in mind when you say "we"? Does this mean that you are also comparing with results from other countries? For instance, are the epidemiological results from United States, from Sweden, from France, all compatible? This is the point behind, I believe, the proposal from Dr. Breslin.

MR. ROCK, United States

Are we talking about dosimeters, comparing dosimeters, is this the question? Or the accuracy of dosimeters?

CHAIRMAN

Intercomparison of results that is the number of WLM as measured by different dosimeters.

MR. ROCK, United States

If everybody understands what they are measuring, and if they've actually analyzed their system accurately, certainly the results should be compatible.

DR. O'RIORDAN, United Kingdom

Mr. Chairman, Dr. Wallauschek asked me to tell you that the Agency tried to set up an intercalibration exercise among laboratories in relation to the measurement of radon and radon daughters in houses and I think since some of the information overlaps, this might be a useful vehicle for further intercomparison.

CHAIRMAN

Excuse me, did you say in houses as opposed to personal dosimetry used for miners?

DR. O'RIORDAN, United Kingdom

I am talking about the instrumentation for measuring radon and radon daughters in houses, but these are essentially area monitors.

DR. WALLAUSCHEK, NEA

I would like to add to this, that this work had started in the framework of our programme on the preparation of radiation protection standards for building materials. If we want to recommend figures to be applied at international level in relation to radon, we should really be sure that such figures are comparable. In order to achieve this, we set up an intercalibration exercise on a very small scale, for the time being, among those countries who are contributing to the preparation of the standard, the United Kingdom, France, Germany, Sweden and two or three other countries. This question will be discussed by the Radiation Protection and Public Health Committee, and in this Committee all NEA countries are represented including the United States, and we are certainly prepared to discuss a widening of the system, either for radon measurement in houses, or even in relation to what we are discussing here.

DR. TANIGUCHI, Canada

Would it be in order to ask how this is being done technically, not in detail but broad enough to understand why because, I am wondering how you would be able to do it for radon daughters. I can see you can send out standards for radon. At what stage is this intercomparison?

DR. O'RIORDAN (UK)

I think it would be done by circulating an instrument, and I might suggest that although it is outside the NEA framework at present, one of the first intercomparisons will be with the Canadian laboratory.

DR. TANIGUCHI, Canada

One final clarification then, you said you would send an instrument around to CANADA? I would like to know whether you are interested in getting a standard reading on the building material rather than on the instrument?

DR. O'RIORDAN (UK)

The instrument will be a common factor. In fact, you will measure radon daughter concentrations in locations chosen to give suitable value and intercompare the readings with the circulating instrument.

DR. BRESLIN, United States

May I ask if what you have heard of the proposal is compatible with your own proposal on systematic programmes for comparison.

DR. O'RIORDAN, United Kingdom

I had not formulated any specific ideas about how this could be done. I am sure there are many approaches thatcould be taken. I would like to offer one comment about mine instrument drawing on our experience and I am certain that other people have had the same experience. I think mine instruments have to be tested both for calibration under more or less controlled conditions in the laboratory but that is only one step. Equally important and certainly a dominant step is to test these instruments for proper operation in the very hostile conditions of a working uranium mine. Doing that on a systematic basis with a number of different kinds of instruments, is not easy but I do think it is necessary and it ought to somehow or other be incorporated in any kind of a test programme.

CHAIRMAN

Any further comment on the presentations or the summary from Mr. Pradel?

MR. ROCK, United States

Like I said in my paper, we are at a technological stage where to delay another five years or ten is ridiculous. A stage is reached where we can actually come up with some measurements that are meaningful and to me it is something that we ought to concentrate on. The measurements may not be absolutely 100%, but can be in a range so far much advanced to what we are using right now that certainly we ought to entertain the fact of trying to put the pressure on to start using these devices. We are using them in the mines, we are testing them in the mines, so to assume that we are at a stage now we are still insecure to the extend that we don't feel it is an advantage is something that I would like to get out of our heads. The longer we delay, we know, as soon as we start accumulating the information, we are still going to have to wait several years before we can use this actually. Some instruments are better than others and no doubt the information is going to improve as we use them. I am perfectly willing to compare instruments and I think we all have the right theory and the right calibration procedures. I am confident in what we are doing, so the thought I would like to leave is that the state of the art is such that we should start using what we have.

CHAIRMAN

The point is well taken, Mr. Rock and it should be of great interest to the representatives from different sectors including from regulatory authorities. However, they would wish to be reasonably confident that equipment is available to provide meaningful information As indicated in the introduction there is a number of such instruments which seem to be giving different information which in turn is used in epidemiological and dose-response studies. It is difficult not to entertain reasonable doubts in the wide application of such information from a legal point of view.

MR. ROCK, United States

I would like to point out that it would be foolhardy on for our part to have worked so hard on this equipment and to have confidence in it and just keep delaying the use of it on the basis that we ought to all get together and make sure that we are doing everything just exactly alike. I think the end-results are not worth the delay and no matter what adjustments are needed, they should be pretty close to each other.

DR. BRESLIN, United States

I just wanted to point out that with regard to confidence level, there is no reason in the world why MESA in the United States and the corresponding agencies in other countries cannot start applying their dosimeters if they choose, I do think that some kind of an intercomparison by each agency or country or user will help to provide better confidence in the equipment if results correspond with the equipment designed by other agencies and users.

MR. PRADEL, France

Je voudrais simplement dire que je suis d'accord avec M. Rock et je pense que nous sommes arrivés à un stade où le plus important, c'est d'utiliser ces appareils, de voir les résultats obtenus, éventuellement de les modifier, de les améliorer. Il ne faut certainement pas attendre et faire beaucoup d'études en laboratoire, maintenant. Nous avons atteint un stade opérationnel, je pense qu'il y a plusieurs appareils qui sont valables, certainement, il n'y a aucune raison, dans tous les domaines on peut avoir plusieurs solutions. Je crois qu'il faut les essayer, puis peut-être dans quelques années on en choisira un parmi ceux qui existent, ou on en conservera plusieurs.

MR. YOURT, Canada

In Ontario, industry and government have been doing this with regard to personal dose sampling and a decision was made to go ahead and use the instruments that are available now, and improve them as we go along. We realize that the one we are using at present may not be the one we use five years from now, but at least we are getting some answers. The other comment I would like to make is that I think we are already trying each other's instruments out. Very recently in our study we endeavoured to use the same arrangements as Mr. Rock is using in Denver. Furthermore, the Atomic Energy Control Board has proposed to test some of the French-CEA instruments. I suggest we should expedite this type of procedure.

CHAIRMAN

In fact, such exchange does exist on a limited scale, but I believe the reference that Dr. Breslin had in mind was to enlarge the scale to more than just a few countries.

Dr. BRESLIN, United States

May I just add one more comment in this regard - I think we should keep in mind, particularly in the area of dosimetry that we should work towards something which miners are happy with, that they will respect and be willing to use because that is a hurdle that we must take care of.

DR. GRAY, Canada

I would just like to say that I would be interested in such a program. In view of the comments that have been made, it could start with the standard instrument and try it out in the mine.

DR. TANIGUCHI, Canada

May I offer just a brief comment on our experience on low intercomparisons on radioactivity. We do national comparisons but we find also very useful the intercomparisons with EPA, Las Vegas, the IAEA, Vienna and the international reference centre in France. I think we would be interested in intercomparison program for radon.

CHAIRMAN

As you can see, the Canadian team is interested.

DR. AHMED, IAEA

Mr. Chairman, I just wanted to add that the IAEA had conducted several programs on intercalibration and intercomparison on various aspects. I can give you the government proposal. I was thinking that the agency may come into the picture in organizing and helping in some way to have intercomparison or intercalibration of monitoring instruments used in mines and mills.

CHAIRMAN

Thank you Dr. Ahmed. This should definitely be kept in mind.

MR. ROCK, United States

I support the use of dosimeters as often and in as many places as possible, and I am most sympathetic with the intercomparison. However, I hope decisions in this direction will not turn off the follow-up work that I feel is necessary to meet some practical demands, namely, lighter instruments, cheaper instruments, such that the use for the instruments can grow, such that operators will find it feasible in the dollar sense to equip more miners and eventually get what we all want to get to, namely better exposure records for all underground miners. I think we should look on this as a step in the right direction, but there are still some hills off in the distance that we still have to climb.

CHAIRMAN

Thank you Mr. Rock. At this point, I would like to invite Dr. Fry to interject a few of the ideas, from his summary Round Table discussion, into this discussion.

DR. FRY, Australia

In the Round Table discussion of personal dosimeters, we looked at a number of questions starting with a fairly basic one, namely, what really are we aiming to measure in order to protect the miner? Our basic aim should be to protect the miner rather than to measure with ever greater and greater refinement exactly what his lung dose might be. That led us to a discussion of risk in general because I believe we all agreed the only acceptable basis we have at the present time for setting what one would call an acceptable risk for miners (we were not suggesting for a moment that we were able to set, decide what was acceptable), but the only way an acceptable risk can be established is by the epidemiological data. Dr. Jacobi who is one of the pioneers in developing lung models and lung dose calculations also agrees with this. I believe that at this particular stage anyway, we are not in a position to set risks on the basis of estimated lung doses. I think the important thing that comes out of this is: if we are relying on the epidemiological data, what we have is a correlation between excess lung cancer cases in uranium miners and exposure, cumulative exposure, estimated on the basis of area monitoring measurements an occupancy factor. That led us to wonder what relationship exists between personal monitoring results and with assessments of exposure calculated on the basis of area monitoring and occupancy factor. One of the general conclusions I think was that people who had used experimentally personal monitors had found invariably that personal monitoring led to a higher estimate of cumulative working level than that which you would get by area monitoring and occupancy factor. That then led us to a note of caution when interpreting the epidemiological data. One must remember that what is correlated is cumulative working level months exposure based on the environmentally measured area monitoring data, and not on personal monitoring data, so there needs to be some attempt at normalizing the epidemiological data to new measurements made with personal monitors. We also discussed perhaps a little irrelevantly at that session the relationship between risk and smoking, but I won't propose to dwell on that now. I might interpose here the question raised in Mr. Bush's paper which I think is an important one. You have a very rich ore body here in Canada, and we have at least one very rich ore body in Australia. The relationship between gallery exposure and permitted working level months exposure is important. I believe in fact that Bill Bush's third formula is the correct one and I believe one can justify this, on the basis of assessing risks, and I understand that this is the approach that the ICRP would adopt. What one should do is to assess the risk associated with the external gallery exposure, add that to the risk associated with the working level exposure, and the sum of those two risks must not exceed the risk associated with a whole body exposure of 5 rems per year. Now, if you put in the assessed risks associated with whole body general exposure, the risk associated with the working level month exposure, you put the numbers in that I have got in my paper and which agrees with Dr. Jacobi's assessment, lo and behold! you come out with the formula that the gamma dose + the working level months exposure must not be greater than 5, and I think that is an important conclusion, and as I say, I was speaking to Dr. Jacobi about this and I think he said that is exactly what an ICRP approach could be. This is important, because I am just looking through the Ham reports last night and I noticed that in the recent results for the mine containing rich ore, 6 lb ore, relatively rich ore, the annual external exposure rate to miners there was about 2.8 rems per year. That means that if we use the Bush formula No. 3, the maximum permissible working overexposure is only going to be 2.2 working level months per year in that situation. Now if one measures that exposure with personal dosimeters, what is going to assess a higher working level month exposure than would have done on the basis of area monitoring, which, as I say, is the basis of the epidemiological data, so it is something to be borne in mind because 2.2 working level months per year might very well be rather difficult to achieve on average in a mine which not only has rich ore and therefore high gamma radiation and very possibly although not necessarily, but very possibly, high radon exposures as well.

We discussed risks and we discussed what it was that we ought to try to measure, and then we got on to the question: is personal monitoring really necessary and I believe that the answer to that was "yes". It was certainly necessary if one is interested in measuring with greater accuracy the actual exposure that an individual miner receives and one of the reasons for doing this is one wishes to have better data for correlation in future epidemiological surveys. Although I guess we all realize that, and we are all hopeful, the epidemiological data is not going to be improved in the future because we are hopeful that the incidence of excess lung cancers in uranium mining will be very, very much reduced as of now. However, because we wish to know as much as we can about the individual exposures of miners then I believe we agreed that personal monitoring was certainly necessary, but, we also agreed that it was not, it could not be, all that one needed to do. You cannot just monitor personally and that is the end of it all, because personal monitoring does not help you much in controlling the mine atmosphere which I sense what you are aiming to do. Now, that of course leads on to the questions about area monitoring which I shall leave to Dr. Jacobi. I believe, however, our conclusion was that both area monitoring and personal monitoring were highly desirable, if not essential. We then addressed ourselves to the question as to whether personal monitors were available, and we know that they are not, because we had, two or three of them, in fact, described here. There are two types: the French type using track etch detectors and the other using TLD, both of course working on the same basic principles. Both types work, both types are still in the developmental stage, the French type is commercially available but I still believe it would be correct to say it was still a developmental instrument, at the present time. I believe that we recommended to ourselves that development should continue on personal monitoring and that we should not limit ourselves necessarily to these two types of monitors which are both monitoring on the same principle. We should try to think of other means of doing this, particularly if it can be done more cheaply. The suggestion was made that perhaps passive personal monitors are possible although I don't believe anybody was able to come up with a suggestion as to how this might be done. I think in conclusion, we decided that when it came to answering the question which many of us will be asked by mine managers in the not-too-distant future, as to exactly what he should buy, we will have to say at the present time we don't know. We know that there are a number of instruments under development and we will be keeping all these experts and will be keeping these techniques under close scrutiny and would be advising him hopefully in the very near future that he could invest his money in one particular type or another. Thank you, Mr. Chairman.

CHAIRMAN

Thank you Dr. Fry. Before asking for comments, may I make a few of my own with regard to your remarks on personal dosimetry being necessary to obtain all the data for epidemiological consideration, but it may take a long time. Indeed this may be so, but nevertheless, studies, retrospective studies are possible and we might be able to get valuable data in this way conditions being equal of course. I can't help thinking that many of the epidemiological data that we have now are in fact by retrospect in so far as they are estimated by assuming that prior to such and such a date, such and such conditions prevailed and therefore the exposure must have been such and such, based on what we know today. These assumptions could be improved by personal dosimeters and I would imagine that in a matter of a few years, we will have a much better understanding of conditions of exposure. My second comment relates to the variability of working level in the mine. The point was made by a number of speakers and it would seem to be an inducement for personal dosimeters. With these two additional thoughts, I would now invite comments from the participants.

MR. ROCK, United States

The main feature here, the main role, the main objective is the personal dosimeter problem and I just cannot think of any reasonable arguments of why we would not want this. Everybody wants to know what is the exposure, and so far we have never been able to come forward and say so many working level months, plus or minus so many per cent. This being the case, I think this is the thing that we are going to stress in the minutes of this meeting and I think the essence of what we are saying is yes we have progressed to the state, although they are not commercially available, where, if we can get the ball rolling, so to speak, if you will accept this, say France can start putting all her people on dosimeters and this would probably be much easier than in the United States because the mines are controlled by the Government, then we see the thing progress into the other country. Certainly I think all the people that are working on these prototype dosimeters, believe it is just a matter of getting them into commercial production. I think all these people are knowledgeable and they know that what they are measuring is correct, and so therefore I am saying that the highlight of this meeting should be this: that we are in a stage and we should use all our means to get these things into practical use.

DR. JACOBI, Federal Republic of Germany

I want to make additional comments on the future development of personal dosimeters. We should not only consider the instrument itself, we should also consider the method to read out the personal dosimeter as I think that is the true difference between the two methods which we have heard here, the French dosimeter which is using the track etch detector, and the other method from the United States which is using the TLD detector. If you want to use personal dosimeters for monitoring, then the automatic and simple instruments to read out these personal dosimeters should be available.

CHAIRMAN

Thank you, Dr. Jacobi. This is a very important point. I might add here that such automatic read out systems for track etch type is possible, has been developed and is available in the laboratory.

DR. JACOBI, Federal Republic of Germany

It is still in development. I think that, in the case of TLD commercial instruments are available. There is the difference, I think.

MR. PRADEL, France

Je crois que le développement des appareils de lecture n'est pas un problème très difficile et nous n'avons pas fait beaucoup d'efforts dans ce domaine, parce qu'il faut d'abord que nous sachions combien de dosimètres on a à lire. Si nous fabriquons seulement 10 ou 100 dosimètres, il n'est pas intéressant de développer des appareils automatiques, cela coûte trop cher, nous n'avons pas les moyens financiers pour l'instant de le faire, et on peut très bien, s'il y en a peu, le faire éventuellement au microscope ou avec des solutions simples. Comme l'a dit M. Zettwoog l'autre jour, on peut avec des photomètres faire la mesure dans le cadre de nos détecteurs. Si on a de grandes séries, il existe déjà des appareils, les appareils américains, il y a un appareil qui s'appelle un quantimètre qui est capable de compter très facilement et automatiquement tous les détecteurs, c'est une question de prix et de nombre. Donc il y a un choix à faire de l'appareil en fonction du nombre de détecteurs que l'on possède.

DR. TANIGUCHI, Canada

Yes, just to follow up on the comment about having to recognize the radon system, I would remind you of the presentation we did on existing automatic systems which can quite readily be adopted to radon daughter monitoring systems. I think in Canada we do have the advantage that radiation personal monitoring is offered by only one national group in contrast to various commercial companies, south of the border, and all these data come in to one central point. I am taking your comments as encouragement that somebody should develop, should adapt automatic TLD read out to the radon daughter collecting system.

DR. ROSS, United States

I always think downstream a little bit, and I picture all these suggestions being pressed into action forthwith, and then I think of the what else in the system is a stumbling block, and using the thing that Dr. Taniguchi just said, the fact that they have a national registry of dose, I am reminded that we in our country do not, and here we are accumulating all these fine data in a population of miners that may switch jobs four times a year and move from state to state, and unless these beautiful data, that we are accumulating are to be other than fragments in many places, I think we need a strong effort toward finding a way to gather them, to keep them, and only then will we protect the miners which is our objective.

CHAIRMAN

Indeed, this is a very important point to the extent, as Dr. Taniguchi has indicated, in Canada it is a must. If there is not any further point, I will invite Mr. Rock to give us his views on the session on area monitoring.

MR. ROCK, United States

Firstly, there were 15 papers from seven different countries. They primarily outlined the studies which had been made, and to a lesser extent, the description of the instruments that they used. Secondly, we had assurance from official agencies and organizations that they had not left us in the lurch, the IAEA and the ICRP. Thirdly, it seems to me that there was recognition that the problem of radon and its daughters exists where radium exists and non-uranium mines such as tin, fluospar, krypton, magnetitite, fluorite, syllabar and many others, will need to be surveyed to provide assurance that radiological hazards do not exist. It really came as somewhat of a surprise to me that many of the papers and much of the discussion did not deal with uranium mines at all. It dealt with the specific problem of radon and radon daughters and its detection in mines. Fourthly, the problem of area monitoring does not necessarily end in the mine. There are problems with structures, and with the emanation of radons from underground and surface mines that must be evaluated. In mentioning this, I intrude slightly on Dr. Jacobi's Round Table discussion, but it seems to fit here. Fifthly, there seemed to be no dearth of ideas, concepts and prototypes for various monitoring devices. Sixth, three means seem to emerge: one, the need for attention by design engineers as opposed to research scientists to pick the lab tested prototype and ruggedise it for use in the mines. It has seemed to me in past years that we have seen many instruments developed in the laboratories and then more or less dropped because they were not rugged enough to be used in the mines. The services of design engineers should have been used to put this into effect. I do feel there is a need for a centralized control for intercalibrating prototype instruments so that clearly superior instruments can be selected for widespread use. People could hand-carry their instruments or their gadgets, to Oakridge, spend a couple of weeks there while they go through the process of calibrating and comparing them with each other. There is an international intercomparison available for whole body counters. It is true they do not carry the whole body counters, but the concept is widely used and I think that this situation cries out for having some place where they can be intercalibrated. Taking it one step further, I think there is a need to control field testing in the mines of the more promising instruments. I thought that a strong case was made for the practical significance of area monitors, constant area monitors for use in mines to identify the relation problems in real time so that counter measures could be taken properly. Now finally, I would like to call attention to the fact that Mr. Bush spoke in Session IV, and I thought that his ideas had been widely poached here by previous surmises, but finally I would mention the concept of measuring exposure to combined risk from radon daughters and gamma radiation. It seems to me that there is more than a modicum of logic in this concept and I predict that we will hear from it again.

CHAIRMAN

Thank you, Mr. Rock. Indeed, a lot of the information that was conveyed to us originates from non-uranium mines and when reviewing the papers, I noticed the wording used in the Italian paper namely that from a radiation protection point of view, the distinction between mines should not be related to uranium mines but the distinction should be based on whether or not there are radiological hazards. Now, I invite questions and comments from the participants.

DR. SHREVE, United States

When thinking of Bill Bush's talk and this combining of hazards, I also recall a thing that did not seem to be mentioned in any of the summary, namely that there are some new radiological hazards that may synergise with radiological hazards, and of course primarily, you know what I want to say, diesel, smoke inhalation. While we look on this thing of combining radiological hazards as complex and one demanding a lot of forethought, there must be still more efforts to really get the total picture which must include these other things and of course uranium in dust on top of it.

DR. JACOBI, F.R. of Germany

I have mentioned during this meeting that probably in the future the limit of exposure in miners should be derived from the basis of the available epidemiological results. These results are coming from mines with exposures not only to radon daughters but also to dust. So this means that the radon only, I think, is not sufficient to measure exposure. But in reality, exposure limit increase with all the agents which are contributing to the levels of known lung cancer.

MR. ROCK, United States

What I want to say is that concentrating strictly on mortality is not the usual case in our health programmes. We should be thinking injury also. It is much more complicated but nevertheless when talk health and other matters, we talk silicosis, we talk about the stages of damage done to the individual, compensation is awarded on the degree of the bazard - here we are talking strictly mortality; so I think that this certainly has to be. Our thinking has to be modified in this to this extent, that we know that other lung diseases or even perhaps the straight radiation diseases themselves can develop strictly from radiation exposure. In any health or safety programme, thinking straight mortality really overlooks a great segment of damage and not just strict human suffering. Somehow we have got to let ourselves be lulled into thinking the problem is not simply those who die. Has anybody got any comment on this who knows more about the epidemiological effects than I do, I wish they would state it. I have often been concerned about this aspect. We are only conning something, we are not conning the guy who has lost his ability to be, to work normally.

CHAIRMAN

If I may just respond to this, Mr. Rock. Indeed, the point you are raising is very real and very appropriate. This morning, in one of the better known newspapers, there was an article about Workmen Compensation relating exactly to what you are saying namely that compensation is awarded on the basis of the most apparent and visible effect, which in this case, is mortality. Although in substance compensation is supposed to be governed by the policy of giving the benefit of the doubt in fact, this does not seem to be the case. This complicated by the fact that it is extremely difficult in some cases to make an objective judgment and perhaps this kind of information might help indeed to assist in making such a judgment. We may still be a long way off, but this may be the first step.

DR. CHAMEAUD, France

Je crois que vous venez d'évoquer là un chapitre de la pathologie minière qui est très très important. Je crois que vous avez raison M. Rock quand vous dites qu'on s'intéresse à la mortalité de cancer du poumon, mais il y a effectivement dans la mine de très nombreux co-facteurs. Alors pour chacun de ces co-facteurs, on donne une limite maximale admissible. Mais l'addition de toutes ces limites maximales admissibles donnent quelquefois des intoxications ou des atteintes du poumon qui sont tout à fait curieuses. On peut avoir, par exemple dans certains coins 10 polluants différents et 10 limites maximales admissibles. Cela devient ridicule. Alors je crois que ce problème doit être étudié très sérieusement, et nous avons commencé à le faire pour notre part, et il y a des co-facteurs qui en expérimentation animale, doivent être étudiés complètement. Je vous ai signalé d'ailleurs dans mon papier, que chez l'animal on constatait, pour des doses relativement faibles, je dis bien relativement faibles de radon, des altérations des alvéoles pulmonaires, de la circulation pulmonaire, qui peuvent grandement altérer les capacités respiratoires. C'est donc un problème très très important qui vient après, bien entendu, la mortalité, mais qui est quand même très invalidante pour les travailleurs.

CHAIRMAN

Time is pressing on. Some of you may wish to leave to visit the laboratory while others may wish to continue. Are there any further questions on the point being discussed, that is on the co-factors.

MR. PRADEL, France

M. Jacobi nous a dit que les enquêtes épidémiologiques avaient été faites en tenant compte des différents risques qui existaient dans les mines. Je crois qu'il faut être très conscients que ces risques peuvent changer et par exemple chez nous, depuis quelques années, nous avons vu apparaître les engins diesel. Il n'y en avait pas au début de l'année, alors si nous avions fait une enquête épidémiologique on n'aurait pas les fumées de diesel. Je crois qu'il faudrait que l'attention des exploitants soit attirée sur ce fait. On peut voir un jour des décisions qui seront prises sans qu'on nous demande notre avis. On pourrait imaginer des moteurs à essence avec du plomb au fond de la mine, par exemple. On peut imaginer des tas de choses comme cela. Je crois qu'il est bon d'attirer l'attention des exploitants sur l'existence de ces co-facteurs.

CHAIRMAN

Merci M. Pradel. I would now call on Dr. Jacobi to summarize the Round Table discussions on area monitoring.

DR. JACOBI, F.R. of Germany

We are all aware that at the present time, the main purpose of area monitoring is the estimation of partial exposure of miners to radon daughters. The papers on this conference and the discussions, have however shown that there is an interpretation of area measurements in terms of individual exposure that can evolve in some cases much earlier. Therefore, I think we all agreed that in the future the development and application of personal dosimeters should be enhanced to reach the same standard of safety which we have reached now in practically all other fields of radiation protection. The quantity which should be measured with such dosimeters should correspond to the quantity for which the first limit is given. The second important aspect of area monitoring is the direct control of the mine atmosphere to determine the main sources and to obtain answers about the distribution of the activity in the mine. Last, but not least, it is very important to optimize the ventilation system and to increase the efficiency. For this purpose, the instruments with a direct readout should be preferred, either the instant working level meters, or the continuously operating instruments with a direct readout. We have also discussed shortly in the Round Table that it might be useful in the future to have in large mines, automatic control systems which not only control the radiation level but also the dust level and ventilation level in characteristic regions of the mine. We have seen that it is necessary to optimize this control and following counter measures to put co-ordination and co-operation between all persons who are involved in, and responsible for the measurement and for the lay out of the ventilation system. We discussed briefly the problems of the measurement of emissions rate from mines. I can mention that in the future we should be getting more quantitative data about this by measuring the set up which might be evolved in the emission of radon daughters from mines. For this purpose, it was not necessary to have instruments which measure continuously the concentration, or which give direct response about the concentration during a long period of time. In closing, it is interesting to note how many instruments are in development and available for the monitoring of radon daughters. Thank you.

CHAIRMAN

Thank you, Dr. Jacobi. With reference to the emission from the mine, I believe the point was discussed and the conclusion reached that it needed to be quantified. Some remarks were made also to the effect that this may not necessarily be a measure of the primary hazards. Although the consequences may not be very significant, it needs to be quantified for such things as the preparation of environmental impact statements. I am concerned that we do not convey the wrong impression. Because we are measuring emission from the mine does not necessarily mean that something very significant is coming out. This at the moment is my understanding of the situation.

MR. YOURT, Canada

I would like to comment on this aspect because it came up yesterday and I did some digging for information that is minimum in scope. Readings taken recently indicated that immediately above a tailings area, standing right on the tailings and holding the filter as close to the surface of tailing as it is possible without getting it dirty, the highest readings were in the order of 0.1 WL and this with no wind. The other question came up about exhaust from upcast airways, and the reading quoted to me was in the order also of 0.1 WL in the order of 200 feet from the exhaust airway. This is as I say from a limited number of readings and I am sure more will be done. In an endeavour to get significant readings, the use of a box was required to get a significant reading on the tailings.

MR. ROCK, United States

I would also like to second this thought. I think that certainly we are going to be asked figures on exhausts from mines. We are going to be asked to explain just what the relationship is in total, following a comparison with the normal environmental impact. But I think it is negligible. That is just one of those academic things, but it is a procedure that we have to go through. This kind of thing is necessary, certainly, but I would not want to overemphasize it or give environmentalists the idea this is of real importance as far as the environment.

DR. JACOBI, F.R. of Germany

I would think, I fear that anyone who mentioned this would come up with this problem and therefore we shall try to resolve this problem before then. If I look at the available data then the maximum permissible dose for the population in the neighbourhood of nuclear power plants, limits are in the range of 10 mrems per year. I use the number from Mr. Yourt about the working level which he measured over tailing which are accessible for the population I think this dose will be considerably higher than the dose limits were for a power plant. In our country this limit is not only for power plants but for all other nuclear plants too.

CHAIRMAN

Perhaps we can summarize this by saying that the instrumentation is available to do the measurements and why not use it. I wonder if I can raise a point Dr. Jacobi, in connection with the need for close inspection over and above the instrumentation and all the equipment that might be available. Some seem to believe that it should not depend entirely on instrumentation and there is always the need to walk around the mine and carry out close inspection of airways and ducts. Is this part of your session?

DR. JACOBI, F.R. of Germany

Mr. Chairman, I have mentioned only the sentence that good cooperation and good coordination should be obtained from all people who are taking measurements and who are going to perform the regulations. That makes it clear I hope.

DR. SNIHS, Sweden

Mr. Chairman, I would like to come back to this question of continuous measurements of the emission of mine sweepers. I think that is a very controversial and difficult question which must be considered very seriously before making a recommendation. I agree that perhaps it would be necessary to make an estimation of the environmental impact of the emission and that can be made by measuring the releases, then using atmospheric models estimate the doses in the environment. That is one thing. The other thing is for continuous measurement of the emission, and you have to compare that with the nuclear power and I think there is difference because these measurements are made to trigger some kind of action, some kind of counter measures if the emission is too high. That is possible in nuclear reactors but I don't know how possible it is with the mine, what can you do to reduce the emission. It can be made over a period of time, perhaps if you change the ventilation system. But a rapid change is I think very difficult to do. So, I think it should be recommended that this question should be considered seriously.

MR. YOURT, Canada

Since I am so close to the source of the emissions, here I feel obliged to make some further comment on this. A lot of money has been expended in revegetation right in this area, and it is particularly difficult here because of the acid condition and that requires a great deal of lime. Revegetation will add to the cover and reduce the emanation but this is not easy, and I say to reduce the emanation, the more you put on top of tailings the more the radon will disintegrate in the covering.

DR. JACOBI, F.R. of Germany

I think we should not recommend in general the measurement of the emission from each mine, but I think for some larger mines, it would be reasonable to get some values to estimate environmental impact statement. I think we shall do it this way.

CHAIRMAN

I also believe it would be "prudent" to do so.

DR. CHAMEAUD, France

Je m'excuse de revenir en arrière, mais enfin je suis un peu surpris qu'on n'ait pas parlé davantage pendant ces journées des poussières à vie longue. M. Pradel a noté tout à l'heure dans son résumé, on a parlé de tous les autres polluants, des échappements de moteur, des gaz de tir, toutes espèces de choses, mais de poussières à vie longue très peu. Personnellement, c'est une question que je pose, j'aimerais savoir si on sait beaucoup de choses sur ces poussières à vie longue, sur par exemple leur granulométrie, sur leur effet radiologique, et si elles sont intéressantes ou si on peut les négliger.

CHAIRMAN

Le point a été soulevé à quelques reprises. En particulier, si la mémoire m'est fidèle, on avait cru qu'elles étaient négligeables dans certains cas, surtout lorsque cela correspond à une dose inférieure à un dixième, ou un tiers de la dose limite. Puisque effectivement ces poussières font aussi partie de la dose, je ne vois pas très bien comment on pourrait les négliger sur la base de cet argument. Il suffirait seulement de subdiviser la dose en dix éléments différents qui chacun serait inférieur à un dixième de la dose totale, donc on négligerait tout. Cela devient un peu absurde. I would invite further comments from the representative on the point that Dr. Chameaud is making namely why haven't we discussed the measurements and the recording of doses from long life dusts such as uranium, thorium and why should it be part of the exposure records?

MR. PRADEL, France

Je pense que ces quantités peuvent être négligeables encore qu'il faudrait être bien sûr de la concentration maximale admissible ce qui n'est pas très clair si on regarde dans l'ICRP. C'est surtout une question de concentration et de teneur des minerais. Si on a un minerai de teneur à 1%, on peut dépasser facilement la concentration maximale admissible en poussière dans les mines, c'est très facile. Donc personnellement, je ne pense pas que ce soit un problème très négligeable.

DR. CHAMEAUD, France

Quand je disais négligeable, ce n'est pas que je considère qu'il est négligeable, mais j'aurais aimé savoir si des études approfondies avaient été faites au sujet des poussières.

DR. JACOBI, F.R. of Germany

      Mr. Chairman, I think that the answer on these questions why we have not dealt here with long life dust particles is very simply because, this meeting is "Radon and Daughters" and I think this may be the subject of discussion in a future meeting.

CHAIRMAN

      Bien que ce soit une réponse valable, je ne crois pas qu'elle satisfasse entièrement la curiosité du docteur Chameaud.

DR. JOHNSON, Canada

      The ICRP in its report from the Task Group Committee will be giving a recommendation of no more than 70 pCi hours per litre for an annual intake of alpha activity in ore dust. That will be coming out in publication in a few months.

CHAIRMAN

      Would you repeat that.

DR. JOHNSON, Canada

      This again is just a draft but there will be a recommendation on ore dust made to the Committee IV, which is not a recommendation of course to the main committee. Correct me if I am wrong, any of these members of the Task Group: the resulting annual operational limit for alpha activity in ore dust is 70 pCi hours per litre. That is an accumulated exposure to concentration multiplied by the number of hours of exposure to it. The quarterly limit is one half that, i.e. 35 pCi hours per litre. Sorry, the quarterly is half the annual limit.

MR. BUSH, Canada

      I would like to remind you that like all the other permissible intakes, that value can only be used if only uranium exists in the air. We still have to consider the additional risks which were suggested in my paper, and I don't know whether we have a problem but certainly we have to find out whether we have one or not.

DR. JACOBI, F.R. of Germany

      I don't want to make a comment on the last point, but I would like to hear something about the method of thoron daughters in mines. I have heard that in many mines thoron content is relatively high and I would be interested to hear something of the fractional activity which has aggravated for so long in these daughters.

MR. BUSH, Canada

      The Interior Department of Health made a measurement two or three weeks ago and I understand they may be .3 working levels of thoron daughters. That was the highest measurement and at the same time the highest measurement of radon daughters was .5. The lowest value of thoron was down around .03. I think these are the only measurements that have been made in the last ten years in Canada and ten years ago they measured similar concentrations.

DR. O'RIORDAN, United Kingdom

      Could I also draw your attention to some measurements one of my colleagues made a few years ago of thoron levels in non-uranium mines, specifically, tin mines, haematite mines and fluospar mines, published in "Public Health visits in 1973, volume 25".

M. PRADEL, France

Je voudrais simplement dire que nous n'avons pas de thorium dans nos mines, donc il n'y a pas de thoron, pratiquement pas.

MR. BUSH, Canada

I have some more comprehensive results here now: in the Denison mines there were 9 samples taken, and the thoron concentrations were from .2 WL to .3 WL for those 9 samples and the radon daughters were .3 WL to .5 WL. In the Rio Algom mines, 6 samples showed from .1 WL to .2 WL of thoron daughters and .2 WL to .5 WL of radon daughters.

MR ROCK, United States

I would just like to say that we are fortunate that ventilation is so much more effective in controlling thoron daughters. If it were not so much more effective, you would have tremendous levels. The fact that you have two or three working levels on radon daughters with roughly half the amount of working levels in thoron daughters is quite significant. If ventilation weren't so effective, we would be in tremendous trouble. We see this of course in boring unventilated areas and we have measured some up in Alaska. We went up there and took some measurements in a mine around 1% uranium and 1% thorium, of course in areas where it was poorly ventilated. We could easily see the relationship between the two, so it is just very fortunate that thoron daughters are so easily controllable.

CHAIRMAN

Thank you Mr. Rock. We have reached a point where we should consider comments and proposals for future activities.

DR. AHMED, IAEA

Thank you, Mr. Chairman. I should like to take this opportunity to ask this meeting for some guidelines on future activities which could be done at international level. Before, I would like to give some background also. I have outlined the activities of the Agency in my intervention yesterday, and I should also like to mention some of the related activities. We had a panel in 1973 under the chairmanship of Prof. Marsa from Rochester on estimating the hazard from airborne contaminants. During this meeting, I have noted what remains to be done in development of methods, principles, instruments and monitoring of radon and radon daughters. I have also noted that there are some considerations to be given to the basis for development of standards for radon and radon daughters to be applied in mines and mills. So I should like to take this opportunity to ask this meeting if they could provide guidelines on what IAEA or for that matter any international organization could do during the coming years in helping in any way in the development of methods, instruments, principles, standards or whatever it is.

CHAIRMAN

Thank you, Dr. Ahmed. Indeed the fact that you are with us for instance is an indication of the great interest and close collaboration of the IAEA in the NEA work. There may be some differences at times in specific objectives and scheduling. As you might have noticed, we are eager to get on with the jobs and to obtain specific solutions to specific problems in the shortest time possible. This may influence the activities but nevertheless, there is no reason why this could not be done in close collaboration with IAEA wherever it is possible. Perhaps you can indicate if some of the concerns expressed here might have already been considered by IAEA.

DR. AHMED, IAEA

Mr. Chairman, I think I already mentioned that the proposal of intercomparison, intercalibration is a good one and IAEA can play a role in collaboration with NEA or any other international organization. This is one I have already noted, but I also wanted to have more ideas from this meeting.

DR. WALLAUSCHEK, NEA

I should like to say on this subject that for many years there has been excellent collaboration with IAEA in all fields in which we are working. I think the two Agencies have complementary objectives. IAEA has done enormous and very useful work through its publication of guidelines and manuals proposed to be followed on a worldwide basis. Our objectives are more practically linked to specific problems. I do not think that the type of meeting we are having here would be organised by IAEA but I feel that it certainly serves in a complementary way to IAEA's activities. When proposals are made by Member countries, either IAEA or NEA could take up certain activities or both could collaborate, for instance, in organising a joint symposium. I hope that you have the impression that the development of knowledge and guidance in relation to topics we are discussing here is done in the best way to serve our needs.

CHAIRMAN

Thank you, Dr. Wallauschek. May we proceed now with what needs to be done instead of how we can do it. Yes, proposals or comments are invited as to where we should go from here. Unless Mr. Yourt you wish to comment further on the IAEA/NEA relationship.

MR. YOURT, Canada

Yes, on one point. I would like to ask if IAEA or NEA could bring any pressure on ICRP to come out with revised thinking. Current thinking dates back to 1959 when the 30 pCi/ℓ per year came out with a factor to take care of attached and unattached daughters. We have never heard any clearcut interpretation of this that mining companies, uranium mining companies, could use. In 1960, we received advice from individual members of ICRP that we should try to get down to 30 instead of a 100; we acted on that without strong urgence from any government organization. Again it appears to me that if we could get some action from ICRP on their thinking, it would be a great help to the uranium mining companies.

CHAIRMAN

Thank you Mr. Yourt. There is a representative of ICRP present here and I am sure he has taken note of your request. I am confident we could add our support to it without seeking a concensus as I believe the point is sufficiently clear.

DR. JACOBI, F.R. of Germany

But I think it might be useful if the two agencies could write an official letter to ICRP on how significant these problems are and that they are very interested to obtain some recommendations as soon as possible.

MR. JOHNSON, Canada

The Chairman of the ICRP is also the Chairman of the Task Group which is considering these questions. I will relay the message.

CHAIRMAN

Can we go back to suggestions or comments for consideration of specific problems.

MR. ROCK, United States

We ought to have some way of keeping aware of progress by others. We ought to have some way of keeping ourselves current in what is going on. Somewhere there should be a central system to communicate with each other and finding out what is new in a useful way.

CHAIRMAN

May I take this opportunity Mr. Rock, to explain that this meeting is part of such a system. It is held under the auspices of NEA as a result of recommendation made by the expert group that met in April 1976 to the Radiation Protection and Public Health Committee. Recommendations from this meeting will be going to that Committee. The group that met in April is still alive and exchanging information within the scope of its mandate. This meeting of experts on the personal dosimetry and area monitoring is another way of facilitating the information exchange. It has been set up by the group in April. The US correspondent is Dr. Cunningham who is with the USNRC and should you feel that there should be another system, it should be discussed.

DR. WALLAUSCHEK, NEA

I would like to add to this, this group may recommend there should be another meeting in two, or three or five years time to look into new developments and to renew our exchange of views on the subject, even next year if you wish. We are very flexible, but I feel next year would be too soon. So what would be your practical proposal for this type of meeting to be repeated?

MR. ROCK, United States

I would say certainly have one in another year. We would certainly be prepared to make a much more thorough presentation of our efforts and where we stand and certainly other people in the room would also have progressed accordingly. It just seems to me that to have a meeting and publish the proceedings and then we all tend to go our own way, this is the normal trend of the way things happen, and it seems to me we ought to be able to preserve our relationship here and keep up the most rewarding process.

CHAIRMAN

Can I just ask Mr. Rock whether your proposal for this future meeting relates to these same subjects?

MR. ROCK, United States

Well, I think certainly we ought to add a few things that we mentioned during this meeting, but certainly we should also expand on other things.

CHAIRMAN

Extension of controls to non-uranium mines and to naturally occurring sources have been mentioned. This could be conveyed to NEA as part of the package.

MR. ZETTWOOG, France

En ce qui nous concerne, nous avons l'intention de commencer une campagne de mesures de comparaison comme celles que je vous ai déjà présentées, portant sur environ 80 dosimètres. Je pense donc qu'à la fin de l'année nous aurons probablement de l'ordre de 800 points de comparaison. Alors cela commence à pouvoir être exploité. A mon avis, le délai serait trop court si on se rencontrait au bout seulement de douze mois. Il faudrait quand même un peu plus de temps et je pense que deux ans serait peut-être plus raisonnable.

CHAIRMAN

La suggestion est en effet raisonnable et une demi-mesure serait douze mois après la publication des proceedings.

DR. WALLAUSCHEK, NEA

Mr. Chairman, can I have a little more advice on what might be the subject of the second meeting. I feel that this meeting has given us the opportunity to find out what is the present state of what is available, what type of measurements we have and other related problems. Should it not be the next step to adopt a more critical approach and look into the real problems and come to more real conclusions. For instance, questions were raised on controls for radon, on the role of area monitoring and personal monitoring - do we need personal monitoring at all - their application in relation to small mines and to big mines, where should the measurements be made. There seems to be a need for the development of a more critical approach. Or should we just continue to explore the subjects we have identified during the week?

DR. SHREVE, United States

I would like to ask if it would not be possible just to confer on radon and radon dosimetry in general? As we have seen some of the instruments already developed have application not only to occupational exposures but also to environmental exposures.

DR. WALLAUSCHEK, NEA

Can I clarify this a little more. Do you mean one day dealing with one subject, e.g. occupational exposure, the next with another subject. I don't see clearly what your proposal is.

DR. SHREVE, United States

It seems we have covered two specific topics during the week. We specifically looked at area monitors, we specifically looked at personal dosimeters. What I am asking is, why not look at just the problems of radon daughter monitoring and measurements in general. At the Round Table discussion, we considered two specific topics, not relating to mines. To have a Round Table discussion specifically related to mines, possibly. I don't know how to make myself any clearer.

DR. WALLAUSCHEK, NEA

I am wondering what is the difference between the meeting we had here and your proposal.

CHAIRMAN

May I attempt to summarize the discussion. My understanding is that you are suggesting that we should meet under the general heading of radon daughter monitoring but have Round Table discussions on the application of radon daughter monitoring for occupational health purpose, for environmental effects control or for whatever might be of interest to the group.

MR. ROCK, United States

What I want to say is we can develop the rapport here. I think we have a very expert group of specialists, and we complement each other. I am confident we could produce papers on special investigations for the next session.

CHAIRMAN

Thank you Mr. Rock. Let it be known that the intent is to meet again in the next 18 months so everybody who wish to, can start preparing for it and the Secretariat can start planning. This should be ample notice.

DR. SNIHS, Sweden

Mr. Chairman, may we comment about the proposal. It seems to me that it should be another meeting on technical problems. I think that at this meeting we had a number of lectures on different personal dosimeters as well as summaries of the experiences in different countries. This will be presented in the proceedings with the discussions. That will be a lot of pages to read and one will be looking for some conclusions on the urgent problems, for instance what kind of instruments shall I use now in our mines, what method shall I use, and when I read the proceedings I shall find a difference of opinions about that. For instance, we have heard today Mr. Rock say that the highlights of these methods should be that personal dosimeters should be used if they exist. Dr. Jacobi said that the monitoring is perhaps not good to determine personal exposure. With all the uncertainties and errors, I have doubts about the conclusions. As I said earlier, even with the laboratory scale testing in mines, I think it is perhaps too early to recognize these dosimeters. They are still perhaps a little too experimental, some of them are too early, and the question which was raised by Dr. Jacobi about the standards is perhaps also a controversial one. What could be a valuable fall-out of this meeting, I think, is along the lines of the proposal by Dr. Wallauschek to have an analysis of the activities and purposes of the measurements and surveillance of dosimetry in mines and who would analyze the meaning of all these measurements, is what we are looking for.

_____?

As indicated by Dr. Fry in the Round Table discussion, the main purpose should be to protect the workers and it follows that one is after the best method to do this taking into account the money and the manpower.

DR. SNIHS, Sweden

I think that personal analysis could be made based on what has been presented here this week concerning the instruments and methods and the practical experience. We should also consider at the same time the given standards and if there is any need for change - that may be perhaps another topic but it should be taken into consideration and analyzed anyway. All these analyses may perhaps lead to some kind of recommendation about reasonable conditions and efforts in this field. Then perhaps we can also make some constructive critique of what has been said this week, suggest different approaches and collect opinion on what has been said here. So I will support the proposal made by Dr. Wallauschek about some kind of follow-up meeting. I don't know if it should be a Round Table meeting or a small expert group meeting, how it should be made.

DR. WALLAUSCHEK, NEA

All these suggestions are most helpful to focus on what we should do in the future in this field. We have started with this meeting, and as Mr. Rock said, this is a good group for such discussions. We should now decide if the problems have been examined and understood to such an extent that this group does not need to meet again. On the other hand, we know that a lot remains to be done in this field, and in view of this, we could follow Mr. Rock's and Mr. Snihs' proposals and meet again in order to find out what developments have occurred in the meantime. In this respect, I should like to know your opinion on whether such a meeting should be organised at a mine or mill site, or it could be held at our Paris headquarters.

MR. ROCK, United States

I do agree that it would be very advantageous to see the various types of mine. I think that the ones with the most problems and they are the ones that most concerns me, are in Grand Junction for instance. Although they may be small, they really place in perspective the whole problems in United States. It would be advantageous to see those things as I feel most people here are really not mining people. One really has to have a feel for this, well it is almost intuition. I hate to use that word but you would certainly see a much cheaper, a much smaller, a much less organized situation, yet these people receive exposure to the daughter products and the mortality rate are such that they affect the whole country and perhaps the whole world through our epidemiological study. Although some people here would be able to realize the problems, I am not sure that the group as a whole would gain much from it. I guess that it is the best I can say.

MR. MONTEZEMOLO, Italy

Thank you Mr. Chairman. I do think that if the United States will organize the next meeting, some people are not interested in visiting mines and mills, there would be always the possibility of organizing a hike along the Colorado River.

MR. PRADEL, France

Ce que je voulais dire, c'est que nous pourrions je crois organiser cette réunion en France si l'OCDE nous en faisait la demande. Je pense que nous aurions l'accord de nos autorités. Nous pourrions organiser la visite de mines et on pourrait soit faire une réunion sur les mines, soit la faire à Paris. Il faut choisir ce que l'on veut voir. Ou bien on veut voir les mines, ou bien on veut voir Paris. On peut peut-être faire deux jours à un endroit et deux jours à l'autre.

CHAIRMAN

We could leave it to the Secretariat to identify a host country. Now let's go back to Dr. Snihs' proposal which seems very compatible with the previous one. If we are to meet again in the next two years or so, the analysis and critical assessment that would have been done in the meantime could be discussed. Unfortunately, for the time being there's no easy answer to your direct questions, except to note that there is a number of systems and instruments that provide coherent information within a given system of application, but the intercomparison with other systems and instruments may not and this is what has to be analyzed. We are faced with a number of separate identities, at least this is my impression, and one would have to decide for himself which is better suited for specific cases. Would you wish to see something done about follow-up in the nearer future than two years?

DR. SNIHS, Sweden

No, not necessarily. My point is only that we should not lose the contacts, the impetus and the results from this meeting. The final conclusions and the follow-up could flow from their analysis.

CHAIRMAN

Therefore, this is compatible and comparable with the previous proposal and it should be combined. If I may summarize, then it is proposed to meet again with two years to discuss the various aspects indicated by Dr. Kerr and yourself.

DR. SNIHS, Sweden

Perhaps we could add an informal suggestion that the things that were discussed here give rise to exchange of correspondence to keep a rapport between some of us and that would help to keep us in a more informed state.

CHAIRMAN

Any further suggestions? on our proposal before I close this session?

DR. JOHNSON, Canada

I think that we do need a specific sort of recommendation to follow this meeting, Mr. Chairman, and to leave it as it is on a possible future discussion is too indefinite, particularly in view of the cost and energy that has been put into it. It seems to me that there are some specific recommendations that should be made that we should ask to be applied to mining right now. It seems to me that following the very fine introductory papers to this meeting, that mining represents a critical step in the nuclear fuel cycle. Environmentatlists are attacking all steps of the fuel cycle and here we are at the very basic step, literally and figuratively. We have problems that our fast breeder technology above ground has coped with, and it's time we coped with those underground. I would be disappointed to leave the meeting without some specific technical recommendations, for the mining industry. For example, by implication only, I think, we have chosen working level as a basic measuring unit. Now, is that the consensus of the meeting - it seems to me that it is but I think we need to specifically say it. We also need to give some specific recommandations about the monitoring that can be done right now. For example, area monitoring is vital in a mine. In the exhaust air within the mine itself, there should be a continuous monitor. Exhaust air from the mine itself should be monitored. We should recommend registration of uranium miners, the application of personal dosimeters as quickly as possible, the surveying of all types of mines for the radon/radon daughter hazards, that the ventilation safety group within the mine do detailed surveys despite the fact that they have area monitoring, that the development and the publication of standardized instrumentation which we can be used as a basis for intercomparison later on, and after we have done recommendation like this, it seems to me then we should agree on meeting again. Thank you very much.

CHAIRMAN

Thank you very much, Dr. Johnson. All you have said is indeed most appropriate and should be emphasized. The item under discussion is with regard to future activity of this group and there is one recommendation which is to meet again in two years from now to discuss radon daughter monitoring in general but with specific application to occupational and environmental aspects. The points you have raised should be noted as recommendations from this meeting on personal dosimetry and area monitoring. Some specific recommendation may be difficult to generalize as for instance whether or not uranium miners should be declared atomic radiation workers because a lot depends on the national legislation. But in any case, we could recommend that the exposure of miners can be measured with the equipment that we have discussed. I wonder if I could contact you again for writing these recommendations.

MR. PRADEL, France

Si l'on voulait faire les recommandations suggérées, il nous faudrait au moins une huitaine de jours. J'ai participé au groupe de travail de l'Agence et au groupe de travail de l'ICRP, et on a essayé de formuler des recommandations, mais il faut un temps extraordinaire. Il faut discuter chaque mot et ce n'est pas avec le temps disponible qu'on peut le faire.

DR. JOHNSON, Canada

I am not sure, whether I really want to respond in detail. I did not have in mind preparing an ICRP type document on the subject. It seems to me that as a result of our meeting, we have some recommendations. We have had enough discussions to make at least these general recommendations without specific detail or any that became legally binding.

CHAIRMAN

This indeed is the intent. With regard to the conference communique at this late hour, the best I can offer is to draft a text by to-morrow morning and to finalize it with the sessions and Round Table chairmen. This should be a simple statement of facts about the meeting with the very broad conclusions we have identified. Thank you very much for your time and collaboration. I declare this session closed.

DR. WALLAUSCHEK, NEA

Before closing our meeting, I should like to remind you that preparatory arrangements were made with only four speakers to present papers and the organisation for the rest of the week was left open. When I arrived at Elliot Lake I had the feeling that the meeting was not well prepared. You will also remember that at our first session we set up our programme of work for the week, and I was most pleased about your very active response in offering so many excellent contributions to our work. I really feel, and you have confirmed this several times, that this week's work has been excellent. I should not terminate without saying that the good results achieved, and I am sure you all agree with me, are due to our very pleasant collaboration with our Canadian friends, in particular with Paul Hamel. He was extremely helpful in assisting in making a lot of the arrangements and in particular in relation to the very important Canadian contribution to our

work. We have received a great deal of help from Mr. Kidd, the
Director of the Elliot Lake Centre, as well as from his secretary and
the secretary of Mr. Hamel. I would like on behalf of you all to
thank them for their efforts. I should not forget our interpreters,
they have worked really hard. In closing this meeting I thank you
again for your very active collaboration and wish you a good return
to your respective home countries - I hope we shall see each other in
two years' time.

List of participants

Liste des participants

AUSTRALIA - AUSTRALIE

FRY, R.M., Director, Environmental and Public Health Unit, Australian Atomic Energy Commission, 45 Beach Street, Coogee, N.S.W.

CANADA

BARDSWICH, W.A., Chief, Environmental Engineer, Ontario Ministry of Natural Resources, Mines Engineering Branch, 260 Cedar Street, Sudbury, Ontario P3B 3X2

BUSH, W.R., Scientific Advisor, Atomic Energy Control Board, Directorate of Licensing, P.O. Box 1046, Ottawa, Ontario K1P 5S9

GRAY, W.M., Dr., Senior Scientist, Mining Research Laboratories, CANMET, EMR, 555 Booth Street, Ottawa, Ontario K1A 0G1

GROGAN, D., Chief, Radiation Dosimetry Division, Radiation Protection Bureau, Department of National Health & Welfare, Brookfield Road, Ottawa, Ontario K1A 1C1

HAM, D.J.M., Dr., School of Graduate Studies, University of Toronto, Toronto, Ontario M5S 1A1

HAMEL, P.E., Director, Research and Coordination Directorate, Atomic Energy Control Board, P.O. Box 1046, Ottawa, Ontario K1P 5S9

HORWOOD, J.L., Research Scientist, Mineral Sciences Laboratories, CANMET, EMR, 555 Booth Street, Ottawa, Ontario K1A 0G1

JOHNSON, H.M., Health Physicist, Radiation and Industrial Safety Section, Whiteshell Nuclear Research Establishment, Atomic Energy of Canada Limited, Pinawa, Manitoba ROE 1L0

JOHNSON, J.R., Medical Research Branch, Chalk River Nuclear Laboratories, Atomic Energy of Canada Limited, Chalk River, Ontario K0J 1J0 (see also ICRP)

KNIGHT, G., Research Scientist, Elliot Lake Laboratory, Mining Research Laboratories, CANMET, E.M.R., P.O. Box 100, Elliot Lake, Ontario P5A 2J6

POMROY, C., Head, Human Monitoring Section, Radiation Medicine Division, Radiation Protection Bureau, Department of National Health & Welfare, Brookfield Road, Ottawa, Ontario K1A 1C1

STOCKER, H., Research and Coordination Directorate, Atomic Energy Control Board, P.O. Box 1046, Ottawa, Ontario K1P 5S9

TANIGUCHI, H., Dr, Chief, Nuclear Safety Division, Radiation Protection Bureau, Department of National Health & Welfare, Brookfield Road, Ottawa, Ontario K1A 1C1

WASHINGTON, R.A., Dr., Research Scientist, Elliot Lake Laboratory, Mining Research Laboratories, P.O. Box 100, Elliot Lake, Ontario P5A 2J6

ZAHARY, G., Health and Safety Programme, Mining Research Laboratories, CANMET, EMR, P.O. Box 100, Elliot Lake, Ontario P5A 2J6

FRANCE

BEAU, P., Dr., Département de Protection - SPS, Centre d'Etudes Nucléaires de Fontenay-aux-Roses, Commissariat à l'Energie Atomique, B.P. n° 6, 92260 Fontenay-aux-Roses

BRESSON, G., Adjoint au Chef du Département de Protection, Centre d'Etudes Nucléaires de Fontenay-aux-Roses, Commissariat à l'Energie Atomique, B.P. n° 6, 92260 Fontenay-aux-Roses

CHAMEAUD, J., Dr., COGEMA, Chef du Service Médical "Branche Mines", Division Minière de la Crouzille, B.P. n° 1, 87640 Razes

PRADEL, J., Chef du Service Technique d'Etudes de Protection et de Pollution Atmosphériques, Département de Protection, Centre d'Etudes Nucléaires de Fontenay-aux-Roses, Commissariat à l'Energie Atomique, B.P. n° 6, 92260 Fontenay-aux-Roses

ZETTWOOG, P., Service Technique d'Etude de la Pollution dans l'Atmosphère et dans les Mines, Département de Protection, Centre d'Etudes Nucléaires de Fontenay-aux-Roses, Commissariat à l'Energie Atomique, B.P. n° 6, 92260 Fontenay-aux-Roses

FEDERAL REPUBLIC OF GERMANY - REPUBLIQUE FEDERALE D'ALLEMAGNE

JACOBI, W., Prof., Gesellschaft für Strahlen- und Umweltforschung mbH., Institut für Strahlenschutz, Ingolstädter Landstr. 1, Post Oberschleissheim, D-8042 Neuherberg b. München

MILDE, W., Uranerzbergbau GmbH & Co. KG., c/o Agnew Lake Mines Ltd., P.O. Box 1970, Espanola, Ontario, Canada

ITALY - ITALIE

BASSIGNANI, S., Responsable du Bureau de Protection Radiologique et de Dosimétrie de l'AGIP Nucléaire, 20122 Corso P. Romana, 68 Milano

BREUER, F., Dr., Division de la Protection de l'Environnement, Direction Centrale de Sécurité et de Protection Sanitaire, Comitato Nazionale per l'Energia Nucleare, Viale Regina Margherita 125, 00198 Rome

CORDERO di MONTEZEMOLO, AGIP Attivita Minerarie SPA, Division des Ressources Energétiques Diverses, Service d'Exploitation Minière, 20097 San Donato Milanese

SCIOCCHETTI, G., Dipartimento Radiazioni, Comitato Nazionale per l'Energia Nucleare, Centro Studi Nucleari Casaccia, Via Anguillarese km 1300, Rome

JAPAN - JAPON

KITAHARA, Y., Senior Staff, Health and Safety Office, Power Reactor and Nuclear Fuel Development Corporation, 9-13, 1-Chome, Akasaka, Minato-ku, Tokyo

KUROKAWA, Y., Dr., Manager, Health and Safety Office, Power Reactor and Nuclear Fuel Development Corporation, 9-13, 1-Chome, Akasaka, Minato-ku, Tokyo

KUROSAWA, R., Prof., Institute of Physics and Technology, Waseda
    University, 17 Kikui-cho, Shinjuku-ku, Tokyo

SWEDEN - SUEDE

AGNEDAL, P.O., AB Atomenergi, Studsvik, Fack, S-611 01 Nyköping 1

EHDVALL, H., National Institute of Radiation Protection, Fack,
    S-104 01 Stockholm 60

SNIHS, J.O., National Institute of Radiation Protection, Fack,
    S-104 01 Stockholm 60

SWITZERLAND - SUISSE

KAUFMANN, E., Head of Physical Section, Swiss National Accident
    Insurance Fund, Fluhmattstrasse 1, 6002 Lucerne

UNITED KINGDOM - ROYAUME-UNI

O'RIORDAN, M.C., National Radiological Protection Board, Harwell,
    Didcot, Oxfordshire OX11 ORQ

UNITED STATES - ETATS-UNIS

BRESLIN, A.J., Dr., Director, Health Protection Engineering Division,
    Health and Safety Laboratory, US Energy Research and Development
    Administration, 376 Hudson Street, New York, N.Y. 10014

KERR, G.D., Dr., Health Physics Division, Oak Ridge National
    Laboratory, P.O. Box X, Oak Ridge, Tennessee 37830

ROCK, R.L., Chief, Radiation Branch, Denver Technical Support
    Center, United States Department of the Interior, Mining
    Enforcement and Safety Administration, P.O. Box 25367, Denver
    Federal Center, Denver, Colorado 80225

ROSS, D., Dr., Division of Safety, Standards & Compliance, U.S.
    Energy Research and Development Administration, Washington,
    D.C. 20545

SHREVE, Jr., J.D., Dr., Technical Specialist, Physical Science and
    Measurement Department, Kerr McGee Corporation, Post Office
    Box 25861, Oklahoma City, Oklahoma 73125

INTERNATIONAL ATOMIC ENERGY AGENCY
AGENCE INTERNATIONALE DE L'ENERGIE ATOMIQUE

AHMED, J-U., Dr., Division of Nuclear Safety and Environmental
    Protection, International Atomic Energy Agency, Kärntnerring 11,
    A-1011 Vienna, Austria

INTERNATIONAL COMMISSION OF RADIOLOGICAL PROTECTION
COMMISSION INTERNATIONALE DE PROTECTION RADIOLOGIQUE

JOHNSON, J.R., Dr., Medical Research Branch, Chalk River Nuclear Laboratories, Atomic Energy of Canada Limited, Chalk River, Ontario K0J 1J0 (see also Canada)

OECD NUCLEAR ENERGY AGENCY
AGENCE DE L'OCDE POUR L'ENERGIE NUCLEAIRE

WALLAUSCHEK, E., Dr., Head, Radiation Protection and Radioactive Waste Management Division, OECD Nuclear Energy Agency, 38 boulevard Suchet, 75016 Paris, France

Some other publications of NEA

ACTIVITY REPORTS

Activity Reports of the OECD Second Activity Report (1973)
Nuclear Energy Agency (NEA)   71 pages (crown 4to)

Third Activity Report (1974)
75 pages (crown 4to)

Fourth Activity Report (1975)
77 pages (crown 4to)

Free on request

Annual Report of the OECD High   Fifteenth Report (1973-1974)
Temperature Reactor Project   85 pages (crown 4to)
(DRAGON)
Sixteenth Report (1974-1975)
99 pages (crown 4to)

Free on request

Annual Reports of the OECD   Fourteenth Report (1973)
Halden Reactor Project   105 pages (crown 4to)

Fifteenth Report (1974)
103 pages (crown 4to)

Sixteenth Report (1975)
70 pages (crown 4to)

SCIENTIFIC AND TECHNICAL CONFERENCE PROCEEDINGS

Application of On-Line Computers to Nuclear Reactors
Proceedings of the Sandefjord Seminar, September 1968
900 pages (crown 4to)
£ 7.5s., $ 20, F 85, FS 78, DM 70

Third Party Liability and Insurance in the Field of Maritime Carriage of Nuclear Substances
Proceedings of the Monaco Symposium, October 1968
529 pages (crown 8vo)
£ 2.12s., $ 7.50, F 34, FS 28.50, DM 22.50

The Physics Problems of Reactor Shielding
Proceedings of the Specialist Meeting, Paris, December 1970
175 pages (crown 4to)
£ 1.75, $ 5, F 23, FS 20, DM 15.60

Magnetohydrodynamic Electrical Power Generation
Proceedings of the Fifth International Conference, München, April 1971
499 pages (crown 4to)
£ 4.88, $ 14, F 65, FS 50, DM 43

Marine Radioecology
Proceedings of the Hamburg Seminar, September 1971
213 pages (crown 8vo)
£ 1.50, $ 4.50, F 20, FS 15.60, DM 13.60

Disposal of Radioactive Waste
Proceedings of the Information Meeting, Paris, 12th-14th April 1972
290 pages (crown 8vo)
£ 2.60, $ 7.75, F 32, FS 25, DM 20

Power from Radioisotopes
Proceedings of the Second International Symposium, Madrid, 29th May-1st June 1972
986 pages (crown 4to)
£ 9, $ 24, F 110, FS 83.50, DM 68.80

The Management of Radioactive Wastes from Fuel Reprocessing
Proceedings of the Paris Symposium, 27th November-1st December 1972
1265 pages (crown 8vo)
£ 12, $ 34, F 140, FS 107, DM 88

The Monitoring of Radioactive Effluents
Proceedings of the Karlsruhe Seminar, 14th-17th May 1974
452 pages (crown 8vo)
£ 4.40, $ 11, F 44

Management of Plutonium-Contaminated Solid Wastes
Proceedings of the Marcoule Seminar, 14th-16th October 1974
248 pages (crown 8vo)
£ 3.80, $ 9.50, F 38

Bituminization of Low and Medium Level Radioactive Wastes
Proceedings of Antwerp Seminar, 18th-19th May 1976
251 pages (crown 8vo)
£ 4.70, $ 10,00, F 42

Personal Dosimetry and Area
Monitoring Suitable for Radon
and Daughter Products

Proceedings of the NEA Specialist
Meeting, Elliot Lake (Canada),
4th-8th October 1976
318 pages (crown 8vo)

SCIENTIFIC AND TECHNICAL REPORTS

Power Reactor Characteristics

September 1966
83 pages (crown 4to)
15s., $ 2.50, F 10, FS 10, DM 8.30

Uranium Resources
(Revised Estimates)

December 1967
27 pages (crown 4to)
Free on request

Prospects for Nuclear Energy in
Western Europe : Illustrative
Power Reactor Programmes

May 1968
47 pages (crown 4to)
17s.6d., $ 2.50, F 10, FS 10,
DM 8.30

Uranium - Production and Short
Term Demand

January 1969
29 pages (crown 4to)
7s., $ 1, F 4, FS 4, DM 3.30

Uranium - Resources, Production
and Demand

September 1970
54 pages (crown 4to)
£ 1, $ 3, F 13, FS 11.50, DM 9.10

Uranium - Resources, Production
and Demand

August 1973
140 pages (crown 4to)
£ 1.76, $ 5, F 20, FS 15.60,
DM 12.50

Uranium - Resources, Production
and Demand, including other
Nuclear Fuel Cycle Data

December 1975
78 pages (crown 4to)
£ 3.10, $ 7.00, F 28

Reprocessing of Spent Nuclear
Fuels in OECD Countries

January 1977
47 pages (crown 4to)
£ 2.50, $ 5,00, F 20

Water Cooled Reactor Safety

May 1970
179 pages (crown 4to)
£ 1.52, $ 4.50, F 20, FS 17.50,
DM 13.60

Glossary of Terms and Symbols in
Thermionic Conversion

1971
112 pages (crown 4to)
£ 1.75, $ 5, F 23, FS 20, DM 15.60

Radioactive Waste Disposal
Operation into the Atlantic 1967

September 1968
74 pages (crown 8vo)
12s., $ 1.80, F 8, FS 7, DM 5.80

Radioactive Waste Management
Practices in Western Europe

1972
126 pages (crown 8vo)
£ 1.15, $ 3.25, F 15, FS 11.70,
DM 10.50

Radioactive Waste Management Practices in Japan	1974 45 pages (crown 8vo) Free on request
Basic Approach for Safety Analysis and Control of Products Containing Radionuclides and Available to the General Public	June 1970 31 pages (crown 8vo) 11s., $ 1.50, F 7, FS 6, DM 4.90
Radiation Protection Standards for Gaseous Tritium Light Devices	1973 23 pages (crown 8vo) Free on request
Radiation Protection Considerations on the Design and Operation of Particle Accelerators	1974 80 pages (crown 8vo) Free on request
Interim Radiation Protection Standards for the Design, Construction, Testing and Control of Radioisotopic Cardiac Pacemakers	1974 54 pages (crown 8vo) £ 1, $ 2.50, F 10
Guidelines for Sea Disposal Packages of Radioactive Waste	November 1974 32 pages (crown 8vo) Free on request
Estimated Population Exposure from Nuclear Power Production and Other Radiation Sources	January 1976 48 pages (crown 8vo) £ 1.60, $ 3.50, F 14

LEGAL PUBLICATIONS

Convention on Third Party Liability in the Field of Nuclear Energy	July 1960, incorporating provisions of Additional Protocol of January 1964 73 pages (crown 4to) Free on request
Nuclear Legislation, Analytical Study : "Organisation and General Regime Governing Nuclear Activities"	1969 230 pages (crown 8vo) £ 2, $ 6, F 24, FS 24, DM 20
Nuclear Legislation, Analytical Study : "Regulations Governing Nuclear Installation and Radiation Protection"	1972 492 pages (crown 8vo) £ 3.70, $ 11, F 45, FS 34.60, DM 29.80
Nuclear Legislation, Analytical Study : "Nuclear Third Party Liability"	Revised Version 1977 in preparation
Nuclear Law Bulletin	Annual Subscription Two issues and supplements £ 2.80, $ 6.25, F 25

OECD SALES AGENTS
DEPOSITAIRES DES PUBLICATIONS DE L'OCDE

ARGENTINA – ARGENTINE
Carlos Hirsch S.R.L.,
Florida 165, BUENOS-AIRES.
☎ 33-1787-2391 Y 30-7122

AUSTRALIA – AUSTRALIE
International B.C.N. Library Suppliers Pty Ltd.,
161 Sturt St., South MELBOURNE, Vic. 3205.
☎ 699-6388
658 Pittwater Road, BROOKVALE NSW 2100.
☎ 938 2267

AUSTRIA – AUTRICHE
Gerold and Co., Graben 31, WIEN 1. ☎ 52.22.35

BELGIUM – BELGIQUE
Librairie des Sciences
Coudenberg 76-78, B 1000 BRUXELLES 1.
☎ 512-05-60

BRAZIL – BRESIL
Mestre Jou S.A., Rua Guaipá 518,
Caixa Postal 24090, 05089 SAO PAULO 10.
☎ 216-1920
Rua Senador Dantas 19 s/205-6, RIO DE JANEIRO GB. ☎ 232-07. 32

CANADA
Renouf Publishing Company Limited
2182 St. Catherine Street West
MONTREAL, Quebec H3H 1M7
☎ (514) 937-3519

DENMARK – DANEMARK
Munksgaards Boghandel
Nørregade 6, 1165 KØBENHAVN K.
☎ (01) 12 69 70

FINLAND – FINLANDE
Akateeminen Kirjakauppa
Keskuskatu 1, 00100 HELSINKI 10. ☎ 625.901

FRANCE
Bureau des Publications de l'OCDE
2 rue André-Pascal, 75775 PARIS CEDEX 16.
☎ 524.81.67
Principaux correspondants :
13602 AIX-EN-PROVENCE : Librairie de l'Université. ☎ 26.18.08
38000 GRENOBLE : B. Arthaud. ☎ 87.25.11

GERMANY – ALLEMAGNE
Verlag Weltarchiv G.m.b.H.
D 2000 HAMBURG 36, Neuer Jungfernstieg 21
☎ 040-35-62-500

GREECE – GRECE
Librairie Kauffmann, 28 rue du Stade,
ATHENES 132. ☎ 322.21.60

HONG-KONG
Government Information Services,
Sales and Publications Office,
Beaconsfield House, 1st floor,
Queen's Road, Central
☎ H-233191

ICELAND – ISLANDE
Snaebjörn Jónsson and Co., h.f.,
Hafnarstraeti 4 and 9, P.O.B. 1131,
REYKJAVIK. ☎ 13133/14281/11936

INDIA – INDE
Oxford Book and Stationery Co. :
NEW DELHI, Scindia House. ☎ 45896
CALCUTTA, 17 Park Street. ☎ 240832

IRELAND – IRLANDE
Eason and Son, 40 Lower O'Connell Street,
P.O.B. 42, DUBLIN 1. ☎ 74 39 35

ISRAEL
Emanuel Brown :
35 Allenby Road, TEL AVIV. ☎ 51049/54082
also at :
9, Shlomzion Hamalka Street, JERUSALEM.
☎ 234807
48 Nahlath Benjamin Street, TEL AVIV.
☎ 53276

ITALY – ITALIE
Libreria Commissionaria Sansoni :
Via Lamarmora 45, 50121 FIRENZE. ☎ 579751
Via Bartolini 29, 20155 MILANO. ☎ 365083
Sous-dépositaires :
Editrice e Libreria Herder,
Piazza Montecitorio 120, 00186 ROMA.
☎ 674628
Libreria Hoepli, Via Hoepli 5, 20121 MILANO.
☎ 865446
Libreria Lattes, Via Garibaldi 3, 10122 TORINO.
☎ 519274
La diffusione delle edizioni OCDE è inoltre assicurata dalle migliori librerie nelle città più importanti.

JAPAN – JAPON
OECD Publications Centre,
Akasaka Park Building,
2-3-4 Akasaka,
Minato-ku
TOKYO 107. ☎ 586-2016

KOREA – COREE
Pan Korea Book Corporation
P.O.Box n° 101 Kwangwhamun, SEOUL
☎ 72-7369

LEBANON – LIBAN
Documenta Scientifica/Redico
Edison Building, Bliss Street,
P.O.Box 5641, BEIRUT. ☎ 354429 – 344425

THE NETHERLANDS – PAYS-BAS
W.P. Van Stockum
Buitenhof 36, DEN HAAG. ☎ 070-65.68.08

NEW ZEALAND – NOUVELLE-ZELANDE
The Publications Manager,
Government Printing Office,
WELLINGTON : Mulgrave Street (Private Bag),
World Trade Centre, Cubacade, Cuba Street,
Rutherford House, Lambton Quay ☎ 737-320
AUCKLAND : Rutland Street (P.O.Box 5344)
☎ 32.919
CHRISTCHURCH : 130 Oxford Tce, (Private Bag)
☎ 50.331
HAMILTON : Barton Street (P.O.Box 857)
☎ 80.103
DUNEDIN : T & G Building, Princes Street
(P.O.Box 1104), ☎ 78.294

NORWAY – NORVEGE
Johan Grundt Tanums Bokhandel,
Karl Johansgate 41/43, OSLO 1. ☎ 02-332980

PAKISTAN
Mirza Book Agency, 65 Shahrah Quaid-E-Azam,
LAHORE 3. ☎ 66839

PHILIPPINES
R.M. Garcia Publishing House,
903 Quezon Blvd. Ext., QUEZON CITY,
P.O. Box 1860 – MANILA. ☎ 99.98.47

PORTUGAL
Livraria Portugal,
Rua do Carmo 70-74. LISBOA 2. ☎ 360582/3

SPAIN – ESPAGNE
Libreria Mundi Prensa
Castelló 37, MADRID-1. ☎ 275.46.55
Libreria Bastinos
Pelayo, 52, BARCELONA 1. ☎ 222.06.00

SWEDEN – SUEDE
Fritzes Kungl. Hovbokhandel,
Fredsgatan 2, 11152 STOCKHOLM 16.
☎ 08/23 89 00

SWITZERLAND – SUISSE
Librairie Payot, 6 rue Grenus, 1211 GENEVE 11.
☎ 022-31.89.50

TAIWAN
National Book Company
84-5 Sing Sung S. Rd., Sec. 3
TAIPEI 107.

TURKEY – TURQUIE
Librairie Hachette,
469 Istiklal Caddesi,
Beyoglu, ISTANBUL, ☎ 44.94.70
et 14 E Ziya Gökalp Caddesi
ANKARA. ☎ 12.10.80

UNITED KINGDOM – ROYAUME-UNI
H.M. Stationery Office, P.O.B. 569, LONDON
SE1 9 NH, ☎ 01-928-6977, Ext. 410
or
49 High Holborn
LONDON WC1V 6HB (personal callers)
Branches at: EDINBURGH, BIRMINGHAM,
BRISTOL, MANCHESTER, CARDIFF,
BELFAST.

UNITED STATES OF AMERICA
OECD Publications Center, Suite 1207,
1750 Pennsylvania Ave, N.W.
WASHINGTON, D.C. 20006. ☎ (202)298-8755

VENEZUELA
Libreria del Este, Avda. F. Miranda 52,
Edificio Galipán, Aptdo. 60 337, CARACAS 106.
☎ 32 23 01/33 26 04/33 24 73

YUGOSLAVIA – YOUGOSLAVIE
Jugoslovenska Knjiga, Terazije 27, P.O.B. 36,
BEOGRAD. ☎ 621-992

Les commandes provenant de pays où l'OCDE n'a pas encore désigné de dépositaire peuvent être adressées à :
OCDE, Bureau des Publications, 2 rue André-Pascal, 75775 Paris CEDEX 16
Orders and inquiries from countries where sales agents have not yet been appointed may be sent to
OECD, Publications Office, 2 rue André-Pascal, 75775 Paris CEDEX 16

PUBLICATIONS DE L'OCDE, 2, rue André-Pascal, 75775 Paris Cedex 16 - No. 38.592 1977

IMPRIMÉ EN FRANCE

NO LONGER THE PROPERTY
OF THE
UNIVERSITY OF R.I. LIBRARY